Michael Schwung

Cooperative Event-Based Control of Mobile Agents

Logos Verlag Berlin

λογος

Bibliografische Information der Deutschen Nationalbibliothek

Die Deutsche Nationalbibliothek verzeichnet diese Publikation in der Deutschen Nationalbibliografie; detaillierte bibliografische Daten sind im Internet über http://dnb.d-nb.de abrufbar.

ISBN 978-3-8325-5470-5

Logos Verlag Berlin GmbH
Georg-Knorr-Str. 4, Geb. 10,
D-12681 Berlin, Germany

Tel.: +49 (0)30 / 42 85 10 90
Fax: +49 (0)30 / 42 85 10 92

http://www.logos-verlag.de

Acknowledgements

This thesis summarises the results of my years as a research associate at the Institute of Automation and Computer Control at Ruhr University Bochum, Germany that would not have been possible without the support of several important persons that I would like to thank here.

First, I would like to thank my supervisor Prof. Dr.-Ing. Jan Lunze, who accepted me as his PhD student even before I completed my master degree. Thanks to his experience and critical judgement I was able to constantly improve both my research results and my teaching abilities. I would also like to thank Prof. Dr.-Ing. Aydin Sezgin to review this thesis. During meetings he gave me insights in the field of communication technology and has taken influence on my research.

Special thanks go to my fellow PhD students Alexander Schwab, Kai Schenk, Philipp Welz, Marc Wissing, Christian Wölfel, Markus Zgorzelski, Sven Bodenburg, Daniel Vey, René Schuh, Melanie Schuh and Andrej Mosebach. They were always willing to discuss my research towards the future of aviation. We became good friends and I am grateful for the moments we have shared. Thanks also go to the backbone of the institute Kerstin Funke, Susanne Malow, Andrea Marschall and Rudolf Pura for their constant technical and administrative support. I would also like to thank my student assistants Benjamin Littek and Dilara Serif Oglu, who contributed to my work. In particular they supported me at the experimental plant MULAN.

I owe a profound gratitude to my family, who have always been there at the right moments to support me through my work. In particular, I have to thank my mother Ursula, my father Klaus and my brother Andreas for the constant support from the beginning of my education until completing this thesis. Finally, I would like to thank my fiancée Larissa, who has always given me the strength and who will always be my motivation. You always show me the really important things in life, you are my why.

Bochum, March 2022 Michael Schwung

Contents

Part II Fundamental methods

Part III Event-based control over an ideal communication network

PART V SIMULATION STUDIES AND EXPERIMENTS

Appendix

Abstract

In the past decade the use of autonomously moving objects called 'agents' has become increasingly popular. In particular, unmanned aerial vehicles (UAVs) are used for a wide range of tasks due to their low operating and maintenance costs. At the same time, the greater number of UAVs increases the risk of collisions. This thesis proposes a method that plans trajectories for agents to enable them to fulfil individual tasks as well as cooperative tasks in groups by ensuring the collision-free movement. The agents are locally controlled and connected over an unreliable communication network that may induce packet losses and transmission delays. Further sensors e.g. for a distance measurement are not used and communication should only be invoked if it is necessary to avoid a collision.

The basic problem occurs for two agents. The first agent is called the stand-on agent, it can change its trajectory at any time without regard to the second agent. This agent is named the give-way agent. It has to ensure the collision avoidance by adapting its trajectory based on local data and communicated information about the current and future movement of the stand-on agent. A control unit for the give-way agent is introduced that has to execute four tasks to ensure the control aims: 1. Estimation of the current network properties. 2. Prediction of the movement of the stand-on agent. 3. Invocation of communication whenever the local data becomes too uncertain. 4. Planning of collision avoiding trajectories.

The aim of this thesis is to provide methods so as to solve the four tasks. To this aim a control unit consisting of four parts is introduced that uses approaches from control theory and communication technology. A delay estimator generates an estimate of the properties of the communication network that vary with the distance between the agents. A prediction unit determines a set that includes the uncertain future movement of the neighbouring agent. An event generator monitors the control aims and invokes communication in an event-based fashion when the local data becomes too uncertain. Furthermore, it decides when it is necessary to change the trajectory of the agent in order to ensure the collision-free movement. A trajectory planning unit provides collision-free trajectories based on Bézier curves. As a result it is proven that collisions are avoided using the proposed method even in the presence of transmission delays and packet losses induced by the communication network. It is shown that the network estimation is suitable for the controller and the set of predicted positions of the nearby agent is feasible.

The main result of the thesis is a novel control method for mobile agents that are connected over an unreliable communication network. Local trajectories are planned for the agents so that they are able to fulfil individual tasks as well as cooperative tasks in a group with a guaranteed collision avoidance. Communication is only utilised whenever it is necessary. The method can be used for different types of agents (e.g. UAVs or cars) with only slight modifications. The proposed methods are tested and evaluated through simulations and experiments with two quadrotors.

Part I

Introduction

Introduction to the event-based control of mobile agents

1

1.1 Purpose of the cooperative control of mobile agents

The use of autonomously moving objects such as unmanned aerial vehicles (UAVs) or cars enables one the execution of individual tasks, e.g. parcel delivery by using drones. Another application example is illustrated in Fig. 1.1. Here, two UAVs are utilised to provide communication to several moving ground objects, which are not able to communicate directly to each other. Due to the large distances between the ground objects two UAVs are required that act as aerial communication relay stations to provide all objects with communication links.

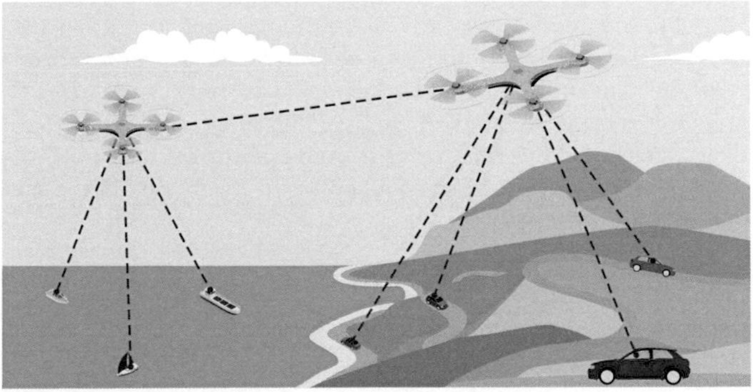

Fig. 1.1: Application example: Communication between ground objects using UAVs as aerial communication relay stations.

This application example illustrates that the autonomous movements of the UAVs need to be coordinated to fulfil their individual tasks. Furthermore, the increasing number of unmanned moving objects raises the risk of collisions. This thesis provides a method for the cooperative control of autonomously moving objects called *agents* in the literature of control theory. The agents are connected over a communication network and should fulfil the following control aims:

1. Collision avoidance between autonomously moving agents.

2. Satisfaction of individual tasks of the agents (e.g. reaching their individual destinations).

As the existing communication networks are increasingly utilised to capacity, with the control method the control aims should be satisfied with a significantly reduced communication exchange between the agents. To this aim the method uses the idea of event-based control from control theory and combines it with models from communication technology.

1.2 Event-based control of mobile agents

The agents to be considered are only able to measure their own positions and speeds locally to save energy. They communicate these data together with their current trajectory to the neighbouring agent. As the agents can change their trajectories at any time (e.g. due to an obstacle on their trajectory), a conflict of the control aims introduced in the preceding section may occur. However, not every change of the trajectories leads directly to a conflict of the control aims. Hence, a continuous communication is not necessary, but it is sufficient to communicate and to change the trajectories of the agents only if the uncertainties with respect to the control aims become too large. For this reason, the idea of event-based control is utilised, where information are only sent when a threshold is violated that indicates the uncertainty of the local information.

The problem statement of this thesis results from a conflict of the control aims. In this case the control method to be developed should plan the trajectories for the agents so as to fulfil the following two objectives:

- Avoid collisions between the agents by maintaining a safety distance between them.
- Maintain a maximum distance between the agents.

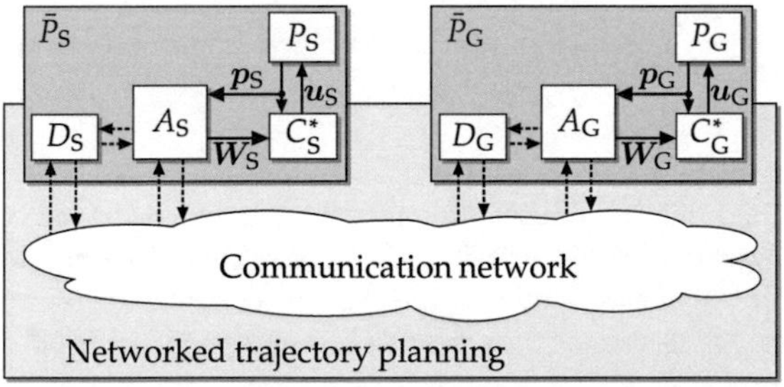

Fig. 1.2: Structure of the networked control system.

In the following, the networked system described in Section 1.1 is transferred into the control engineering structure, which is depicted in Fig. 1.2. In this thesis the control system consists of two mobile agents that are assigned with the functions *stand-on*[1] (agent $\bar{P}_{\mathrm{S}}$) and *give-way*[1] (agent $\bar{P}_{\mathrm{G}}$). The assignment of the tasks to the agents depends on the current situation as illustrated in Section 1.3.2. The stand-on agent $\bar{P}_{\mathrm{S}}$ and the give-way agent $\bar{P}_{\mathrm{G}}$ consist of the physical objects P_{S} and P_{G}, which are independently controlled by local flatness-based two-degrees-of-freedom controllers C^{*}_{S} and C^{*}_{G}. They generate the control inputs $\boldsymbol{u}_{\mathrm{S}}(t)$ and $\boldsymbol{u}_{\mathrm{G}}(t)$ and make the objects follow their trajectories generated by the corresponding event-based control units A_{S} and A_{G} even in the presence of external disturbances and model uncertainties. These control units use the locally measured positions $\boldsymbol{p}_{\mathrm{S}}(t)$ and $\boldsymbol{p}_{\mathrm{G}}(t)$ of the agents to generate the matrices $\boldsymbol{W}_{\mathrm{S}}(t)$ and $\boldsymbol{W}_{\mathrm{G}}(t)$, which consist of the trajectories and their first four derivatives and are able to communicate over the communication network if necessary.

The communication network is assumed to be unreliable in the sense that data packets may be received delayed or lost. The delay estimators D_{S} and D_{G} contain a Markov model that provides an estimate of the current properties of the network to take them into account for the trajectory planning.

Remark. The controllers C^{*}_{S} and C^{*}_{G} are designed using conventional control engineering methods. The design of the controllers depends on the specific type of the object to be controlled. First, the thesis focusses on the design of the event-based control units A_{S} and A_{G}, the controllers are derived in Chapter 10 for a quadrotor as the demonstration example.

In order to achieve both objectives stated above in a distributed fashion the control units A_{S} and A_{G} of the agents execute the following four tasks in an event-based fashion:

1. Network estimation: Estimate the current quality-of-service (QoS) properties of the communication channel, which vary with the distance between the agents.

2. Prediction: Estimate the future positions of the neighbouring agent.

3. Communication: Invoke communication by incorporating the estimated QoS properties of the channel only when the local data becomes too uncertain indicated by a threshold.

4. Trajectory planning: Plan the trajectory so as to fulfil the control aims. Modify the trajectory whenever a violation of the control aims threatens due to a changed situation.

The aim of this thesis is to elaborate methods to solve the four tasks. Particular attention is laid on the fact that communication between the agents is only invoked if it is absolutely necessary. The collision avoidance should be guaranteed even in the presence of

[1]The term is motivated by the International Regulations for Preventing Collisions at Sea 1972 (COLREGs) for maritime navigation [95].

transmission delays or packet losses induced by the communication network. In addition, the method should be able to control different types of agents (e.g. UAVs, cars) with only small modifications.

1.3 Motivation and application fields

1.3.1 *Motivation for the event-based control of mobile agents*

Practical motivation. Autonomous moving agents especially UAVs are used in many different applications due to their ease of deployment, low maintenance cost, high mobility and ability to hover. Such vehicles are utilised, for example, for providing wireless coverage, real-time monitoring of road traffic, remote sensing or search and rescue operations. Often an UAV is deployed to perform various tasks and some tasks require the cooperation of several UAVs in a formation. Hence, it is appropriate to develop a method with which both, a single agent or a group of agents is able to execute different types of tasks. Furthermore, the increase of UAVs in air traffic raises the risk of collisions. As no pilot is on board, the method to be developed has to ensure the collision-free movement of the agents while fulfilling their tasks.

In order to ensure the collision-free movement and to fulfil the individual tasks of the agents, the positions of nearby agents must be known continuously. In common practice, these positions are measured by using on-board sensors or cameras, which causes the following difficulties:

- Increase of the energy consumption of the flying agents and thus reduction of the flight time.
- Position and distance measurement is limited to a certain area around the agents so that approaching agents might be identified too late for collision avoidance.

For these reasons, it is appropriate to connect the UAVs with each other over a communication network that is used to transmit the required information between the agents. As a continuous data exchange increases the energy consumption by transmitters and receivers on the agents and raises the network load, it is reasonable to develop the method in a way that information is only transmitted at certain time instants when new information is necessary. To this aim, communication should only be invoked if appropriate signals violate a threshold.

Theoretical motivation. Achieving the control aims by using only locally measured data and communicated information is a challenging task since the required information may be outdated or missing. The control aims of the networked system have to be split appropriately into local control aims of the individual agents.

For the stabilisation of a system by a control loop that is closed using a communication network the classical event-based control [36, 88] can be used. It aims at reducing the communication effort within a control loop by closing the feedback only at event time instants if the control error violates a threshold. The idea of the event-based control is transferred in this thesis to a higher abstraction level for the trajectory planning of the agents. This approach is challenging due to the following four aspects:

- As communication is invoked aperiodically, the fundamental assumption of discrete-time systems and sampled-data control that the communication time instants or the sampling time instants are equidistant, is violated. Hence, for the event-based communication scheme new models and methods for the description and the analysis of the networked system have to be found.

- As the event-based control method is not used to close a control loop, but it is applied for a trajectory planning, the theory of the event-based feedback control cannot be used. New event-based methods for the trajectory planning and the monitoring of the control aims need to be developed. In contrast to the event-based feedback control where a single event type is sufficient as a clock for the invocation of communication, now five different types of events need to be utilised for communication, trajectory planning and the estimation of the properties of the unreliable communication network.

- For the event-based feedback control events are generated based on current information. In order to plan smooth trajectories for the agents the event-based method to be developed needs to be able to generate events at future time instants. To this aim it uses communicated information that contain the future trajectories of the agents.

- The estimated transmission delay of the network is a statistical quantity. Based on these statistics, deterministic actions have to be executed by the agents to guarantee the control aims. A new method needs to be designed so that deterministic control actions can be generated based on uncertain information.

1.3.2 *Application scenarios*

The event-based control of mobile agents is a concept that can be applied to different types of agents (e.g. UAVs or cars) that have to fulfil tasks individually or in a group by following locally planned trajectories. This section introduces two scenarios in which the method is applied to two UAVs to evaluate the method in simulations and experiments.

Communication relay over aerial base stations. The first scenario considers the communication relay over aerial base stations for the 5G communication [52, 53, 178]. Here, data packets are forwarded over UAVs to ensure requirements on the quality of the communication link. The required channel quality highly depends on the distance $s(t)$ between the agents, which leads to two objectives that need to be fulfilled:

1. Keep a safety distance $\underline{s}$ to ensure a collision-free movement of the agents.
2. Keep a maximum distance $\bar{s}$ between the agents in order to fulfil the requirements on the channel quality.

The thesis focuses on the satisfaction of the requirement $\underline{s} \leq s(t) \leq \bar{s}$ that results from the two objectives. To this aim appropriate trajectories are planned and the agents are controlled along these trajectories. The communication relay is not further considered. The scenario to be considered is illustrated in Fig. 1.3.

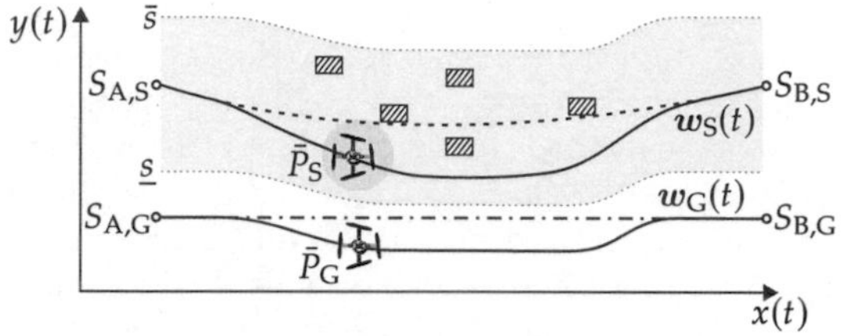

Fig. 1.3: Illustration of the first scenario.

Both agents move on trajectories from their start points $S_{A,S}$, $S_{A,G}$ to their end points $S_{B,S}$, $S_{B,G}$ on which they should satisfy the requirement $\underline{s} \leq s(t) \leq \bar{s}$ on the current distance $s(t)$ for collision avoidance and the communication relay. One of the agents faces obstacles on its trajectory and changes its trajectory from the dotted line to the solid line to avoid a collision. In this situation, the UAVs act with the following functional associations. The first agent is the stand-on agent $\bar{P}_S$, because it stays on its new trajectory to avoid the obstacles (Fig. 1.3). As a reaction, the second agent is the give-way agent $\bar{P}_G$, because it has to change its locally planned trajectory so as to keep the distance between the UAVs inside the required interval $s(t) \in [\underline{s}, \bar{s}]$.

The problem is formally defined in the next section, where the solution steps are also briefly described.

Formation control of UAVs. In the second scenario a formation of two UAVs should be kept within a tolerance range. A formation is a geometric arrangement of agents that maintain a fixed distance to each other. The formation of UAVs is often utilised for the surveillance of areas [40, 76, 164]. The scenario is illustrated in Fig. 1.4 where the agents move on circular trajectories to enable the experimental evaluation of the scenario within the limited space at the test bed 'MULAN' at the Ruhr University Bochum. Moreover, the generation of circular formations is of relevance for various use cases as in [82, 94, 120, 179]. In this thesis the agents should build a formation that satisfies the following objectives:

1. The agents should be located opposite to one another within a tolerance range defined by γ.
2. The agents should move in different heights z_l and z_h.

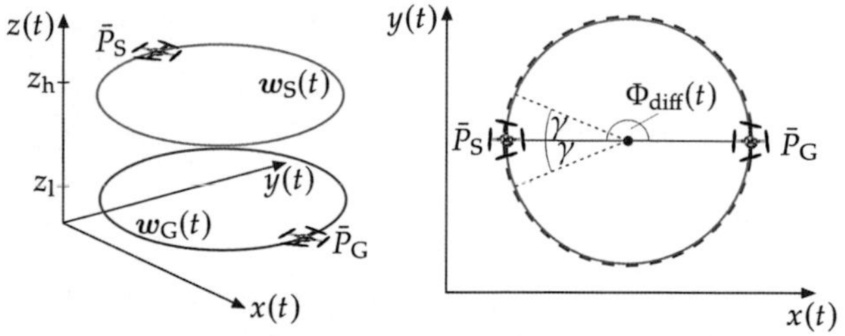

Fig. 1.4: Illustration of the second scenario.

Remark. The agents are located opposite to one another when the angle $\Phi_\mathrm{diff}(t)$ between them satisfies $\Phi_\mathrm{diff}(t) = 180°$. The objectives are similar to the objectives in the first scenario. By considering the distance between the agents, the objectives can also be defined by the distances $\underline{s}$ and $\bar{s}$ of the first scenario. The utilisation of the angle is more convenient to specify the control aims.

As in the first scenario one agent acting as the stand-on agent $\bar{P}_\mathrm{S}$ can change its trajectory $w_\mathrm{S}(t)$ at any time by changing the speed $v_\mathrm{S}(t)$ on the circular path or by changing the height $z_\mathrm{S}(t)$ of its movement. Then, the other agent acting as the give-way agent $\bar{P}_\mathrm{G}$ has to adapt its trajectory $w_\mathrm{G}(t)$ accordingly to keep the objectives satisfied. The problem is formally given in the next section.

Further application scenarios. There are many other use cases for which the method of this thesis can be applied:

- Coordinated search and rescue missions [141, 142, 144].
- Maintenance works on offshore wind farms [55, 77].
- Cooperative movement of heavy loads [96, 162].
- Parcel delivering via drones or autonomous vehicles [91, 155].
- Formations of agents, even formations of satellites [136, 143].
- Merging of vehicle platoons [34, 101].

All application fields have in common that the control aims have to be fulfilled by agents individually or in a group. Furthermore, the control aims can be fulfilled by planning suitable trajectories for the agents, where the agents have to fulfil conditions $\underline{s}$ and $\bar{s}$ on the distance between them.

This means the method enables one a broad range of applications for unmanned mobile agents by planning appropriate trajectories and maintaining application-specific distances $\underline{s}$ and $\bar{s}$.

1.4 Problem formulation and fundamental questions

1.4.1 *Control problem*

The problem is defined for the following two movement scenarios of the agents. First, the agents are allowed to move freely in the 3D space, called *general movement* of the agents. Second, the agents are limited to circular trajectories in two different heights, named *circular movement* of the agents.

Problem statement for agents with a general movement. A networked control system that consists of two agents, which move on trajectories in the 3D space and should satisfy individual tasks as well as cooperative tasks has to fulfil the following requirements:

(A1) **Trajectory planning and monitoring:** The trajectories $\boldsymbol{w}_l(t)$ of the agents $\bar{P}_l$, $(l \in \{S, G\})$ should bring them from their start points $S_{A,l}$ to their end points $S_{B,l}$. The separation $s(t) = ||\boldsymbol{p}_G(t) - \boldsymbol{p}_S(t)||$, given by the euclidean vector norm defined in (2.1) on p. 29 should satisfy

$$\underline{s} \leq s(t), \quad \forall t \tag{1.1}$$

$$s(t) \leq \bar{s} \tag{1.2}$$

where $\underline{s}$ is a safety distance to avoid collisions between the agents and $\bar{s}$ is a maximum separation to achieve specific control aims (e.g. communication relay).

(A2) **Trajectory following:** The agents $\bar{P}_l$, $(l \in \{S, G\})$ should follow given trajectories $\boldsymbol{w}_l(t)$ exactly if there are no external disturbances:

$$\boldsymbol{p}_l(t) = \boldsymbol{w}_l(t), \quad l \in \{S, G\}, \quad \forall t.$$

(A3) **Disturbance compensation:** External disturbances (e.g. wind) should be compensated by the agents to ensure a smooth flight behaviour.

The control method is supposed to respect the following constraints:

(C1) **Network topology:** There is no coordinator or forwarding unit in the network but all control and estimation algorithms have to work with local information.

(C2) **Communication:** The communication link between the agents should only be invoked if the uncertainty of the local information exceeds a threshold.

(C3) **Computation:** The computation power onboard the agents is limited and causes bounded computational delays.

Problem statement for agents with a circular movement. If the agents move on circular trajectories, the control aims and the constraints are the same as above, but due to the restriction of the direction of the movement, specific methods can be developed. As the circular trajectories are represented in the cylindrical coordinate system, it is more convenient to specify the control aims based on the angle difference $\Phi_{\mathrm{diff}}(t)$ between the agents as well as on the height difference $z_{\mathrm{diff}}(t)$. Therefore, for a circular movement of the agents, requirements (1.1) and (1.2) are replaced by:

$$\Phi_{\mathrm{diff}}(t) \in \left[\underline{s}, \bar{s}\right], \quad \forall t \tag{1.3}$$

with

$$\underline{s} = 180^\circ - \gamma \tag{1.4}$$

$$\bar{s} = 180^\circ + \gamma \tag{1.5}$$

and

$$z_{\mathrm{S}}(t), z_{\mathrm{G}}(t) \in [z_{\mathrm{l}}, z_{\mathrm{h}}], \tag{1.6}$$

which causes the height difference

$$z_{\mathrm{diff}}(t) = |z_{\mathrm{S}}(t) - z_{\mathrm{G}}(t)| = \bar{z} \quad \text{with} \quad \bar{z} = |z_{\mathrm{h}} - z_{\mathrm{l}}|. \tag{1.7}$$

Relation (1.3) claims the agents to be located opposite to one another within a tolerance range specified by γ. Equation (1.7) demands that the agents keep a height difference to one another specified by the two allowed heights z_{l} and z_{h} given in (1.6). This requirement should hold whenever the agents move on a constant height, but not during height changes of the agents.

The thesis focusses on the main challenge to develop a method for the planning of trajectories for the agents so as to satisfy control aim (A1). In Chapter 10 a flatness-based two-degrees-of-freedom controller is derived with which control aims (A2) and (A3) are satisfied.

1.4.2 Decomposition of the control problem and steps of solution

Allocation of the control aims of the networked control system into local aims of the agents. In order to develop a method that is executed locally by the agents and that leads to the fulfilment of the control aims (A1) – (A3) by respecting the constraints (C1) – (C3), the control aims of the networked control system have to be subdivided into local control aims of an individual agent. Satisfaction of the local control aims leads to the fulfilment of the control aims of the overall system.

In order to satisfy control aim (A1) the agents have to fulfil their individual control aims, which are named (S1) for the stand-on agent and (G1) for the give-way agent. As the two agents are assigned with different functions, their local aims differ:

(S1) **Trajectory planning and monitoring:** The trajectory $w_S(t)$ of the stand-on agent should bring it from its start point $S_{A,S}$ to its end point $S_{B,S}$. It should equal a desired trajectory $w_{S,d}(t)$ to fulfil its individual task and to avoid collisions with obstacles:

$$w_S(t) = w_{S,d}(t), \quad \forall t.$$

(G1) **Trajectory planning and monitoring:** The trajectory $w_G(t)$ of the give-way agent should bring it from its start point $S_{A,G}$ to its end point $S_{B,G}$. It should be planned to fulfil the following requirements:

- The trajectory $w_G(t)$ should coincide with a desired trajectory $w_{G,d}(t)$ to fulfil its individual task and to avoid collisions with obstacles

$$w_G(t) = w_{G,d}(t), \quad \forall t.$$

- The separation $s(t)$ fulfilling (1.1) and (1.2) should be guaranteed.

If the trajectory is changed at a time instant t_c to fulfil (1.1) or (1.2), the agent should return to the trajectory $w_{G,d}(t)$ as soon as it is possible at a time instant

$$t_{c+1} = \arg\min_{t>t_c} \left\{ \underline{s} \le s(t) \wedge s(t) \le \bar{s} \right\}.$$

The control aims (A2) and (A3) correspond unchanged to the local aims of the agents. The control aims are renamed to (S2), (G2) and (S3), (G3) for consistency. Likewise, the individual agents are subject to the constraints (C1) – (C3).

Remark. The control aim (G1) can be partly contradictory in the sense that the give-way agent might not be able to follow the desired trajectory $w_{G,d}(t)$ and to comply with the conditions (1.1) and (1.2) at the same time. As the conditions (1.1) and (1.2) are of superior priority, the give-way agent has to neglect following the trajectory $w_{G,d}(t)$.

Analysis of the local control aims and fundamental questions. An analysis of the local control aims leads to the fact that the stand-on agent can change its trajectory at any time. As communication is invoked only at event time instants using an unreliable network that induces transmission delays and packet losses the following questions arise that need to be addressed by the event-based method:

- How can the time delay of the information transmission be estimated?
- At which time instant communication needs to be invoked with respect to time delays and packet losses caused by the unreliable communication network?
- Which information does the individual agent need to plan appropriate trajectories locally?
- How can deterministic control actions be generated based on uncertain information?
- How does the trajectory need to be planned to satisfy different control aims?

The solution to these issues enables the agents to fulfil the control aims.

Main problems. Considering the analysis of the local control aims the problems are identified, which need to be solved by the agents to fulfil their control aims. In the thesis it is focussed on the problems that need to be solved by the agent acting as the give-way agent, because it is responsible to fulfil the requirements (1.1) and (1.2) of control aim (A1) or the requirements (1.3) and (1.6) of control aim (A1) for a circular movement. The stand-on agent just has to follow its planned trajectory. Four problems are determined, which have to be solved by the give-way agent:

- An estimation method needs to be developed to derive an estimate of the current QoS properties of the communication channel, because the channel quality varies with the relative movement of the agents to one another.
- A method has to be found to estimate the future positions of the stand-on agent, because the communicated trajectory becomes uncertain over time.
- A method needs to be derived to determine the uncertainty of the local data in order to decide when it is necessary to invoke communication and when to change the trajectory. To this aim appropriate event thresholds have to be found with respect to the uncertain QoS properties of the network and the dynamics of the agents.
- A trajectory planning method has to be developed to get trajectories that respect the dynamics of the agents and fulfil the control aims.

The problems are given in detail below. The method uses two notations $(t_k, \tilde{t}_i)$ to state two different time instants at which events are generated by the event-based control units and a new estimate of the properties of the communication channel is derived by the delay estimators. The notation is stated in detail in Chapter 8. Furthermore, the distance between the agents is denoted by $s(t)$ for the control aims and by $d(t)$ in the context of estimating the properties of the network.

Problem 1.1. (Time delay estimation)

Given:	• Parametrised Markov model describing the channel properties. • Maximum distance $\max \mathrm{dist}(w_{\mathrm{G}}(t), \mathcal{P}_{\mathrm{S}}(t, t_{\mathrm{r},k}, \tau_k))$ between the trajectory of the give-way agent and the possible positions of the stand-on agent defined in Problem 1.2 in a time interval $t \in [\tilde{t}_i, \tilde{t}_{i+1}]$.
Find:	Statistical estimate $\tilde{\tau}_{\max}(d(\tilde{t}_i))$ of the mean time delay $\tau_k = \tau_{\mathrm{c}} + \tau_{\mathrm{n}}$ in a time interval $[\tilde{t}_i, \tilde{t}_{i+1}]$ consisting of the computation time τ_{c} of the agents and the transmission delay τ_{n} of the network.
Boundary conditions:	Channel statistics are constant if the agents remain in a distance of at most $\bar{d}$ to one another (Assumption 8.1 on p. 175).

Problem 1.2. (Position estimation)

Given:	• Communicated position $p_{\mathrm{S}}(t_{\mathrm{c},k})$ and speed $v_{\mathrm{S}}(t_{\mathrm{c},k})$ of the stand-on agent at the communication time instant $t_{\mathrm{c},k}$. • Communicated trajectory $w_{\mathrm{S}}(t)$, $(t_{\mathrm{c},k} \leq t \leq t_{\mathrm{end}})$ of the stand-on agent. • Estimated time delay $\tilde{\tau}_{\max}(d(\tilde{t}_i))$. • Mean time delay τ_k. • Time instant $t_{\mathrm{r},k}$ of reception of information.
Find:	A set $\mathcal{P}_{\mathrm{S}}(t, t_{\mathrm{r},k}, \tau_k)$ that includes the future positions of the stand-on agent such that $$p_{\mathrm{S}}(t) \in \mathcal{P}_{\mathrm{S}}(t, t_{\mathrm{r},k}, \tau_k), \quad t \geq t_k.$$
Boundary conditions:	• System and actuator limitations given by (1.8), (1.9) on p. 17. • Maximum speed (1.10) or (1.11) on p. 17.

Problem 1.3. (Uncertainty estimation of the local data and invocation of communication)

Given:
- Trajectory $w_{\mathrm{G}}(t)$, $(t_{\mathrm{c},k} \leq t \leq t_{\mathrm{end}})$ of the give-way agent.
- Inclusion $\mathcal{P}_{\mathrm{S}}(t, t_{\mathrm{r},k}, \tau_k)$ of positions of the stand-on agent.
- Estimated time delay $\tilde{\tau}_{\max}(d(\tilde{t}_i))$.

Find: Event threshold $\bar{e}_{\mathrm{G}}$ to determine communication time instants $t_{\mathrm{c},k}$ and event time instants t_k. Communication must be invoked, so that the required data can be received at t_k, $(t_k > t_{\mathrm{c},k})$ under consideration of $\tilde{\tau}_{\max}(d(\tilde{t}_i))$.

Boundary conditions: Adherence of a minimum time span between two consecutive events.

Problem 1.4. (Trajectory planning)

Given:
- Communicated trajectory $w_{\mathrm{S}}(t)$, $(t_{\mathrm{c},k} \leq t \leq t_{\mathrm{end}})$ of the stand-on agent.
- Trajectory $w_{\mathrm{G}}(t)$, $(t_{\mathrm{c},k} \leq t \leq t_{\mathrm{end}})$ of the give-way agent.

Find: Trajectory for the give-way agent so that control aim (G1) remains fulfilled.

Boundary conditions:
- System and actuator limitations given by (1.8), (1.9) on p. 17.
- Maximum speed (1.10) or (1.11) on p. 17.

Remark. In the case the agents are connected by an ideal communication network that does not induce any transmission delays or packet losses, communicated information is always received by the agents instantaneously $(t_{\mathrm{c},k} = t_k)$ and the give-way agent does not need to solve Problem 1.1.

Structure of the event-based control unit A. The four problems are solved by the agents using a delay estimator D and an event-based control unit A. In Fig. 1.5 the basic structure of the units A_{G} and D_{G} of the give-way agent is shown as the give-way agent is responsible to satisfy control aim (A1). The structure of the control units A_{S} and D_{S} of the stand-on agent is identical, but they execute different tasks. Solid arrows depict continuous signal transmissions while the dashed arrows represent event-based signal transfers.

The delay estimator D_{G} and the control unit A_{G}, which consists of three parts, execute the following tasks to solve the Problems 1.1 – 1.4. The estimator D_{G} generates the statistical estimate $\tilde{\tau}_{\max}(d(\tilde{t}_i))$ of the time delay using a two-state Markov model based

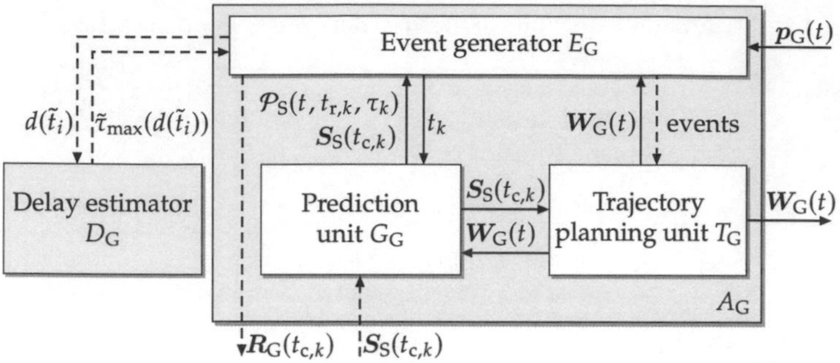

Fig. 1.5: Structure of the event-based control unit A_G with the delay estimator D_G of the agent $\bar{P}_G$.

on the maximum distance $d(\tilde{t}_i)$ between the agents (Problem 1.1). The *prediction unit* G_G receives the communicated data $S_S(t_{c,k})$ introduced on p. 129 of the stand-on agent and generates an inclusion $\mathcal{P}_S(t, t_{r,k}, \tau_k)$ of the position of the stand-on agent between two consecutive event time instants t_k, t_{k+1} (Problem 1.2). The *event generator* E_G supervises the fulfilment of the control aims. It uses the estimated time delay $\tilde{\tau}_{max}(d(\tilde{t}_i))$, the current position $p_G(t)$ of the give-way agent and the matrix $W_G(t)$, which contains its trajectory to invoke communication by generating an event and by sending a request $R_G(t_{c,k})$ introduced on p. 195 (Problem 1.3). It triggers the time delay estimation whenever the current estimate becomes too uncertain. Furthermore, it decides when it is necessary to change the trajectory of the give-way agent by generating an event. The *trajectory planning unit* T_G changes the trajectory appropriately in order to keep control aim (G1) fulfilled (Problem 1.4).

1.4.3 *Main assumptions*

This paragraph states the assumptions that apply throughout the thesis. It should be noted that Assumptions 1.2 and 1.3 concern the possible movements of the UAVs, where Assumption 1.2 relates to the general movement of the agents and Assumption 1.3 applies to the circular movement. In this thesis the method is applied to the control of quadrotors. Hence, these assumptions are made with respect to the dynamics of the quadrotors.

Assumptions regarding the properties of the communication network are made in Chapters 6 and 8.

Assumption 1.1. The agents are only able to measure their local data as position, speed and acceleration.

Assumption 1.2. The dynamics of the quadrotors are restricted by

- state limitations represented by the angles $\phi(t)$ and $\vartheta(t)$ around the axes of the quadrotor

$$\begin{aligned} \underline{\phi} \leq \phi(t) \leq \bar{\phi} \\ \underline{\vartheta} \leq \vartheta(t) \leq \bar{\vartheta}. \end{aligned} \tag{1.8}$$

- actuator limitations of the rotor speeds n_i

$$\underline{n} \leq n_i \leq \bar{n}, \quad i = 1, \ldots, 4. \tag{1.9}$$

The speeds $v_G(t)$ and $v_S(t)$ of the give-way agent and the stand-on agent are bounded:

$$||v_G(t)|| = \left\| \frac{\mathrm{d}\, w_G(t)}{\mathrm{d}\, t} \right\| \leq v_{G,\max} \quad \text{and} \quad ||v_S(t)|| \leq v_{S,\max} = v_{\max}. \tag{1.10}$$

In case the give-way agent has to plan a reactive trajectory introduced in Section 7.4, it is able to increase its speed so that

$$v_{S,\max} < v_{G,\max} = 1.5\, v_{\max}$$

holds.

Assumption 1.3. The quadrotors move on the circular path constantly in one direction of rotation (mathematically positive or negative direction) and the circular paths are only planned in the two different heights z_l and z_h. The speeds $v_S(t)$ and $v_G(t)$ of $\bar{P}_S$ and $\bar{P}_G$ are bounded:

$$\begin{aligned} v_{S,\min} \leq ||v_S(t)|| \leq v_{S,\max} \\ v_{G,\min} \leq ||v_G(t)|| \leq v_{G,\max} \end{aligned} \tag{1.11}$$

with $v_{G,\max} > v_{S,\max}$ and $v_{G,\min} < v_{S,\min}$. The difference of the maximum speeds and the difference of the minimum speeds is identical:

$$|v_{S,\min} - v_{G,\min}| = |v_{S,\max} - v_{G,\max}|.$$

Assumption 1.4. The trajectories of the quadrotors are planned using piecewise Bézier curves introduced in the next chapter. According to Assumption 1.2 the conditions on the start points and the end points of the trajectories should satisfy:

$$\begin{aligned} &w(\bar{t}_i) = 0, \quad ||\dot{w}(\bar{t}_i)|| \leq v_{\max}, \quad ||\dot{w}(\bar{t}_{i+1})|| \leq v_{\max}, \\ &\ddot{w}(\bar{t}_i) = \ddot{w}(\bar{t}_{i+1}) = w^{(3)}(\bar{t}_i) = w^{(3)}(\bar{t}_{i+1}) = w^{(4)}(\bar{t}_i) = w^{(4)}(\bar{t}_{i+1}) = 0, \end{aligned} \tag{1.12}$$

which can be easily achieved by a coordinate shift.

1.5 Literature review

The proposed method combines approaches from the fields of communication technology and control theory. Communication technology provides models of wireless channels for different application fields that depend on the environment the agents are moving in (e.g. urban or rural environment). With these models it is possible to describe the quality-of-service (QoS) parameters of the channels. Control theory contributes methods concerning the position estimation of moving agents, the event-based control of feedback systems, the event-based communication and the trajectory planning for moving agents for various application cases.

This literature review gives an overview over the relevant methods from both fields that are utilised to develop the event-based control method for mobile agents in this thesis.

Modelling of wireless communication channels. Networked control systems have become increasingly popular, because they can be applied in many areas (e.g. connected vehicles, networked UAV operations). The classical control theory has been focussed on the control of these systems under the assumption that the communication network has ideal properties in the sense that it does not induce transmission delays or packet losses. However, this assumption is not valid for systems that use a wireless network with inherent imperfect channel properties or a limited bandwidth as in this thesis. Hence, the channel properties need to be considered for the control with an appropriate model. There are many approaches to model wireless channels for different applications in the literature [103, 163].

Channel models for UAV applications as in this thesis can be subdivided by the propagation scenario into air-to-air channels and air-to-ground channels. The air-to-air propagation is an important aspect for the communication between UAVs. The characteristics of these channels depend mainly on the environmental conditions, UAV flight direction, the existence of a line-of-sight (LOS) path between the UAVs and ground reflections [32, 81, 97, 170]. The models focus either on the large-scale statistics such as path loss and shadowing [49, 151, 174] or on the small-scale statistics that are modelled with various models such as Rayleigh, Rice or Nakagami [31, 102, 150, 182]. These fading amplitude statistics are important to determine the random behaviour of fading channels using the first order statistics given by a cumulative distribution function (CDF) and a probability density function (PDF). The air-to-ground propagation models the communication between an UAV and ground stations and is not further considered in this thesis.

Furthermore, analytical channel models are utilised to model the propagation behaviour of a channel under certain assumptions and parameters. The models are derived with three approaches: deterministic, stochastic and geometry-based [67, 123, 131].

An approach to derive the QoS parameters of a channel, which can be utilised for the control algorithms, is the use of the effective capacity theory. Here, a link layer model is stated in which it is possible to derive the required data rate of a channel to satisfy given QoS parameters of a physical channel. This method was extended to different source rates and a packetised communication in [168, 169].

In this thesis the Rice fading model is utilised to represent the channel statistics. It is used by the delay estimator to generate an estimate of the current transmission delay of the channel.

Position estimation approaches. The position estimation of moving agents especially of UAVs is a field of growing interest. The motivation is the fact that complex tasks can be solved by the cooperation of several agents. However, this requires that the individual agent is aware of the relative positions and orientations of neighbouring agents to determine the distances to these agents. To this aim, the agent should be localised in the same coordinate frame. In this thesis the future positions of the stand-on agent are estimated, because its future trajectory is uncertain. The positions of the stand-on agent need to be known to ensure the collision avoidance.

Many approaches for a position estimation use image-based algorithms, which process the data from onboard cameras or incorporate measurements of GPS sensors [35, 59, 104–106, 128, 171]. In addition, commercial drones utilise real-time kinematics [90, 159]. This is a method of precisely determining position coordinates using satellite navigation techniques. Different approaches for the position estimation have been proposed in [48, 71, 118], where a Kalman filter is applied to estimate the position of agents by analysing the received signal strength from the onboard communication module, the angle of arrival or the homography of a series of images. In [160] the possible positions of ground moving agents have been included in a circular set that expands over time.

A different field of application is to estimate the future positions of agents in a disturbed environment. For collision avoidance it is necessary that an agent is aware of the set of possible future positions of nearby agents in order to be able to plan guaranteed collision-free trajectories. An approach is to analyse the past movement to get a reliable estimate of the future position and path of an agent [38]. Different approaches make use of a known trajectory of an agent and estimate the future positions to be in a tube around this trajectory when the agent is affected by external disturbances [175, 181].

In this thesis the idea from [160] is used. The approach is extended to estimate the position of an agent in the 3D space and the conservatism of the estimate is reduced compared to [160]. Furthermore, with the method in this thesis the future positions of an agent can be estimated.

Event-based control and communication. Networked control systems are control systems, where the components (sensors, actuators and controllers) are connected by a communication network. As the network might be used by several components of a system at the same time it is reasonable to reduce the amount of communicated data as far as this is possible without threatening the control aims. The control approach in this thesis utilises a communication scheme where information between the agents is only exchanged if it is necessary.

At an early stage periodic control schemes (also called time triggered control schemes) are developed due to the ease of design and analysis [62]. Here, information is communicated with a fixed sampling period, which is often chosen to be small in order to guarantee the desired system performance. In order to mitigate the unnecessary waste of network capabilities the event-based control scheme (or event-triggered control scheme) was introduced in [36, 37, 88]. By using this approach control tasks are executed after an event is generated when an appropriately chosen threshold is violated instead of elapsing a time span [87]. The reduction of resource utilisation motivates the wide range of applications of the event-based communication in state-feedback control [86, 133, 158], output-feedback control [63, 64] and formation control and consensus of multi-agent systems [70, 148, 166].

The event-based control approaches can be classified into three categories: event-triggered sampling schemes [44, 63, 64, 74, 117, 156], self-triggered sampling schemes [61, 87, 124] and discrete event-triggered communication schemes [58, 69, 78]. In the event-triggered sampling the state of a system is constantly supervised. An event is generated and a new control input is determined whenever a predefined threshold is violated. The threshold is state-dependent and can be given by a fixed value or by a function. In multi-agent systems where global information may be unavailable for each local system decentralised event-triggered control schemes are proposed that work with locally measured data and communicated information [63, 156]. Using the self-triggered sampling scheme the next event time instant is determined by only evaluating previously received data and utilising knowledge about the dynamics of the system. The state of the system to be controlled is not constantly measured. Due to the over-approximation of the state of the system, often more events are generated compared to the event-triggered control scheme. Both control schemes should guarantee a positive minimum inter-event time span (MIET). Otherwise Zeno behaviour occurs that is generating an infinite number of events in a finite period of time, which renders the control approaches to be infeasible [44]. For several event-triggered control schemes no positive MIET can be guaranteed in the presence of arbitrarily small disturbances [44]. In these cases a discrete event-triggered communication scheme can be applied. Here, the state of the system is sampled periodically while the event condition is evaluated only at a sampling time instant, which excludes Zeno behaviour.

As there are many communication imperfections in networked control systems, such as transmission delays, packet losses, communication constraints or quantisation effects, the event-triggered control scheme is extended with many approaches to cope with these uncertainties. Often a bound on the transmission delay has been stated so that the event-based state-feedback control loop is stable despite using the delayed information or machine learning techniques have been applied to compensate the random delays [79, 111, 176]. In [80, 115] communication protocols have been proposed that use acknowledgement (ACK) messages to deal with delays and packet losses. The idea is to send an ACK message after the reception of a data packet. If the sending agent does not receive an ACK message from all other agents, it resends its data to those agents, which did not send the ACK message.

Most of the approaches in the literature have focussed only on some aspects of the communication imperfections while ignoring the others. In [89] a hybrid system framework has been stated that incorporates communication constraints, varying transmission intervals and varying delays to guarantee stability of a networked control system based on Lyapunov functions.

This thesis utilises the idea of the event-triggered sampling to invoke communication and a change of the trajectory of the agent. In addition the method to be developed is able to generate events at future time instants by an evaluation of the communicated trajectories.

Motion planning and path following. Motion planning is an indispensable part for autonomously moving agents. Motion planning is responsible for coordinating the overall system by interacting with human operators and making decisions in environments with obstacles. Hence, the motion planning is often referred to as the highest layer of an autonomous system. As the method is applied to quadrotors in this thesis, this literature review overviews motion planning approaches for UAVs. However, the basic principles can also be applied to other autonomous agents. The planning problem can be roughly divided into two steps. First, a discrete path must be found and second, a trajectory is generated, which is often planned using optimisation methods.

In order to find suitable paths, methods have been proposed ranging from sampling-based [107] to searching-based [113]. For applying these methods a configuration space has to be defined [108]. It is a set of possible transformations that can be performed by the agents. Using the configuration space, motion planning problems that vary in geometry and kinematics can be solved by the same planning algorithms.

The two most common sampling-based methods are the probabilistic road-map (PRM) [149] and the rapidly-exploring random tree (RRT) [109]. The PRM method picks random samples from the free space of the agent and uses a local planning unit to connect the new samples to its nearby configurations. It results a graph model, where the motion planning problem is solved using graph search algorithms. The RRT method consists of a tree which root is the start configuration. Each tree node is a configuration in the free space. The RRT grows by iteratively adding edges by random configurations if the connection between the random configuration and the nearest tree node is feasible. The random configuration becomes a node and an edge is added between it and the nearest tree node. Once the tree has extended to a configuration near the target configuration, it is terminated and a path is immediately generated. The RRT method efficiently finds a suitable path. However, it does not find the optimal path but converges to the first suitable solution. By extending the method, it is also possible to find an optimal solution [100].

Searching-based methods discretise the configuration space and convert the path finding problem to a graph searching problem. Hence, the widely used algorithms as the breadth-first search or the depth-first search can be applied [57]. While the breadth-first search provides an optimal solution if the graph is a uniform-weighted graph, with the depth-first search an optimal solution cannot be guaranteed [173]. Furthermore, the Dijkstra algorithm can be applied to derive the shortest path and the A^* algorithm is used to find a least-cost path [57].

As most path finding algorithms generate a path without time information, the path needs to be parametrised in time to obtain a trajectory. This is the main objective of the trajectory generation. The trajectory generation problem is formulated as minimising an objective function such as the control cost while fulfilling constraints on the dynamics of the agent. The methods can be categorised as hard-constraint [127, 138] and soft-constraint [146, 183]. The hard-constraint methods treat all constraints on the trajectory in the same way. In contrast, the soft-constraint methods penalise the constraints in the objective function so as to obtain better results especially in noisy environments.

In order to enable a quadrotor to follow the planned paths, several methods are developed. One of the most common control techniques for UAVs is the feedback linearisation. The aim of this control approach is to linearise the system in a certain operation point by applying a non-linear inversion of the system. The non-linearities are erased and the linear control theory can be applied [33, 47]. Another approach is based on the Lyapunov theory. Adherence of the Lyapunov stability condition leads to the convergence of the controller [56]. Furthermore, model predictive control (MPC) can be used for the path following. Here, the control problem is transformed into an optimisation problem [50, 68] and solved iteratively for a time horizon. In [130] a vector field has been utilised to make a quadrotor follow its path.

In this thesis the motion planning is used to determine collision-free trajectories with respect to static obstacles. In Chapter 10 a controller is provided that ensures the path following, which is necessary for guaranteeing the collision avoidance.

Trajectory planning for different applications. Trajectories for moving agents are planned to meet specific control aims. In this thesis trajectories are planned in order to achieve a collision-free movement and to provide wireless coverage for moving ground objects. This literature review gives a brief overview of approaches to satisfy the aforementioned two control aims.

The problem of collision avoidance has been addressed with several approaches that can be roughly subdivided into two categories: reactive methods and proactive methods. The reactive methods react on a detected collision and change the movement of an agent to avoid the collision. Some of the methods do only change the speed or the acceleration of the agent along its path that remains unchanged. Such methods use artificial potential fields [112, 121] or velocity obstacles [54, 126]. Another approach is to use a sampling-based check of the feasibility of the current trajectory and to change it when a collision is detected [114]. In terms of sensory requirements vision-based approaches achieve good results, where the line-of-sight angle and the time-to-collision between two agents is taken into account [122]. In contrast, the proactive methods detect a collision in advance and plan the trajectories so that they are collision-free. This requires the complete knowledge of all trajectories of nearby agents and obstacles or an appropriate bound on the uncertainty of the trajectory. Collision-free trajectories can be planned based on Bézier curves [125, 154] or by utilising the receding horizon based method of the MPC [39]. In [30] collision-free paths have been planned with respect to communication constraints.

The approach presented in thesis combines both the proactive and the reactive method. If a collision is detected in the future, the trajectory of the give-way agent is changed in advance for a collision avoidance. In contrast to the methods described above, the method is able to change the trajectory at runtime. If a collision is imminent the method reacts instantaneously with a rapid change of the trajectory to avoid the collision.

In recent years, there has been an increasing interest in using UAVs to provide wireless coverage for 5G communication or in emergency cases where the UAVs act as aerial relay stations [46]. An approach is to optimise the trajectory of an UAV that is capable to move with high speeds to maximise the end-to-end throughput [140, 177]. In [52, 93] an optimal position for an aerial base station has been derived to maximise the coverage range of a single UAV.

The method developed in this thesis does not aim at achieving the largest coverage but at providing moving ground objects communication with a specified channel quality.

Formation control of agents. In a second scenario the event-based control method is applied to keep a formation of agents. The problem of formation control of UAVs has been widely investigated. Different approaches in the literature include the leader-follower control [83, 85], the behaviour-based control [110] and the virtual structure method [180]. For the leader-follower control, the formation is obtained with a controller that depends on a single leader state. The formation with the behaviour-based approach is obtained by weighting the control inputs of several agents. For the virtual structure method the formation of agents is considered as a single agent in the virtual structure. This limits the field of application as only the movement of one agent can be controlled. The mentioned approaches can be categorised into position-based, displacement-based and distance-based control considering the sensing capabilities and the amount of information exchange between the agents. In [99] an event-based communication scheme has been proposed for the formation control of UAVs. In [53] a circular trajectory has been planned for an UAV acting as an aerial communication relay station. The method aims at reducing the energy for the flight manoeuvres and the communication relay.

In this thesis two quadrotors build a formation while moving on circular trajectories.

1.6 Contribution of this thesis

The approach presented in this thesis aims at avoiding collisions and keeping a maximum separation between two agents. It uses only local data and communicated information over an unreliable network, where communication is only invoked if it is necessary for the satisfaction of the control aims. The control method combines approaches from both, control theory and communication technology, to improve the control results. The main contributions of the thesis with respect to developing an event-based control method for the agents are as follows.

Trajectory planning. The agents follow locally planned trajectories to fulfil the control aims. As the agents should be able to fulfil different tasks, different trajectories need to be planned to fulfil them. In the literature often trajectory planning methods are stated that are able to fulfil one specified control task (e.g. guarantee collision avoidance) but they are not able to fulfil various tasks. Hence, this thesis develops a trajectory planning method that consists of two parts: First, in the planning task part the boundary conditions are specified for a trajectory to fulfil the control aims. Second, the planning algorithm part determines the trajectory with respect to the planning task and the dynamics of the agent. This approach enables the agents to fulfil different control tasks by only changing the trajectory planning task, while the planning algorithm remains unchanged.

The main result is a trajectory planning method that generates trajectories for the agents so that they are able to fulfil different control tasks.

Event-based communication. The communication between the agents is event-based in the sense that information is only transmitted when the uncertainty about the locally available data becomes too large. For the communication scheme the idea is utilised that data only needs to be sent when it is necessary as performed in the event-based feedback control. In the literature the event-based control is utilised to stabilise systems where the feedback loop is closed over a communication network. It aims at reducing the network load. For the control of the mobile agents a method is developed that realises the idea of the event-based control on a higher abstraction level to use it for a trajectory planning. Communication is invoked whenever a threshold is violated that represents the uncertainty about the local data. Furthermore, the method utilises thresholds to decide when the movements of the agents need to be changed so as to fulfil the control aims. The thresholds are determined with respect to the dynamics of the agents and consider the estimate of the current time delay of an information transfer.

The main result is an event-based method that invokes communication only when the uncertainty about the local data becomes too large and that invokes a change of the trajectories of the agents only when it is necessary. It takes the estimated time delay into account to generate appropriate control actions and generates events at future time instants in order to ensure the control aims.

Estimation of positions of an agent. For the determination when communication has to be invoked or the trajectory of the give-way agent needs to be changed, the distance between the agents has to be known. This means the position of the neighbouring agent must be known continuously.

In contrast to most approaches in the literature no sensors for a distance measurement are used in this thesis and communication takes place only at discrete time instants. Additionally the trajectory of the neighbouring agent is uncertain, which requires a prediction of the possible positions of the neighbouring agent between two event time instants. Hence, the thesis presents a method that generates an ellipsoidal inclusion of all possible future positions of a nearby agent between two event time instants. The inclusion is only based on the position and the speed of the agent at the last event time instant and is less conservative compared to the approach in [160] since it incorporates the dynamics of the agents.

The main result is an estimation method that provides an inclusion of future positions of an agent with respect to its dynamics.

Estimation of network properties. Communication is invoked with an event-based scheme only at discrete time instants when an agent needs new information. This requires that information is received immediately by the agent. However, using an unreliable communication network causes transmission delays and packet losses, so that information may be received late by an agent or it is even lost. Furthermore, the limited computational power onboard the agents results in computational delays when processing the data. Hence, these time delays must be taken into account when determining the event time instants.

In the literature of communication technology the approaches focus often only on the modelling of a wireless channel, while in the literature of control theory the results are often achieved by using fixed upper bounds on the time delay, which makes the results conservative. Therefore, this thesis uses a channel model from the literature of communication technology that represents the properties of the network. An event-based method is introduced that determines the transmission delay induced by the network using the channel model. Furthermore, it generates an estimate of the current time delay consisting of the transmission delay and the computational delay caused by the limited computational power of the agents. This variable estimate that depends on the movement of the agents is used for determining the next communication time instants. In order to obtain an accurate estimate of the time delay over time the method updates the channel model depending on the movement of the agents.

The main result is a method that estimates the current time delay of an information transfer. It is used for the invocation of communication in the presence of these delays.

1.7 Structure of this thesis

The thesis consists of five parts. Part I introduces the control problem and the mathematical basics for the event-based control method. Part II gives the basic methods needed for the event-based control of the agents. In Part III, these methods are applied for the control of agents that are connected over an ideal communication network. This part focusses on the description of the parts of the control units. In Part IV the methods are extended to ensure the control aims for agents communicating over an unreliable network. The focus of this part is to show the impact of the methods from communication technology on the control methods. Part V states the simulation results and the experimental results for an application of the method to two quadrotors.

Chapter 2 introduces the notation, which is used throughout the thesis. The mathematical basics of Bézier curves that are used for the trajectory planning are stated. Relevant algorithms are given that are utilised to determine the distance between two Bézier curves. Two demonstration examples are described, which are used as running examples in order to illustrate the parts of the control method.

Chapter 3 presents the trajectory planning method. Different methods are given to plan trajectories in the cartesian coordinate system and to plan circular trajectories in the cylindrical coordinate system. Furthermore, the planning of trajectories to change the speed on the circular path or the height of the circular movement are stated.

Chapter 4 proposes the prediction method for both, a general movement of the agents in the cartesian coordinate system and a circular movement in the cylindrical coordinate system. Furthermore, the basics of the event generation for the control of the agents are given.

Chapter 5 describes first the communication between the agents over an ideal network and second over an unreliable network. An overview over two common channel models from communication technology to represent the properties of a network is given. A Markov model of the wireless channel is presented, which is suitable for UAV applications.

Chapter 6 proposes the tasks of the event-based control units of the agents for an ideal communication network. The tasks of the control unit A_S of the stand-on agent and the tasks of the control unit A_G of the give-way agent are summarised.

Chapter 7 presents the parts of the control unit A_G of the give-way agent. The operating principles of the prediction unit G_G, the event generator E_G and the trajectory planning unit T_G are described. The event-based communication flow is given and a minimum inter event time span is derived. The method is summarised by an algorithm.

Chapter 8 describes the consequences on the event-based control method when information between the agents is transmitted over an unreliable network. The idea of handling the uncertainties arising from the network is motivated. Furthermore, the extended structure of the event-based control units A_S and A_G together with the delay estimators D_S and D_G is proposed.

Chapter 9 introduces the operating principle of the delay estimator and gives the extensions of the parts of the control units to cope with time delays and packet losses are given. The communication flow over an unreliable network is analysed and the method is summarised by an algorithm.

Chapter 10 introduces the quadrotor as the demonstration example. The quadrotor dynamics are stated and its control is derived. For the experiments the hardware of the physical quadrotor and the onboard software are introduced.

Chapter 11 presents the simulation results of three different scenarios. The analysis concentrates on three effects: time delays, packet losses and the possibility to fulfil the control aims even in the presence of packet losses.

Chapter 12 describes the experimental test bed and the results of the experiments with two quadrotors connected by an ideal communication network. In the scenario, the UAVs have to keep a formation while moving on circular trajectories. The analysis concentrates on the effects that are caused by the experimental environment. Furthermore, the event-based control method is applied to two robots in the same scenario to show that the control method is not limited to a single type of agents.

Chapter 13 summarises and concludes this thesis and states an outlook about possible future research directions.

Preliminaries

2

This chapter summarises the general notations used in this thesis in Section 2.1. A detailed list of symbols can be found in Appendix C. In Section 2.2 the notation of the position of an agent is introduced. The mathematical fundamentals for Bézier curves are stated in Section 2.3. In Section 2.4 the gradient descent optimisation method is presented. In Section 2.5 unmanned aerial vehicles (UAVs) and cars are presented, which serve as demonstration examples throughout this thesis

2.1 Notation

Throughout this thesis, scalars are represented by italic letters ($a \in \mathbb{R}$), vectors by bold italic letters ($\boldsymbol{x} \in \mathbb{R}^n$) and matrices by upper case bold italic letters ($\boldsymbol{A} \in \mathbb{R}^{n \times n}$). The entire set of real numbers is denoted by $\mathbb{R}$. $\boldsymbol{I}$ states the identity matrix of appropriate dimension. The i-th element of the j-th column of the matrix $\boldsymbol{A} \in \mathbb{R}^{n \times m}$ is given by $(\boldsymbol{A}_{ij})$ and the i-th entry of a vector $\boldsymbol{x}$ by $(x)_i$, respectively. $\mathbb{1} = \left(1 \dots 1\right)^{\mathrm{T}}$ is the one vector and $\boldsymbol{0} = \left(0 \dots 0\right)^{\mathrm{T}}$ the zero vector. The matrix $\boldsymbol{A} \in \mathbb{R}^{r \times c}$ with the entries a_{ij}, $i = 1, 2, \dots, r$ and $j = 1, 2, \dots, c$ is denoted by $\boldsymbol{A} = (a_{ij})$. The dimension of a vector $\boldsymbol{x}$ is denoted by $\dim(\boldsymbol{x})$ and $\mathrm{rank}(\boldsymbol{A})$ represents the rank of a matrix $\boldsymbol{A} \in \mathbb{R}^{n \times m}$. The transpose of a vector $\boldsymbol{x}$ or a matrix $\boldsymbol{A}$ is denoted by $\boldsymbol{x}^{\mathrm{T}}$ or $\boldsymbol{A}^{\mathrm{T}}$, respectively.

The i-th eigenvalue of $\boldsymbol{A} \in \mathbb{R}^{n \times n}$ is represented by $\lambda_i(\boldsymbol{A})$. The largest eigenvalue of $\boldsymbol{A} \in \mathbb{R}^{n \times n}$ is defined as

$$\lambda_{\mathrm{max}}(\boldsymbol{A}) = \max \left\{ \lambda_i(\boldsymbol{A}),\ i = 1, 2, \dots, n \right\}$$

while

$$\lambda_{\mathrm{min}}(\boldsymbol{A}) = \min \left\{ \lambda_i(\boldsymbol{A}),\ i = 1, 2, \dots, n \right\}$$

defines the smallest eigenvalue of $\boldsymbol{A} \in \mathbb{R}^{n \times n}$.

Sets are represented by calligraphic letters ($\mathcal{P}$), where their cardinality or size are denoted by $|\mathcal{P}|$. The absolute value of a scalar a is symbolised by $|a|$. $||\boldsymbol{x}||$ denotes the euclidean vector norm of $\boldsymbol{x} \in \mathbb{R}^n$ and $||\boldsymbol{A}||$ the compatible spectral norm of the matrix $\boldsymbol{A} \in \mathbb{R}^{n \times n}$, which are defined by

$$||\boldsymbol{x}|| = \sqrt{\sum_{i=1}^{N} (\boldsymbol{x})_i^2} \quad \text{and} \quad ||\boldsymbol{A}|| = \sqrt{\lambda_{\mathrm{max}}(\boldsymbol{A}^{\mathrm{T}}\boldsymbol{A})}, \tag{2.1}$$

respectively. Moreover, $||\boldsymbol{x}(t)||$ represents the euclidean vector norm at time instant t. $\text{diag}(\lambda_i)$ is a diagonal matrix with the diagonal entries $\lambda_1, \lambda_2, \ldots, \lambda_N$.

A continuous-time linear time-invariant system is given by

$$\begin{aligned}\dot{\boldsymbol{x}}(t) &= \boldsymbol{A}\boldsymbol{x}(t) + \boldsymbol{B}\boldsymbol{u}(t) + \boldsymbol{E}\boldsymbol{d}(t), \quad \boldsymbol{x}(0) = \boldsymbol{x}_0 \\ \boldsymbol{y}(t) &= \boldsymbol{C}\boldsymbol{x}(t)\end{aligned}$$

where $\boldsymbol{x} \in \mathbb{R}^n$ denotes the state of the system with the initial value $\boldsymbol{x}_0$, the input $\boldsymbol{u} \in \mathbb{R}^m$ and the measured output $\boldsymbol{y} \in \mathbb{R}^r$. $\boldsymbol{d} \in \mathbb{R}^l$ represents exogenous disturbances. $\boldsymbol{A} \in \mathbb{R}^{n \times n}$, $\boldsymbol{B} \in \mathbb{R}^{n \times m}$, $\boldsymbol{E} \in \mathbb{R}^{n \times l}$ and $\boldsymbol{C} \in \mathbb{R}^{r \times n}$ are real matrices.

2.2 Specification of the position of an agent

Position of an agent in the cartesian coordinate system. In order to specify the position of an agent for a general movement the cartesian coordinate system is used. The position is denoted by

$$\boldsymbol{p}(t) = \begin{pmatrix} x(t) & y(t) & z(t) \end{pmatrix}^{\mathrm{T}}. \tag{2.2}$$

Position of an agent in the cylindrical coordinate system. For a circular movement of an agent the cylindrical coordinate system is used. The position (2.2) is transformed into cylindrical coordinates by

$$\begin{aligned}x(t) &= r\,\cos(\omega t) \\ y(t) &= r\,\sin(\omega t) \\ z(t) &= z(t).\end{aligned}$$

Angle between two agents moving on circular paths. In order to ease the calculation of the angle $\measuredangle(\boldsymbol{p}_1(t), \boldsymbol{p}_2(t))$ enclosed between two agents with positions $\boldsymbol{p}_1(t)$ and $\boldsymbol{p}_2(t)$, the phase $\Phi(\boldsymbol{p}(t))$ of the position $\boldsymbol{p}(t)$ of an agent given by (2.2) that moves on a circular path in the xy-plane is specified. For a circle with the centre $\boldsymbol{p}_{\mathrm{m}} = (x_{\mathrm{m}} \quad y_{\mathrm{m}} \quad z_{\mathrm{m}})^{\mathrm{T}}$ the phase $\Phi(\boldsymbol{p}(t))$ is defined as

$$\Phi(\boldsymbol{p}(t)) := \begin{cases} \arctan\left(\frac{y'(t)}{x'(t)}\right) & for\ x'(t) > 0,\ y'(t) \in \mathbb{R} \\ \arctan\left(\frac{y'(t)}{x'(t)}\right) + 180^\circ & for\ x'(t) < 0,\ y'(t) > 0 \\ \arctan\left(\frac{y'(t)}{x'(t)}\right) - 180^\circ & for\ x'(t) < 0,\ y'(t) < 0 \\ +90^\circ & for\ x'(t) = 0,\ y'(t) > 0 \\ -90^\circ & for\ x'(t) = 0,\ y'(t) < 0 \\ +180^\circ & for\ x'(t) < 0,\ y'(t) = 0 \end{cases} \tag{2.3}$$

with the relative position

$$p'(t) = \begin{pmatrix} x'(t) \\ y'(t) \\ z'(t) \end{pmatrix} = \begin{pmatrix} x(t) - x_{\mathrm{m}} \\ y(t) - y_{\mathrm{m}} \\ z(t) - z_{\mathrm{m}} \end{pmatrix}$$

in relation to the centre of the circle. The phase of an agent corresponds to the angle of the cylindrical coordinates when the centre of the circle is at the origin of the coordinate system. Using the modulo operator defined as

$$\mathrm{mod}\,(a, m) = a - \left\lfloor \frac{a}{m} \right\rfloor \cdot m \tag{2.4}$$

the angle enclosed between two agents is determined as

$$\Phi_{\mathrm{diff}}(t) = \measuredangle(\boldsymbol{p}_1(t), \boldsymbol{p}_2(t)) = \mathrm{mod}\,(\Phi_1(t) - \Phi_2(t), 360°)$$

where $\Phi_1(t) = \Phi(\boldsymbol{p}_1(t))$ and $\Phi_2(t) = \Phi(\boldsymbol{p}_2(t))$ holds for simplicity and state the phases of the two agents, respectively.

2.3 Bézier curves

2.3.1 *Definition and properties*

Motivation. Numerically simple parametric curves are described using a parameter representation in the monomial base given in (2.15) on p. 36. However, in general the coefficients $\boldsymbol{a}_i$ of the curve do not provide much information about the course of the curve. Just $\boldsymbol{a}_0$ (start point of the curve) and $\boldsymbol{a}_1$ (tangential vector of the curve at $\boldsymbol{a}_0$) have concrete geometric interpretations. In contrast, the representation of the polynomials in the following Bernstein base allows to make predictions about the course of the curve on the basis of the coefficients. Hence, Bézier curves are used in this thesis for the trajectory planning of the agents.

Bernstein base on the unit interval $[0, 1]$. Bézier curves in the three-dimensional space of degree $m = n - 1$ are defined over the interval $u \in [0, 1]$ by their n control points $\boldsymbol{b}_k \in \mathbb{R}^3$ as [66]

$$\boldsymbol{v}(u) = \begin{pmatrix} v_{\mathrm{x}}(u) \\ v_{\mathrm{y}}(u) \\ v_{\mathrm{z}}(u) \end{pmatrix} = \sum_{k=0}^{m} \boldsymbol{b}_k \tilde{B}_k^m(u), \quad u \in [0, 1], \tag{2.5}$$

where $\tilde{B}_k^m(u)$, $(k = 0, 1, \ldots, m)$ are the Bernstein polynomials of order m given by

$$\tilde{B}_k^m(u) = \binom{m}{k} u^k (1-u)^{m-k}, \; u \in [0, 1], \; k = 0, 1, \ldots, m. \tag{2.6}$$

The Bernstein polynomials of order $m = 9$ are exemplarily shown in Fig. 2.1. The control points are given by the vector

$$\boldsymbol{b}_k = \begin{pmatrix} b_{k,x} \\ b_{k,y} \\ b_{k,z} \end{pmatrix}.$$

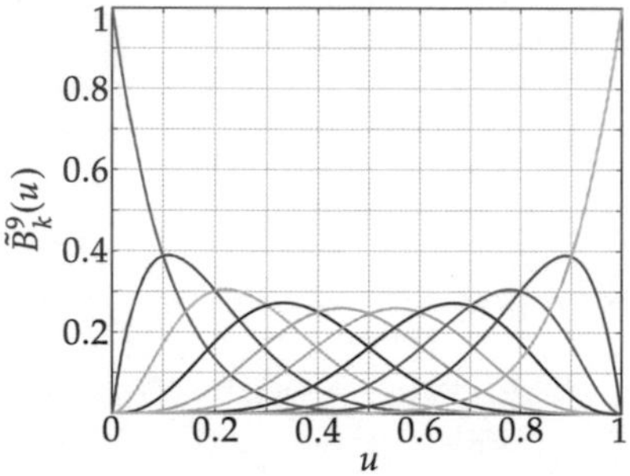

Fig. 2.1: Bernstein polynomials of order $m = 9$, $k = 0, 1, \ldots, 9$ over the interval $[0, 1]$.

The trajectories are obtained by a superposition of the Bernstein polynomials, which are weighted by the control points. Thus, the shape of the Bézier curve of order $m = 9$ is determined by the position of $n = 10$ control points. The polygonal chain defined by the control points is named control polygon. As

$$B_0^m(0) = B_m^m(1) = 1 \tag{2.7}$$

holds, the start point and the end point of the control polygon correspond to the start point and the end point of the Bézier curve, respectively. This property is called the *interpolation property*.

Properties of a Bézier curve. Bézier curves have the following important properties for path planning [66]:

- Any point on the curve is a weighted average of its control points following from the Binomial theorem:
$$\sum_{k=0}^{m} \tilde{B}_k^m(u) = 1, \quad u \in [0, 1].$$
- Following from the previous property the curve lies within the convex hull formed by all control points.
- $\boldsymbol{b}_0$ and $\boldsymbol{b}_m$ fix the start point and the end point of the curve, as shown in Fig. 2.2, for the x-direction. That is
$$\boldsymbol{v}(0) = \boldsymbol{b}_0, \quad \boldsymbol{v}(1) = \boldsymbol{b}_m.$$

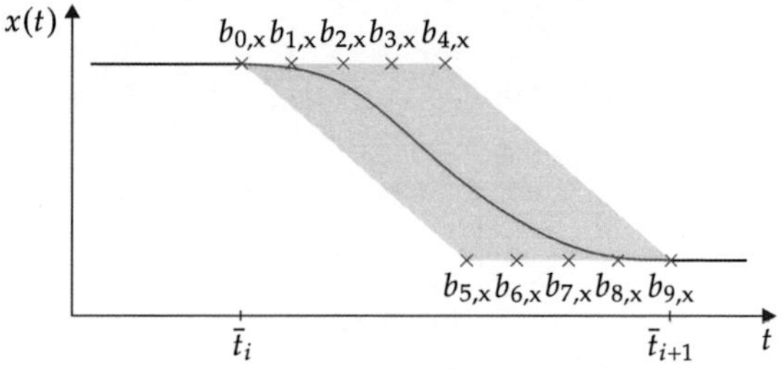

Fig. 2.2: Control points of a Bézier curve.

- The Bézier curve has the property of the affine invariance. As the curve is a convex combination, as it can be seen in eqn. (2.5), invariance holds for affine transformations. Hence, the curve can be scaled or shifted in space by scaling or shifting the control polygon. Here, the control polygon is the closed section of the control points.

- Due to its representation by eqn. (2.5) the curve is parametrised by n control points $\boldsymbol{b}_k$. As the Bernstein polynomials $\tilde{B}_k^m(u)$ are known to each agent and do not change, for communicating the curve, it is sufficient to send the control points $\boldsymbol{b}_k$ together with the parameter interval $[0, 1]$.

- The derivative of an m-th order Bézier curve results in an $(m-1)$–st order Bézier curve:

$$\frac{\mathrm{d}}{\mathrm{d}t}\boldsymbol{v}(u) = \frac{\mathrm{d}}{\mathrm{d}t}\left(\sum_{k=0}^{m} \boldsymbol{b}_k\, \tilde{B}_k^m(u)\right) = m\sum_{k=0}^{m-1}(\boldsymbol{b}_{k+1} - \boldsymbol{b}_k)\tilde{B}_k^{m-1}(u).$$

- The addition or subtraction of m-th order Bézier curves $\boldsymbol{v}_1(u)$, $\boldsymbol{v}_2(u)$ results in another m-th order Bézier curve $\boldsymbol{v}_3(u)$ with control points given by the sum or difference of the control points of the original curves:

$$\boldsymbol{v}_3(u) = \boldsymbol{v}_1(u) \pm \boldsymbol{v}_2(u) = \sum_{k=0}^{m}(\boldsymbol{b}_{1,k} \pm \boldsymbol{b}_{2,k})\, \tilde{B}_k^m(u),\ u \in [0, 1] \tag{2.8}$$

  Bézier curves of different orders $m \neq l$ can be added or subtracted by first applying the degree elevation algorithm stated on p. 37 until $m = l$ holds.

- The multiplication of two Bézier curves $\boldsymbol{v}(u)$ and $\boldsymbol{w}(u)$ over the same interval $[0, 1]$ with orders m and l results in another Bézier curve with order $m + l$ over the interval $[0, 1]$.

- Following from the previous two properties for any m-th order curve $\boldsymbol{v}(u)$ the square of its norm $||\boldsymbol{v}(u)||^2$ can be written as a curve of order $2m$:

$$v_{\mathrm{sq}}(u) = ||\boldsymbol{v}(u)||^2 = \sum_{k=0}^{2m} \boldsymbol{b}_{\mathrm{sq},k} \tilde{B}_k^m(u), \quad u \in [0,1]. \tag{2.9}$$

  The control points $\boldsymbol{b}_{\mathrm{sq},k}$ result from a multiplication and an addition of the control points $\boldsymbol{b}_k$ of the curve $\boldsymbol{v}(u)$.

Bernstein base on the interval $[\bar{t}_i, \bar{t}_{i+1}]$. In addition to the positioning of the control points, the change of the parameter interval represents a further degree of freedom to influence the trajectory. The parameter interval can be interpreted as the transition time interval required for an agent to move along the trajectory from the start point to the end point of the trajectory. Hence, a prolongation of the transition time interval $t_{\mathrm{t}} = \bar{t}_{i+1} - \bar{t}_i$ can lead to the satisfaction of the actuator limitations of an agent.

The properties of the Bernstein base of degree m on the interval $[0,1]$ can be transformed to the base

$$B_k^m(t) = \frac{1}{(\bar{t}_{i+1} - \bar{t}_i)^m} \binom{m}{k} (t - \bar{t}_i)^k (\bar{t}_{i+1} - t)^{m-k}, \; t \in [\bar{t}_i, \bar{t}_{i+1}] \tag{2.10}$$

defined on the arbitrary interval $[\bar{t}_i, \bar{t}_{i+1}]$, leading to a Bézier curve over this interval

$$\boldsymbol{w}(t) = \sum_{k=0}^{m} \boldsymbol{b}_k B_k^m(t), \quad t \in [\bar{t}_i, \bar{t}_{i+1}]. \tag{2.11}$$

With $u \in [0,1]$, $t \in [\bar{t}_i, \bar{t}_{i+1}]$ and the affine parameter transformation

$$t = \bar{t}_i + (\bar{t}_{i+1} - \bar{t}_i)u$$

it can be concluded with (2.5), (2.6), (2.10) and (2.11) that

$$\boldsymbol{v}(u) = \sum_{k=0}^{m} \boldsymbol{b}_k \tilde{B}_k^m(u) = \sum_{k=0}^{m} \boldsymbol{b}_k B_k^m\left(\frac{t - \bar{t}_i}{\bar{t}_{i+1} - \bar{t}_i}\right) = \sum_{k=0}^{m} \boldsymbol{b}_k B_k^m(t) = \boldsymbol{w}(t).$$

Remark. The Bézier curve is independent of the used parameter interval as it can be seen in Fig 2.3 exemplarily for the x-direction. However, this is not valid for the derivatives of the Bézier curve. As it can be seen in eqn. (2.12), the length of the parameter interval is included in the calculation of the derivative of the curve.

Derivatives of a Bézier curve on the interval $[\bar{t}_i, \bar{t}_{i+1}]$. The derivatives of the Bézier curve (2.11) are determined by using the derivatives of the Bernstein polynomials:

$$\frac{\mathrm{d}^j}{\mathrm{d}\,t^j} B_k^m(t) = \frac{1}{(\bar{t}_{i+1} - \bar{t}_i)^j} \frac{m!}{(m-j)!} \sum_{i=0}^{j} (-1)^i \binom{j}{i} B_{k-j+i}^{m-j}(t)$$

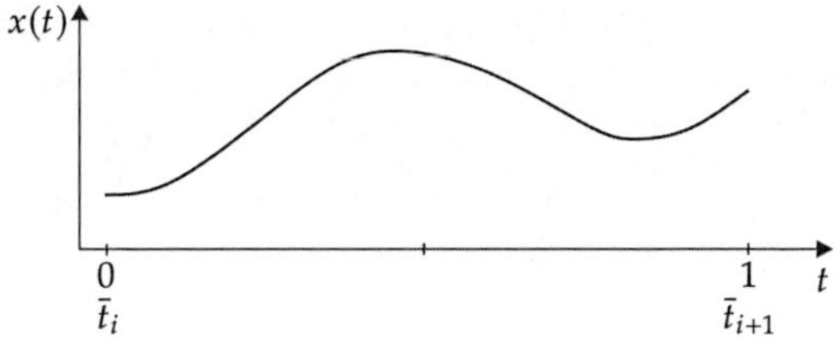

Fig. 2.3: Bézier curve over different parameter intervals $[0, 1]$ and $[\bar{t}_i, \bar{t}_{i+1}]$.

Then, the derivatives of (2.11) are given by

$$\begin{aligned}\frac{\mathrm{d}^j}{\mathrm{d}\,t^j}\boldsymbol{w}(t) &= \frac{\mathrm{d}^j}{\mathrm{d}t^j}\left(\sum_{k=0}^{m} \boldsymbol{b}_k\, B_k^m(t)\right) = \sum_{k=0}^{m} \Delta^j \boldsymbol{b}_k \frac{\mathrm{d}^j}{\mathrm{d}\,t^j} B_k^m(t) \\ &= \frac{1}{(\bar{t}_{i+1} - \bar{t}_i)^j} \frac{m!}{(m-j)!} \sum_{k=0}^{m-j} \Delta^j \boldsymbol{b}_k B_k^{m-j}(t) \end{aligned} \tag{2.12}$$

with

$$\Delta^j \boldsymbol{b}_k = \sum_{i=0}^{j} \binom{j}{i} (-1)^i \boldsymbol{b}_{k+j-i}.$$

Piecewise Bézier curves. The trajectories $\boldsymbol{w}(t)$ of the agents are planned as piecewise Bézier curves. Each trajectory is composed of several Bézier curves over appropriate intervals, which connect a sequence of intermediate points assigned to a sequence of time instants

$$\bar{t} = \{\bar{t}_1 = t_{\text{start}}, \ldots, \bar{t}_i, \bar{t}_{i+1}, \ldots, \bar{t}_N = t_{\text{end}}\}. \tag{2.13}$$

The entire trajectory of the UAV defined over the interval $t \in [t_{\text{start}}, t_{\text{end}}]$ consists of trajectories $\boldsymbol{w}^i(t)$ in the intervals $t \in [\bar{t}_i, \bar{t}_{i+1}] \subset [t_{\text{start}}, t_{\text{end}}]$ given by

$$\boldsymbol{w}^i(t) = \boldsymbol{w}^i\left(\frac{t - \bar{t}_i}{\bar{t}_{i+1} - \bar{t}_i}\right) = \sum_{k=0}^{m} \boldsymbol{b}_k^i B_k^m\left(\frac{t - \bar{t}_i}{\bar{t}_{i+1} - \bar{t}_i}\right). \tag{2.14}$$

The end time instant $\bar{t}_{i+1}$ of the trajectory $\boldsymbol{w}^i(t)$ corresponds to the start time instant of the next trajectory $\boldsymbol{w}^{i+1}(t)$.

The trajectories are desired to be C^k– continuous depending on the relative degree $r = k$ of the agents to achieve a smooth transition between two trajectories $\boldsymbol{w}^i(t)$, $\boldsymbol{w}^{i+1}(t)$. Two trajectories $\boldsymbol{w}^i(t)$ and $\boldsymbol{w}^{i+1}(t)$ are said to be C^k– continuous at $\bar{t}_i$ if

$$\begin{aligned}
\boldsymbol{w}^i(\bar{t}_i) &= \boldsymbol{w}^{i+1}(\bar{t}_i) \\
\dot{\boldsymbol{w}}^i(\bar{t}_i) &= \dot{\boldsymbol{w}}^{i+1}(\bar{t}_i) \\
&\vdots \\
\boldsymbol{w}^{(k)i}(\bar{t}_i) &= \boldsymbol{w}^{(k)i+1}(\bar{t}_i).
\end{aligned}$$

2.3.2 *Transformation of a curve from the Bernstein base into the monomial base*

Parametric curves in the monomial base. The monomial base is the most commonly used base for the theoretical representation of properties of parametric curves. However, for practical calculations it is often neither the most appropriate nor the most numerically stable base. Polynomials in the three dimensional space are described in the monomial base by

$$\boldsymbol{w}(t) = \begin{pmatrix} w_x(t) \\ w_y(t) \\ w_z(t) \end{pmatrix} = \sum_{i=0}^{m} \boldsymbol{a}_i \cdot t^i = \sum_{i=0}^{m} \boldsymbol{a}_i \cdot M_i(t),\ t \in [\bar{t}_i, \bar{t}_{i+1}],\ \boldsymbol{a}_i \in \mathbb{R}^3. \tag{2.15}$$

The polynomials $M_i(t) = t^i$, $(i = 0, \ldots, m)$ are called monomials of degree i and form a base of the vector space of the polynomials of degree m over $[\bar{t}_i, \bar{t}_{i+1}]$.

Remark. In this thesis polynomials in the monomial base are required for the feedforward controllers of the controlled agents to generate their control inputs.

Transformation of a curve. For the transformation of a curve from the Bernstein base into the monomial base the polynomial (2.15) is rewritten. For a single dimension it can also be represented in the form

$$w(t) = \begin{pmatrix} a_0 & a_1 & \ldots & a_m \end{pmatrix} \begin{pmatrix} t^0 \\ t^1 \\ \vdots \\ t^m \end{pmatrix}, \quad a_i \in \mathbb{R} \tag{2.16}$$

for each dimension x, y and z separately. The Bézier curve (2.11) can be written for the dimensions x, y and z separately by

$$w(t) = \begin{pmatrix} b_0 & b_1 & \ldots & b_m \end{pmatrix} \begin{pmatrix} B_0^m(t) \\ B_1^m(t) \\ \vdots \\ B_m^m(t) \end{pmatrix}, \quad b_k \in \mathbb{R}.$$

For the transformation the vector consisting of the Bernstein polynomials in the above equation is substituted by a transformation matrix $\boldsymbol{C}$ and a vector consisting of the monomials t^i, $(i = 0, 1, \ldots, m)$ as

$$w(t) = \begin{pmatrix} b_0 & b_1 & \ldots & b_m \end{pmatrix} \boldsymbol{C} \begin{pmatrix} t^0 \\ t^1 \\ \vdots \\ t^m \end{pmatrix}. \tag{2.17}$$

The coefficients a_i of the polynomials in the monomial base for each dimension x, y and z can be computed with the control points of the Bézier curve by comparing eqs. (2.16) and (2.17) as

$$\begin{pmatrix} a_0 & a_1 & \ldots & a_m \end{pmatrix} \begin{pmatrix} t^0 \\ t^1 \\ \vdots \\ t^m \end{pmatrix} = \begin{pmatrix} b_0 & b_1 & \ldots & b_m \end{pmatrix} \boldsymbol{C} \begin{pmatrix} t^0 \\ t^1 \\ \vdots \\ t^m \end{pmatrix}.$$

The transformation matrix $\boldsymbol{C}$ is a $(m+1) \times (m+1)$ matrix, where the inverse of $\boldsymbol{C}$ exists. The elements of $\boldsymbol{C}$ are given by

$$c_{k,i} = \begin{cases} (-1)^{i-k} \begin{pmatrix} m \\ k-1 \end{pmatrix} \begin{pmatrix} m-k+1 \\ i-k \end{pmatrix}, & i \geq k \\ 0, & \text{others.} \end{cases} \tag{2.18}$$

2.3.3 *Algorithms*

De Casteljau. The de Casteljau algorithm [66] is used to subdivide a Bézier curve on the interval $u \in [0, 1]$ into two independent curves at a chosen point $u = u_\mathrm{d}$ between 0 and 1. Application of the algorithm yields two Bézier curves

$$\boldsymbol{v}(u) = \begin{cases} \boldsymbol{v}_1\left(\frac{u}{u_\mathrm{d}}\right), & u \in [0, u_\mathrm{d}] \\ \boldsymbol{v}_2\left(\frac{u-u_\mathrm{d}}{1-u_\mathrm{d}}\right), & u \in [u_\mathrm{d}, 1], \end{cases}$$

which represent the initial curve $\boldsymbol{v}(u)$. By applying the de Casteljau algorithm, the piecewise trajectories of the agents can be divided into consistent intervals so that the algorithms for determining the minimum and maximum distances between the agents, stated in the following paragraphs, are applicable.

Degree elevation. With the repeated application of the degree elevation algorithm [66] any l-th order Bézier curve can be written as an m-th order Bézier curve with $m > l$.

Given the control points $\tilde{b}_k$, $(k = 0, 1, \dots, l)$ the algorithm finds the control points b_i, $(i = 0, 1, \dots, m)$ such that

$$v(u) = \sum_{k=0}^{l} \tilde{b}_k \tilde{B}_k^l(u) = \sum_{i=0}^{m} b_i \tilde{B}_i^m(u), \quad u \in [0, 1],$$

in which $m > l$ holds.

Minimum distance calculation. The minimum distance between two Bézier curves $v_1(u_1)$ and $v_2(u_2)$ with $u_1, u_2 \in [0, 1]$ can be efficiently calculated with the algorithm presented in [51] as

$$d_{\min} = \min_{u_1, u_2 \in [0,1]} ||v_1(u_1) - v_2(u_2)||, \quad u_{\mathrm{d,min}} = \arg\min_{u_1, u_2 \in [0,1]} ||v_1(u_1) - v_2(u_2)||.$$

The algorithm has a much faster convergence compared to root-finding based approaches and uses the GJK algorithm introduced in [73] and its efficient implementation given in [42].

Maximum distance calculation. In order to calculate the maximum distance between two Bézier curves $v_1(u_1)$ and $v_2(u_2)$ with $u_1, u_2 \in [0, 1]$, first the difference $v_{\mathrm{diff}}(u)$ between the curves is determined with (2.8). After that the square of the norm $v_{\mathrm{sq,diff}}(u) = ||v_{\mathrm{diff}}(u)||^2$ is computed with (2.9). Then the distance between $v_{\mathrm{sq,diff}}(u)$ and a constant curve with value $E_{\max}^2$ is considered to calculate the maximum distance between the curves $v_1(u_1)$ and $v_2(u_2)$. $E_{\max}$ is appropriately chosen, so that it is ensured that the value is always above the curve $v_{\mathrm{sq,diff}}(u)$. Using the minimum distance algorithm [51] as

$$u_{\mathrm{d,max}} = \arg\min_{u \in [0,1]} ||v_{\mathrm{sq,diff}}(u) - E_{\max}^2||, \quad u \in [0, 1]$$

the parameter $u_{\mathrm{d,max}}$ can be found at which the distance between the curves $v_1(u_1)$ and $v_2(u_2)$ is maximum. Inserting $u_{\mathrm{d,max}}$ in the difference $v_{\mathrm{diff}}(u)$ the value of the maximum distance

$$d_{\max} = ||v_{\mathrm{diff}}(u_{\mathrm{d,max}})||$$

is obtained.

2.4 Gradient descent optimisation method

In this thesis the gradient descent optimisation method [45] is used for solving an optimisation problem. Algorithm 1 summarises the steps of the approach. For a convex function $f(x)$ with its domain D_f the result of the algorithm is the value $x_{\min}$, for which the function $f(x)$ is minimum. Hence, $x_{\min}$ defines the global minimum of $f(x)$. Δx states the direction of the descent, $x_0 \in D_f$ is a fixed start point on the function $f(x)$ from which the algorithm starts, $\nabla f(x)$ is the gradient of $f(x)$ and s defines the step size. The index k indicates the iteration step.

Algorithm 1 Gradient descent method

Given: Convex function $f(x)$.
Start point $x_0 \in D_f$.
Step size s.

1: Set $x_{k+1} = x_k - s \cdot \nabla f(x_k)$.

2: If $f(x_{k+1}) < f(x_k)$: go to step 1.
Otherwise set $x_k = x_{\min}$.

Result: Optimum $x_{\min} = \arg\min_x f(x)$.

Applying Algorithm 1 to the exemplary convex function, shown in Fig. 2.4, illustrates the method. Starting from a fixed start point $x_0 \in D_f$ the algorithm performs a descent towards the negative gradient $-\nabla f(x_0)$. The step size s determines the convergence speed of the algorithm and the achievable accuracy. A large step size leads to a fast convergence but a low accuracy.

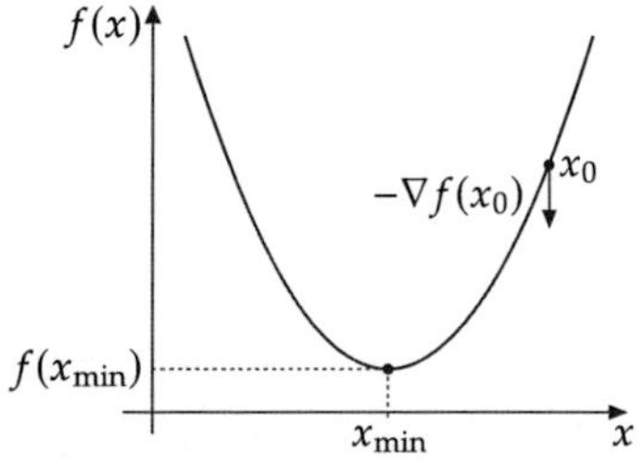

Fig. 2.4: Illustration of the gradient descent method.

2.5 Demonstration examples

2.5.1 *Unmanned aerial vehicle*

Figure 2.5 shows the structure of an UAV, which is used as a demonstration example throughout this thesis in order to illustrate the concepts. It is utilised in simulations and experiments, which results are presented in Chapters 11 and 12. The dynamics of a multirotor UAV are described by considering the UAV in two different coordinate systems as shown in Fig. 2.5 (left). First, it is a three dimensional earth-fixed coordinate frame I and second, it is a three dimensional body-fixed coordinate frame M. Hence, the UAV can be modelled by a translational and a rotational subsystem. The dynamics of the UAV are given in this section. The derivation of the model and the description of its parameters are presented in detail in Chapter 10.

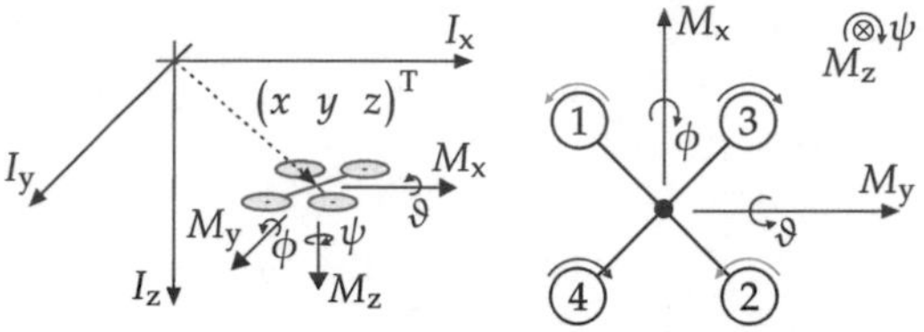

Fig. 2.5: Coordinates of the quadrotor.

The rotational subsystem models the attitude of the UAV, which is the rotation of the body-fixed frame M against the earth-fixed frame I given by the Euler angles $\phi(t)$, $\vartheta(t)$, $\psi(t)$ where $\phi(t)$ is the roll angle, $\vartheta(t)$ is the pitch angle and $\psi(t)$ is the yaw angle. With the state

$$\boldsymbol{x}_{\mathrm{r}}(t) = \begin{pmatrix} \dot{\phi}(t) & \dot{\vartheta}(t) & \dot{\psi}(t) & \phi(t) & \vartheta(t) & \psi(t) \end{pmatrix}^{\mathrm{T}}$$

containing the Euler angles and the angular speeds $\dot{\phi}(t)$, $\dot{\vartheta}(t)$, $\dot{\psi}(t)$ the nonlinear state space model is given by

$$\dot{\boldsymbol{x}}_{\mathrm{r}}(t) = \begin{pmatrix} \frac{(J_y-J_z)\dot{\vartheta}\dot{\psi}}{J_x} + \frac{4\pi^2 c_{\mathrm{th}}\alpha l(-\hat{u}_1+\hat{u}_2+\hat{u}_3-\hat{u}_4)}{J_x} - \frac{J_r\Omega(t)\dot{\vartheta}}{J_x} - \frac{c_\phi\dot{\phi}}{J_x} \\ \frac{(J_z-J_x)\dot{\phi}\dot{\psi}}{J_y} + \frac{4\pi^2 c_{\mathrm{th}}\alpha l(\hat{u}_1+\hat{u}_2-\hat{u}_3-\hat{u}_4)}{J_y} - \frac{J_r\Omega(t)\dot{\phi}}{J_y} - \frac{c_\vartheta\dot{\vartheta}}{J_y} \\ \frac{(J_x-J_y)\dot{\phi}\dot{\vartheta}}{J_z} + \frac{4\pi^2 c_{\mathrm{th}} c_{\mathrm{yaw}}(\hat{u}_1-\hat{u}_2+\hat{u}_3-\hat{u}_4)}{J_z} - \frac{c_\psi\dot{\psi}}{J_z} \\ \dot{\phi} \\ \dot{\vartheta} \\ \dot{\psi} \end{pmatrix} \tag{2.19}$$

$$\boldsymbol{y}_{\mathrm{r}}(t) = (\boldsymbol{0}_{3\times 3}\ \boldsymbol{I}_3)\,\boldsymbol{x}_{\mathrm{r}}(t), \qquad \boldsymbol{x}_{\mathrm{r}}(0) = \boldsymbol{x}_{\mathrm{r}0}$$

The translational subsystem describes the $(x(t), y(t), z(t))$ position of the centre of gravity (COG) of the UAV in I. It is assumed that the COG is in the origin of M and the quadrotor is modelled as a point mass. Using the state

$$\boldsymbol{x}_{\mathrm{t}}(t) = \begin{pmatrix} \dot{x}(t) & \dot{y}(t) & \dot{z}(t) & x(t) & y(t) & z(t) \end{pmatrix}^{\mathrm{T}},$$

which is composed of the position and the speeds $\dot{x}(t)$, $\dot{y}(t)$, $\dot{z}(t)$, the nonlinear model is given by

$$\dot{\boldsymbol{x}}_\mathrm{t}(t) = \begin{pmatrix} -\frac{F_\mathrm{z}(t)}{m}\left(\cos(\phi)\sin(\vartheta)\cos(\psi) + \sin(\phi)\sin(\psi)\right) - \frac{c_\mathrm{x}}{m}\dot{x} \\ -\frac{F_\mathrm{z}(t)}{m}\left(\cos(\phi)\sin(\vartheta)\sin(\psi) - \sin(\phi)\cos(\psi)\right) - \frac{c_\mathrm{y}}{m}\dot{y} \\ g - \frac{F_\mathrm{z}(t)}{m}\left(\cos(\phi)\cos(\vartheta)\right) - \frac{c_\mathrm{z}}{m}\dot{z} \\ \dot{x} \\ \dot{y} \\ \dot{z} \end{pmatrix} \tag{2.20}$$

$$\boldsymbol{y}_\mathrm{t}(t) = \left(\boldsymbol{0}_{3\times 3}\ \boldsymbol{I}_3\right)\boldsymbol{x}_\mathrm{t}(t), \qquad \boldsymbol{x}_\mathrm{t}(0) = \boldsymbol{x}_{\mathrm{t}0}.$$

In Chapter 10 a flatness-based two-degrees-of-freedom is derived for the UAV model (2.19), (2.20). Here, also the relative degree

$$\boldsymbol{r} = \begin{pmatrix}4 & 4 & 2 & 2\end{pmatrix}^\mathrm{T} \tag{2.21}$$

of the system is determined, which requires the planning of trajectories of order $m = 2\,r + 1 = 9$. Although a trajectory of order $m = 5$ is sufficient for the z-component, a trajectory of order $m = 9$ will be determined for consistency with the x-component and the y-component. The yaw angle is not necessary for an appropriate trajectory planning and is set to $\psi(t) = 0$ constantly.

2.5.2 *Car on a motorway*

In order to illustrate that the concepts are not limited to an application to UAVs, they are applied to cars, which are assumed to move on a motorway. The control method is evaluated in experiments with two cars that move on circular trajectories, as presented in Chapter 12.3.4. Throughout this thesis the cars are represented by robots, shown in Fig. 2.6, with the nonlinear state space model derived in [167]:

$$\begin{gathered}\begin{pmatrix}\dot{x}(t)\\ \dot{y}(t)\\ \dot{\phi}(t)\end{pmatrix} = \begin{pmatrix}v(t)\cos(\phi(t))\\ v(t)\sin(\phi(t))\\ \omega(t)\end{pmatrix} \\ \boldsymbol{y}(t) = \begin{pmatrix}x(t)\\ y(t)\\ \phi(t)\end{pmatrix}, \qquad \begin{pmatrix}x(0)\\ y(0)\\ \phi(0)\end{pmatrix} = \begin{pmatrix}x_0\\ y_0\\ \phi_0\end{pmatrix}\end{gathered} \tag{2.22}$$

The position of the robot in the xy-plane is denoted by $x(t)$, $y(t)$ and $\phi(t)$ is the orientation angle. The translational speed $v(t)$ and the rotational speed $\omega(t)$ are the control inputs.

In [152] a flatness-based controller is derived for the robot model (2.22). The relative degree $\boldsymbol{r} = \begin{pmatrix}2 & 2 & 2\end{pmatrix}^\mathrm{T}$ of the system is obtained, which necessitates the planning of trajectories of order $m = 5$.

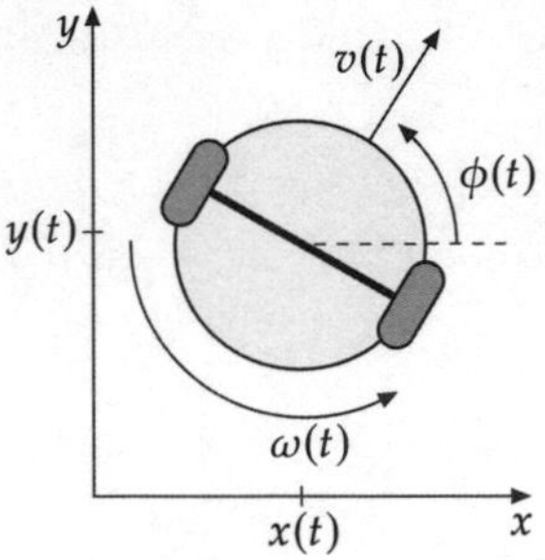

Fig. 2.6: Description of a mobile robot.

Part II

Fundamental methods

Trajectory planning 3

This chapter presents a trajectory planning method using piecewise Bézier curves. The method is designed in a general way so that different trajectories can be planned by only changing the planning task. Section 3.1 states the planning problem. In Section 3.2, the planning method for trajectories for a general movement using cartesian coordinates is given. Section 3.3 presents the method for planning circular trajectories using cylindrical coordinates. In Section 3.4 a method for the determination of trajectories for a change of the speed and the height of the circular movement is stated.

3.1 Planning problem

This chapter deals with the trajectory planning for the mobile agents. The idea of determining the trajectories is to develop a general method, which is able to generate various trajectories by only changing the specific planning task. The planning procedure remains unchanged. For the planning, piecewise Bézier curves, introduced in Section 2.3.1, are used that have to satisfy the limitations (1.8) – (1.10) on p. 17 and that have to fulfil conditions on the start point and the end point of the trajectory to meet the desired C^k – continuity.

Bézier curves are completely defined by their control points, which depend on boundary conditions on the trajectory and the transition time interval. Hence, the determination of the locations of the control points and the transition time interval is presented so that the resulting trajectory satisfies the requirements of the specific planning task. The general planning problem is as follows:

Given:	Planning task consisting of conditions on the start point and the end point of the trajectory and the transition time interval.
Find:	Trajectory $w(t)$ that satisfies the given requirements.
Boundary conditions:	System and actuator limitations (1.8) – (1.10) of the agents.

The method derived in this chapter to solve the planning problem is used by the event-based control unit to solve Problem 1.4. The trajectory $w(t)$ that enables the agents to fulfil their individual control tasks consists of several partial trajectories $\boldsymbol{w}^i(t)$ that correspond to piecewise Bézier curves. The determination of the control points for the partial trajectories $\boldsymbol{w}^i(t)$ is presented for the application to quadrotors in order to obtain closed expressions for the results. However, taking into account the dynamics of any other type of agent, trajectories for these agents can also be determined with small modifications of the presented method. The dynamics of an agent for which the trajectories are generated must be taken into account in two points in the method: On the one hand, the dynamics of the agent influences the order of the trajectory by the relative degree r. On the other hand, the transition time interval is determined using an optimisation problem in which the dynamics of the agent are incorporated as a constraint.

For the quadrotor application trajectories of order $m = 9$ need to be planned due to the relative degree (2.21), which is derived in Chapter 10. Hence, $n = 10$ control points have to be determined, which leads to the requirement of C^4 – continuity.

3.2 General planning method

3.2.1 *Determination of the control points*

For the quadrotor application the trajectories are desired to be C^4 – continuous to achieve a smooth transition between two trajectories $\boldsymbol{w}^i(t)$, $\boldsymbol{w}^{i+1}(t)$ of the piecewise trajectory $\boldsymbol{w}(t)$. A part of the piecewise trajectory is denoted by

$$\boldsymbol{w}^i(t) = \begin{pmatrix} w^i_{\mathrm{x}}(t) & w^i_{\mathrm{y}}(t) & w^i_{\mathrm{z}}(t) \end{pmatrix}^{\mathrm{T}} . \tag{3.1}$$

In order to plan trajectories that fulfil the condition, boundary conditions on the start point and the end point of each curve need to be fulfilled. More precisely, the conditions at the end point of one trajectory have to match the conditions at the start point of the next trajectory. These conditions are summarised in the matrices

$$\boldsymbol{W}^i(\bar{t}_i) = \begin{pmatrix} \boldsymbol{w}^{i\mathrm{T}}(\bar{t}_i) & w^i_{\psi}(\bar{t}_i) \\ \dot{\boldsymbol{w}}^{i\mathrm{T}}(\bar{t}_i) & \dot{w}^i_{\psi}(\bar{t}_i) \\ \ddot{\boldsymbol{w}}^{i\mathrm{T}}(\bar{t}_i) & \ddot{w}^i_{\psi}(\bar{t}_i) \\ \boldsymbol{w}^{(3)i\mathrm{T}}(\bar{t}_i) & w^{(3)i}_{\psi}(\bar{t}_i) \\ \boldsymbol{w}^{(4)i\mathrm{T}}(\bar{t}_i) & w^{(4)i}_{\psi}(\bar{t}_i) \end{pmatrix}, \quad \boldsymbol{W}^i(\bar{t}_{i+1}) = \begin{pmatrix} \boldsymbol{w}^{i\mathrm{T}}(\bar{t}_{i+1}) & w^i_{\psi}(\bar{t}_{i+1}) \\ \dot{\boldsymbol{w}}^{i\mathrm{T}}(\bar{t}_{i+1}) & \dot{w}^i_{\psi}(\bar{t}_{i+1}) \\ \ddot{\boldsymbol{w}}^{i\mathrm{T}}(\bar{t}_{i+1}) & \ddot{w}^i_{\psi}(\bar{t}_{i+1}) \\ \boldsymbol{w}^{(3)i\mathrm{T}}(\bar{t}_{i+1}) & w^{(3)i}_{\psi}(\bar{t}_{i+1}) \\ \boldsymbol{w}^{(4)i\mathrm{T}}(\bar{t}_{i+1}) & w^{(4)i}_{\psi}(\bar{t}_{i+1}) \end{pmatrix} \tag{3.2}$$

for the start point and the end point of the trajectory. They consist of the desired positions in the x, y and z-direction, conditions on the first four derivatives of $\boldsymbol{w}^i(t)$ at these points and the yaw angle $\psi(t)$ and its first four derivatives. The resulting piece of the whole trajectory for the quadrotors is given by

$$\boldsymbol{w}^i_{\mathrm{qc}}(t) = \begin{pmatrix} w^i_{\mathrm{qc,x}}(t) & w^i_{\mathrm{qc,y}}(t) & w^i_{\mathrm{qc,z}}(t) & w^i_{\mathrm{qc},\psi}(t) \end{pmatrix}^{\mathrm{T}} .$$

In the remainder of this chapter for the planning of the trajectories, the yaw angle is assumed to be $w^i_\psi(t) = 0$ without loss of generality. Hence, the trajectory of the agent is only planned for the x, y and z dimension and denoted by (3.1) to simplify the notation.

The positions of the control points must be chosen so as to fulfil the boundary conditions given by (3.2). Five control points at the start point and five control points at the end point of each trajectory are required to meet the conditions.

By a computation of the first four derivatives of the Bézier curve (2.11), the positions of the control points can be determined. By utilising the interpolation property of a Bézier curve (2.7) and applying eqn. (2.12) the first four derivatives of the curve at the start point are given by

$$
\begin{aligned}
\boldsymbol{w}^i(\bar{t}_i) &= \boldsymbol{b}^i_0, \\
\dot{\boldsymbol{w}}^i(\bar{t}_i) &= \tfrac{m}{\bar{t}_{i+1}-\bar{t}_i}(\boldsymbol{b}^i_1 - \boldsymbol{b}^i_0), \\
\ddot{\boldsymbol{w}}^i(\bar{t}_i) &= \tfrac{m(m-1)}{(\bar{t}_{i+1}-\bar{t}_i)^2}(\boldsymbol{b}^i_2 - 2\,\boldsymbol{b}^i_1 + \boldsymbol{b}^i_0), \\
\boldsymbol{w}^{(3)i}(\bar{t}_i) &= \tfrac{m(m-1)(m-2)}{(\bar{t}_{i+1}-\bar{t}_i)^3}(\boldsymbol{b}^i_3 - 3\,\boldsymbol{b}^i_2 + 3\,\boldsymbol{b}^i_1 - \boldsymbol{b}^i_0), \\
\boldsymbol{w}^{(4)i}(\bar{t}_i) &= \tfrac{m(m-1)(m-2)(m-3)}{(\bar{t}_{i+1}-\bar{t}_i)^4}(\boldsymbol{b}^i_4 - 4\,\boldsymbol{b}^i_3 + 6\,\boldsymbol{b}^i_2 - 4\,\boldsymbol{b}^i_1 + \boldsymbol{b}^i_0).
\end{aligned}
$$

In matrix notation the derivatives of the curve for the x-direction can be written as

$$
\underbrace{\begin{pmatrix} 1 & 0 & 0 & 0 & 0 \\ -1 & 1 & 0 & 0 & 0 \\ 1 & -2 & 1 & 0 & 0 \\ -1 & 3 & -3 & 1 & 0 \\ 1 & -4 & 6 & -4 & 1 \end{pmatrix}}_{\boldsymbol{T}_\mathrm{s}}
\underbrace{\begin{pmatrix} b^i_{0,\mathrm{x}} \\ b^i_{1,\mathrm{x}} \\ b^i_{2,\mathrm{x}} \\ b^i_{3,\mathrm{x}} \\ b^i_{4,\mathrm{x}} \end{pmatrix}}_{\boldsymbol{b}^i_\mathrm{s}}
=
\underbrace{\begin{pmatrix} w^i_\mathrm{x}(\bar{t}_i) \\ \dot{w}^i_\mathrm{x}(\bar{t}_i)\frac{\bar{t}_{i+1}-\bar{t}_i}{m} \\ \ddot{w}^i_\mathrm{x}(\bar{t}_i)\frac{(\bar{t}_{i+1}-\bar{t}_i)^2}{m(m-1)} \\ w^{(3)i}_\mathrm{x}(\bar{t}_i)\frac{(\bar{t}_{i+1}-\bar{t}_i)^3}{m(m-1)(m-2)} \\ w^{(4)i}_\mathrm{x}(\bar{t}_i)\frac{(\bar{t}_{i+1}-\bar{t}_i)^4}{m(m-1)(m-2)(m-3)} \end{pmatrix}}_{\boldsymbol{m}_\mathrm{s}} . \tag{3.3}
$$

The matrix representations for the y-direction and z-direction are identical. At the end point of the trajectory the first four derivatives of the curve can be again computed by using eqn. (2.12) as

$$
\begin{aligned}
\boldsymbol{w}^i(\bar{t}_{i+1}) &= \boldsymbol{b}^i_m, \\
\dot{\boldsymbol{w}}^i(\bar{t}_{i+1}) &= \tfrac{m}{\bar{t}_{i+1}-\bar{t}_i}(\boldsymbol{b}^i_m - \boldsymbol{b}^i_{m-1}), \\
\ddot{\boldsymbol{w}}^i(\bar{t}_{i+1}) &= \tfrac{m(m-1)}{(\bar{t}_{i+1}-\bar{t}_i)^2}(\boldsymbol{b}^i_m - 2\,\boldsymbol{b}^i_{m-1} + \boldsymbol{b}^i_{m-2}), \\
\boldsymbol{w}^{(3)i}(\bar{t}_{i+1}) &= \tfrac{m(m-1)(m-2)}{(\bar{t}_{i+1}-\bar{t}_i)^3}(\boldsymbol{b}^i_m - 3\,\boldsymbol{b}^i_{m-1} + 3\,\boldsymbol{b}^i_{m-2} - \boldsymbol{b}^i_{m-3}), \\
\boldsymbol{w}^{(4)i}(\bar{t}_{i+1}) &= \tfrac{m(m-1)(m-2)(m-3)}{(\bar{t}_{i+1}-\bar{t}_i)^4}(\boldsymbol{b}^i_m - 4\,\boldsymbol{b}^i_{m-1} + 6\,\boldsymbol{b}^i_{m-2} - 4\,\boldsymbol{b}^i_{m-3} + \boldsymbol{b}^i_{m-4}).
\end{aligned}
$$

Then, the system of equations for the x-direction can be stated as

$$\underbrace{\begin{pmatrix} 0 & 0 & 0 & 0 & 1 \\ 0 & 0 & 0 & -1 & 1 \\ 0 & 0 & 1 & -2 & 1 \\ 0 & -1 & 3 & -3 & 1 \\ 1 & -4 & 6 & -4 & 1 \end{pmatrix}}_{\boldsymbol{T}_\mathrm{e}} \underbrace{\begin{pmatrix} b^i_{m-4,\mathrm{x}} \\ b^i_{m-3,\mathrm{x}} \\ b^i_{m-2,\mathrm{x}} \\ b^i_{m-1,\mathrm{x}} \\ b^i_{m,\mathrm{x}} \end{pmatrix}}_{\boldsymbol{b}^i_\mathrm{e}} = \underbrace{\begin{pmatrix} w^i_\mathrm{x}(\bar{t}_{i+1}) \\ \dot{w}^i_\mathrm{x}(\bar{t}_{i+1})\frac{\bar{t}_{i+1}-\bar{t}_i}{m} \\ \ddot{w}^i_\mathrm{x}(\bar{t}_{i+1})\frac{(\bar{t}_{i+1}-\bar{t}_i)^2}{m(m-1)} \\ w^{(3)i}_\mathrm{x}(\bar{t}_{i+1})\frac{(\bar{t}_{i+1}-\bar{t}_i)^3}{m(m-1)(m-2)} \\ w^{(4)i}_\mathrm{x}(\bar{t}_{i+1})\frac{(\bar{t}_{i+1}-\bar{t}_i)^4}{m(m-1)(m-2)(m-3)} \end{pmatrix}}_{\boldsymbol{m}_\mathrm{e}}. \tag{3.4}$$

As for the derivatives of the curve at the start point, the matrix representations for the y-direction and z-direction at the end point are identical. For determining the control points, eqs. (3.3) and (3.4) can be summarised as

$$\begin{pmatrix} \boldsymbol{T}_\mathrm{s} & 0 \\ 0 & \boldsymbol{T}_\mathrm{e} \end{pmatrix} \begin{pmatrix} \boldsymbol{b}^i_\mathrm{s} \\ \boldsymbol{b}^i_\mathrm{e} \end{pmatrix} = \begin{pmatrix} \boldsymbol{m}_\mathrm{s} \\ \boldsymbol{m}_\mathrm{e} \end{pmatrix}. \tag{3.5}$$

By a left division the control points are given by

$$\begin{pmatrix} \boldsymbol{b}^i_\mathrm{s} \\ \boldsymbol{b}^i_\mathrm{e} \end{pmatrix} = \begin{pmatrix} \boldsymbol{T}_\mathrm{s} & 0 \\ 0 & \boldsymbol{T}_\mathrm{e} \end{pmatrix}^{-1} \begin{pmatrix} \boldsymbol{m}_\mathrm{s} \\ \boldsymbol{m}_\mathrm{e} \end{pmatrix}. \tag{3.6}$$

Due to the structure of the matrices $\boldsymbol{T}_\mathrm{s}$ and $\boldsymbol{T}_\mathrm{e}$, the determinants of the matrices are given by

$$\det(\boldsymbol{T}_\mathrm{s}) = \det(\boldsymbol{T}_\mathrm{e}) = 1.$$

This implies that the inverses $\boldsymbol{T}_\mathrm{s}^{-1}$ and $\boldsymbol{T}_\mathrm{e}^{-1}$ of the matrices exist. As a result, the control points for a Bézier curve of order $m = 9$ for the x-dimension are given by

$$\begin{pmatrix} b^i_{0,\mathrm{x}} \\ b^i_{1,\mathrm{x}} \\ b^i_{2,\mathrm{x}} \\ b^i_{3,\mathrm{x}} \\ b^i_{4,\mathrm{x}} \\ b^i_{5,\mathrm{x}} \\ b^i_{6,\mathrm{x}} \\ b^i_{7,\mathrm{x}} \\ b^i_{8,\mathrm{x}} \\ b^i_{9,\mathrm{x}} \end{pmatrix} = \begin{pmatrix} w^i_\mathrm{x}(\bar{t}_i) \\ w^i_\mathrm{x}(\bar{t}_i) + \frac{t^i_\mathrm{t}}{9}\dot{w}^i_\mathrm{x}(\bar{t}_i) \\ w^i_\mathrm{x}(\bar{t}_i) + \frac{2t^i_\mathrm{t}}{9}\dot{w}^i_\mathrm{x}(\bar{t}_i) + \frac{t^{i2}_\mathrm{t}}{72}\ddot{w}^i_\mathrm{x}(\bar{t}_i) \\ w^i_\mathrm{x}(\bar{t}_i) + \frac{t^i_\mathrm{t}}{3}\dot{w}^i_\mathrm{x}(\bar{t}_i) + \frac{t^{i2}_\mathrm{t}}{24}\ddot{w}^i_\mathrm{x}(\bar{t}_i) + \frac{t^{i3}_\mathrm{t}}{504}w^{(3)i}_\mathrm{x}(\bar{t}_i) \\ w^i_\mathrm{x}(\bar{t}_i) + \frac{4t^i_\mathrm{t}}{9}\dot{w}^i_\mathrm{x}(\bar{t}_i) + \frac{t^{i2}_\mathrm{t}}{12}\ddot{w}^i_\mathrm{x}(\bar{t}_i) + \frac{t^{i3}_\mathrm{t}}{126}w^{(3)i}_\mathrm{x}(\bar{t}_i) + \frac{t^{i4}_\mathrm{t}}{3024}w^{(4)i}_\mathrm{x}(\bar{t}_i) \\ w^i_\mathrm{x}(\bar{t}_{i+1}) - \frac{4t^i_\mathrm{t}}{9}\dot{w}^i_\mathrm{x}(\bar{t}_{i+1}) + \frac{t^{i2}_\mathrm{t}}{12}\ddot{w}^i_\mathrm{x}(\bar{t}_{i+1}) - \frac{t^{i3}_\mathrm{t}}{126}w^{(3)i}_\mathrm{x}(\bar{t}_{i+1}) + \frac{t^{i4}_\mathrm{t}}{3024}w^{(4)i}_\mathrm{x}(\bar{t}_{i+1}) \\ w^i_\mathrm{x}(\bar{t}_{i+1}) - \frac{t^i_\mathrm{t}}{3}\dot{w}^i_\mathrm{x}(\bar{t}_{i+1}) + \frac{t^{i2}_\mathrm{t}}{24}\ddot{w}^i_\mathrm{x}(\bar{t}_{i+1}) - \frac{t^{i3}_\mathrm{t}}{504}w^{(3)i}_\mathrm{x}(\bar{t}_{i+1}) \\ w^i_\mathrm{x}(\bar{t}_{i+1}) - \frac{2t^i_\mathrm{t}}{9}\dot{w}^i_\mathrm{x}(\bar{t}_{i+1}) + \frac{t^{i2}_\mathrm{t}}{72}\ddot{w}^i_\mathrm{x}(\bar{t}_{i+1}) \\ w^i_\mathrm{x}(\bar{t}_{i+1}) - \frac{t^i_\mathrm{t}}{9}\dot{w}^i_\mathrm{x}(\bar{t}_{i+1}) \\ w^i_\mathrm{x}(\bar{t}_{i+1}) \end{pmatrix} \tag{3.7}$$

with t_t^i stating the parameter interval given in (3.8). The control points for the y-dimension and the z-dimension are obtained in the same way by using the conditions $w_\mathrm{y}^i(\bar{t}_i)$, $w_\mathrm{y}^i(\bar{t}_{i+1})$ and $w_\mathrm{z}^i(\bar{t}_i)$, $w_\mathrm{z}^i(\bar{t}_{i+1})$ and their derivatives included in (3.2). The three components together lead to the required control point vector $\boldsymbol{b}_k^i$.

Remark. The determination of the control points of a trajectory for the robot model (2.22) on p. 41 is carried out in the same way as described in this section. According to Section 2.5.2 for the robot, trajectories of order $m = 5$ are sufficient to meet the conditions on the position, velocity and acceleration. Hence, three control points at the start point and three control points at the end point of the trajectory need to be calculated. Trajectories are determined for the x dimension and the y dimension. However, the robots are also able to follow trajectories of order $m = 9$.

3.2.2 *Default transition time interval*

The positions of the control points are not only specified by the conditions on the start point and the end point of the trajectory, but they are influenced by the parameter interval as well, as it can be seen in eqn. (3.7). This interval can be interpreted as the transition time interval

$$t_\mathrm{t}^i = \bar{t}_{i+1} - \bar{t}_i, \tag{3.8}$$

which is the time span the agents require to move along their trajectories from the start point to the end point. Hence, the transition time interval is an additional degree of freedom to plan the trajectory. Two different types of transition time intervals are used for the trajectory planning. First, a default transition time interval is used for the planning. Second, the transition time interval can be minimised with respect to the agents dynamics to exploit the dynamical limitations. Both agents use a default transition time interval for planning their trajectories. In addition, the give-way agent uses for the generation of a reactive trajectory introduced in Section 7.4 a minimised transition time interval.

A specified default transition time interval $t_\mathrm{t,d}$ can be used to determine the locations of the control points of all pieces of the piecewise Bézier curve as

$$t_\mathrm{t}^i = t_\mathrm{t,d}. \tag{3.9}$$

Generating the trajectory with (3.9) causes the agents to move from the start point to the end point within the time span $t_\mathrm{t,d}$ by fulfilling the limitations (1.8) – (1.10) on p. 17 and without exploiting their dynamical limitations. The time span $t_\mathrm{t,d}$ is chosen to be appropriate for the agents to fulfil their individual tasks.

3.2.3 *Minimisation of the transition time interval*

Formulation of the optimisation problem. With the second method of determining the transition time interval, the agents should reach the end points of their trajectories in a minimum time span by fully exploiting their dynamical limitations (1.8) – (1.10). This aim can be expressed as an optimisation problem. The minimisation of the transition time interval is carried out for the quadrotor used in this thesis. In the optimisation problem the model of the quadrotor is used to derive the constraints, which can be found in Chapter 10. The minimisation of the transition time interval can also be applied to other types of moving agents by replacing the quadrotor model by a model of the agent to be applied in the optimisation problem.

For the optimisation problem the start time instant of the trajectory is set to $\bar{t}_i = 0$, which leads to the transition time interval $t_{\mathrm{t}}^i = \bar{t}_{i+1}$ used as the decision variable. This time shift can be applied without loss of generality due to the property of the affine invariance of Bézier curves as described in Section 2.3.1 and simplifies the optimisation problem. In the following the constraints of the optimisation problem are given.

First, the conditions $\boldsymbol{W}^i(\bar{t}_i)$ and $\boldsymbol{W}^i(\bar{t}_{i+1})$ at the start point and at the end point of the trajectory given in (3.2) need to be considered.

Second, the planned trajectory need to be taken into account. With (3.2) and eqn. (3.6) the planned trajectory (3.19) can be specified as a function of the transition time interval $\bar{t}_{i+1}$ as

$$\boldsymbol{w}^i(t,\bar{t}_{i+1}) = \begin{pmatrix} w_{\mathrm{x}}^i(t,\bar{t}_{i+1}) & w_{\mathrm{y}}^i(t,\bar{t}_{i+1}) & w_{\mathrm{z}}^i(t,\bar{t}_{i+1}) \end{pmatrix}^{\mathrm{T}}. \tag{3.10}$$

The trajectory (3.10) is given in the monomial base defined in (2.15) due to a transformation of (3.10) with (2.18).

Third, the limitations (1.8) – (1.10) need to be considered. The model of the quadrotor is used in order to derive the control input $\boldsymbol{u}(t,\bar{t}_{i+1})$, the angles $\phi(t,\bar{t}_{i+1})$, $\vartheta(t,\bar{t}_{i+1})$ and the speed $||\boldsymbol{v}(t,\bar{t}_{i+1})||$ along the trajectory (3.10). In order to reduce the system complexity, the linear model (10.8), (10.10) on p. 232 of the quadrotor is used to calculate the aforementioned values. As described in Chapter 10, the quadrotor is controlled by a flatness-based two-degrees-of-freedom controller. With the reference trajectory (3.10), the feedforward controller law determined in (10.20) on p. 240 for the partially linearised model (10.6), (10.7) on p. 231 can be specified for the linear model of the quadrotor by rearranging the system dynamics (10.8), (10.10) as

$$\tilde{\boldsymbol{u}}_{\mathrm{ff,l}}(t,\bar{t}_{i+1}) = \begin{pmatrix} \alpha_\phi(t,\bar{t}_{i+1}) \\ \alpha_\vartheta(t,\bar{t}_{i+1}) \\ \alpha_\psi(t,\bar{t}_{i+1}) \\ a_{\mathrm{z}}(t,\bar{t}_{i+1}) \end{pmatrix} = \begin{pmatrix} \frac{1}{g} w_{\mathrm{y}}^{(4)i}(t,\bar{t}_{i+1}) \\ -\frac{1}{g} w_{\mathrm{x}}^{(4)i}(t,\bar{t}_{i+1}) \\ \ddot{w}_\psi^i(t,\bar{t}_{i+1}) \\ g - \ddot{w}_{\mathrm{z}}^i(t,\bar{t}_{i+1}) \end{pmatrix}.$$

As stated in (10.12) on p. 233 the control input needs to be transformed to ease the feedback controller design. The transformed control input is given by

$$\boldsymbol{u}_{\mathrm{ff,l}}(t,\bar{t}_{i+1}) = \tilde{\boldsymbol{B}}_{\mathrm{r}}^{-1}\tilde{\boldsymbol{u}}_{\mathrm{ff,l}}(t,\bar{t}_{i+1}) = \tilde{\boldsymbol{B}}_{\mathrm{r}}^{-1}\begin{pmatrix} \frac{1}{g}w_{\mathrm{y}}^{(4)i}(t,\bar{t}_{i+1}) \\ -\frac{1}{g}w_{\mathrm{x}}^{(4)i}(t,\bar{t}_{i+1}) \\ \ddot{w}_{\psi}^{i}(t,\bar{t}_{i+1}) \\ g-\ddot{w}_{\mathrm{z}}^{i}(t,\bar{t}_{i+1}) \end{pmatrix} \tag{3.11}$$

with $\tilde{\boldsymbol{B}}_{\mathrm{r}}^{-1}$ as the inverse of the input matrix $\tilde{\boldsymbol{B}}_{\mathrm{r}}$ given in (10.9) on p. 232. The angles $\phi(t,\bar{t}_{i+1})$, $\vartheta(t,\bar{t}_{i+1})$ and the speed $||\boldsymbol{v}(t,\bar{t}_{i+1})||$ along the planned trajectory are obtained by rearranging eqs. (10.8), (10.10) as

$$\begin{aligned} \phi(t,\bar{t}_{i+1}) &= \tfrac{1}{g}\ddot{w}_{\mathrm{y}}^{i}(t,\bar{t}_{i+1}) \\ \vartheta(t,\bar{t}_{i+1}) &= -\tfrac{1}{g}\ddot{w}_{\mathrm{x}}^{i}(t,\bar{t}_{i+1}) \\ v(t,\bar{t}_{i+1}) &= ||\boldsymbol{v}(t,\bar{t}_{i+1})|| = \sqrt{\dot{w}_{\mathrm{x}}^{i2}(t,\bar{t}_{i+1}) + \dot{w}_{\mathrm{y}}^{i2}(t,\bar{t}_{i+1}) + \dot{w}_{\mathrm{z}}^{i2}(t,\bar{t}_{i+1})}. \end{aligned} \tag{3.12}$$

As described in Chapter 10, model uncertainties and disturbances are compensated by a feedback controller. For this reason, a control input reserve with $\rho = 0\ldots 1$, $\kappa = 1-\rho$ should be maintained for the trajectory planning.

With the limitations (1.8) – (1.10) and eqs. (3.11), (3.12) the constraints of the optimisation problem are given by

$$\begin{aligned} \underline{u} + \rho\,\bar{u} \le u_j(t,\bar{t}_{i+1}) &\le \bar{u} - \rho\,\bar{u}, \quad j = 1,\ldots,4,\ \forall t \in [0,\bar{t}_{i+1}] \\ \kappa\,\underline{\phi} \le \phi(t,\bar{t}_{i+1}) &\le \kappa\,\bar{\phi} \\ \kappa\,\underline{\vartheta} < \vartheta(t,\bar{t}_{i+1}) &< \kappa\,\bar{\vartheta} \\ -v_{\max} \le v(t,\bar{t}_{i+1}) &\le v_{\max} \end{aligned} \tag{3.13}$$

in which $u_j(t,\bar{t}_{i+1})$, $(j = 1,\ldots,4)$ represents the single control inputs of $\boldsymbol{u}_{\mathrm{ff,l}}(t,\bar{t}_{i+1})$ for simplicity. The optimisation problem is described by

$$\begin{aligned} \min_{\bar{t}_{i+1}} \quad & J = \bar{t}_{i+1} \\ \text{s.t.} \quad & \boldsymbol{W}^i(\bar{t}_i),\ \boldsymbol{W}^i(\bar{t}_{i+1}) \text{ as (3.2)}, \\ & \text{(3.6), (2.18), (3.10) in the form (2.15)}, \\ & \text{(3.11) – (3.13)}. \end{aligned} \tag{3.14}$$

Simplification of the constraints. The constraints of the optimisation problem concern functions of time t, which make the evaluation of the constraints complex. In order to avoid the evaluation of the entire time course, the extrema

$$\begin{aligned}
\underline{u}_j(\bar{t}_{i+1}) &= \min_{t\in[0,\bar{t}_{i+1}]} u_j(t),\ j=1,\dots,4, & \bar{u}_j(\bar{t}_{i+1}) &= \max_{t\in[0,\bar{t}_{i+1}]} u_j(t),\\
\underline{\phi}(\bar{t}_{i+1}) &= \min_{t\in[0,\bar{t}_{i+1}]} \phi(t), & \bar{\phi}(\bar{t}_{i+1}) &= \max_{t\in[0,\bar{t}_{i+1}]} \phi(t),\\
\underline{\vartheta}(\bar{t}_{i+1}) &= \min_{t\in[0,\bar{t}_{i+1}]} \vartheta(t), & \bar{\vartheta}(\bar{t}_{i+1}) &= \max_{t\in[0,\bar{t}_{i+1}]} \vartheta(t),\\
\underline{v}(\bar{t}_{i+1}) &= \min_{t\in[0,\bar{t}_{i+1}]} ||\boldsymbol{v}(t)||, & \bar{v}(\bar{t}_{i+1}) &= \max_{t\in[0,\bar{t}_{i+1}]} ||\boldsymbol{v}(t)||,
\end{aligned} \tag{3.15}$$

together with the initial values $u_j(0)$, $\phi(0)$, $\vartheta(0)$, $v(0)$ and the final values $u_j(\bar{t}_{i+1})$, $\phi(\bar{t}_{i+1})$, $\vartheta(\bar{t}_{i+1})$, $v(\bar{t}_{i+1})$ are considered. An analysis of these points also provides a solution to the optimisation problem. As the constraints are given by polynomials in the monomial base, the extrema can be derived as the solutions of

$$\frac{\mathrm{d}\,u_j(t)}{\mathrm{d}\,t} = 0,\ \frac{\mathrm{d}\,\phi(t)}{\mathrm{d}\,t} = 0,\ \frac{\mathrm{d}\,\vartheta(t)}{\mathrm{d}\,t} = 0 \text{ and } \frac{\mathrm{d}\,v(t)}{\mathrm{d}\,t} = 0.$$

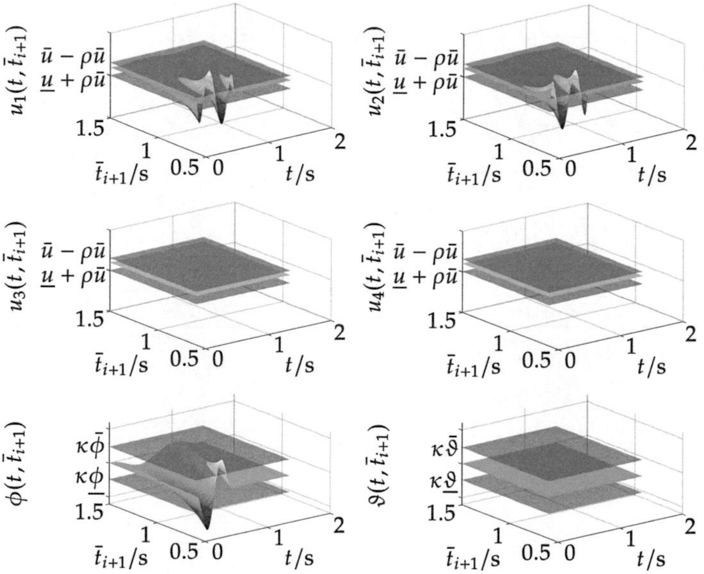

Fig. 3.1: Exemplary courses of the functions $u_j(t,\bar{t}_{i+1})$, $\phi(t,\bar{t}_{i+1})$ and $\vartheta(t,\bar{t}_{i+1})$ in dependence on t and $\bar{t}_{i+1}$.

The approach to evaluate only the extrema of the constraints reduces the function space by the dimension of time. Figure 3.1 illustrates the course of the constraints $u_j(t, \bar{t}_{i+1})$, $(j = 1, \dots, 4)$, $\phi(t, \bar{t}_{i+1})$ and $\vartheta(t, \bar{t}_{i+1})$ of the optimisation problem (3.14) for an exemplary movement of a quadrotor between a start point and an end point. The speed $v(t, \bar{t}_{i+1})$ is similar in course. Each of the constraints spans a plane, which is bounded by the limitations given in (3.13).

In contrast, in Fig. 3.2 the courses of the functions $\underline{u}_j(\bar{t}_{i+1})$, $\bar{u}_j(\bar{t}_{i+1})$, $(j = 1, \dots, 4)$, $\underline{\phi}(\bar{t}_{i+1})$, $\bar{\phi}(\bar{t}_{i+1})$, $\underline{\vartheta}(\bar{t}_{i+1})$ and $\bar{\vartheta}(\bar{t}_{i+1})$ are shown for the same movement of the quadrotor. The simplification leads to a two-dimensional representation of the constraints. Hence, the evaluation of the constraints reduces to the verification whether the values (3.15) are inside the limitations. The smallest value of $\bar{t}_{i+1}$ for which all extrema are inside the limitations corresponds to the minimum transition time interval.

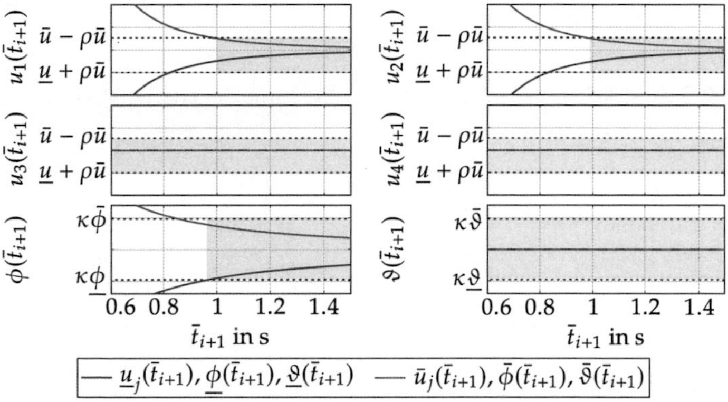

Fig. 3.2: Exemplary courses of the functions $\underline{u}_j(\bar{t}_{i+1})$, $\bar{u}_j(\bar{t}_{i+1})$, $\underline{\phi}(\bar{t}_{i+1})$, $\bar{\phi}(\bar{t}_{i+1})$, $\underline{\vartheta}(\bar{t}_{i+1})$ and $\bar{\vartheta}(\bar{t}_{i+1})$ in dependence on $\bar{t}_{i+1}$.

By combining the extrema with the initial values and the final values in vectors as

$$\begin{aligned}
\hat{\boldsymbol{u}}_j(\bar{t}_{i+1}) &= \begin{pmatrix} \underline{u}_j(\bar{t}_{i+1}) & \bar{u}_j(\bar{t}_{i+1}) & u_j(0) & u_j(\bar{t}_{i+1}) \end{pmatrix}^{\mathrm{T}}, \quad j = 1, \dots, 4 \\
\hat{\boldsymbol{\phi}}(\bar{t}_{i+1}) &= \begin{pmatrix} \underline{\phi}(\bar{t}_{i+1}) & \bar{\phi}(\bar{t}_{i+1}) & \phi(0) & \phi(\bar{t}_{i+1}) \end{pmatrix}^{\mathrm{T}} \\
\hat{\boldsymbol{\vartheta}}(\bar{t}_{i+1}) &= \begin{pmatrix} \underline{\vartheta}(\bar{t}_{i+1}) & \bar{\vartheta}(\bar{t}_{i+1}) & \vartheta(0) & \vartheta(\bar{t}_{i+1}) \end{pmatrix}^{\mathrm{T}} \\
\hat{\boldsymbol{v}}(\bar{t}_{i+1}) &= \begin{pmatrix} \underline{v}(\bar{t}_{i+1}) & \bar{v}(\bar{t}_{i+1}) & v(0) & v(\bar{t}_{i+1}) \end{pmatrix}^{\mathrm{T}}
\end{aligned} \tag{3.16}$$

the constraints (3.13) of the optimisation problem (3.14) can be written as

$$\begin{aligned} \underline{u} + \rho\,\bar{u} \leq \hat{\boldsymbol{u}}_j(\bar{t}_{i+1}) &\leq \bar{u} - \rho\,\bar{u}, \quad j = 1,\dots,4,\ \forall t \in [0, \bar{t}_{i+1}] \\ \kappa\,\underline{\phi} \leq \hat{\phi}(\bar{t}_{i+1}) &\leq \kappa\,\bar{\phi} \\ \kappa\,\underline{\vartheta} \leq \hat{\vartheta}(\bar{t}_{i+1}) &\leq \kappa\,\bar{\vartheta} \\ -v_{\max} \leq \hat{v}(\bar{t}_{i+1}) &\leq v_{\max} \end{aligned} \tag{3.17}$$

in which the constraints must be fulfilled element-wise. With a replacement of the constraints (3.13) by (3.17), the optimisation problem (3.14) can be written as

$$\begin{aligned} \min_{\bar{t}_{i+1}} \quad & J = \bar{t}_{i+1} \\ \text{s.t.} \quad & \boldsymbol{W}^i(\bar{t}_i),\ \boldsymbol{W}^i(\bar{t}_{i+1}) \text{ as (3.2)}, \\ & \text{(3.6), (2.18), (3.10) in the form (2.15)}, \\ & \text{(3.11), (3.12), (3.15) – (3.17)}. \end{aligned} \tag{3.18}$$

The result of the optimisation problem is a transition time interval with which the locations of the control points are determined according to (3.6). The resulting trajectory leads the agent from the start point to the end point in the minimum time span while satisfying all given constraints.

After solving the optimisation problem as described in the next section, the planned trajectory can be used with a time shift by $\bar{t}_i$ to the desired start time instant.

Properties of the optimisation problem. In this paragraph the convexity of the optimisation problem (3.18) should be stated. The qualitative course of the function to be minimised is shown in Fig. 3.3, where the constraints are illustrated as the shaded area. Convexity is defined as follows.

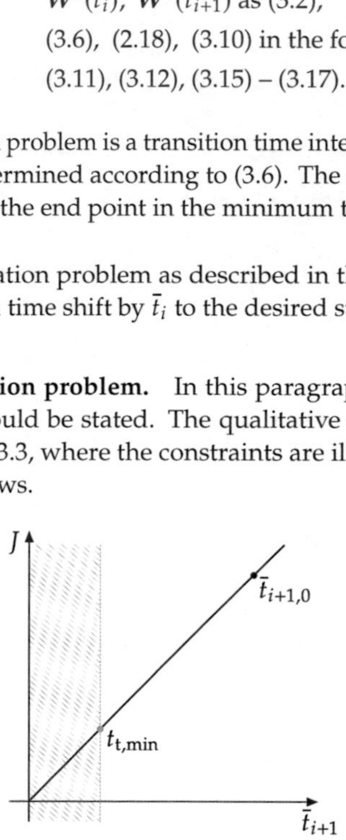

Fig. 3.3: Qualitative course of the cost function with constraints.

Definition 3.1. (Convexity) [45] A function $f : \mathbb{R}^n \rightarrow \mathbb{R}$ is convex if the domain of the function D_f is a convex set and if for all $x_1, x_2 \in D_f$ and α with $0 \leq \alpha \leq 1$ the equation

$$f(\alpha\, x_1 + (1-\alpha)\, x_2) \leq \alpha\, f(x_1) + (1-\alpha) f(x_2)$$

holds.

As the cost function of the optimisation problem (3.18) is a linear function $J = \bar{t}_{i+1}$, which are convex [45] the equation

$$J(\alpha\, \bar{t}_{i+1,1} + (1-\alpha)\, \bar{t}_{i+1,2}) = \alpha\, J(\bar{t}_{i+1,1}) + (1-\alpha) J(\bar{t}_{i+1,2}), \quad 0 \leq \alpha \leq 1,\ \bar{t}_{i+1,1}, \bar{t}_{i+1,2} \in D_J$$

holds. This means that for two arbitrary $\bar{t}_{i+1,1}$, $\bar{t}_{i+1,2}$ on the line $J = \bar{t}_{i+1}$, all points between $\bar{t}_{i+1,1}$, $\bar{t}_{i+1,2}$ are on the line as well. Hence, the cost function is convex.

For the optimisation problem to be convex, the constraints have to be convex as well. In order to find a proof of the convexity with respect to the constraints according to Definition 3.1 analytical expressions for the functions $\underline{u}_j(\bar{t}_{i+1})$, $\bar{u}_j(\bar{t}_{i+1})$, $(j = 1, \ldots, 4)$, $\underline{\phi}(\bar{t}_{i+1})$, $\bar{\phi}(\bar{t}_{i+1})$, $\underline{\vartheta}(\bar{t}_{i+1})$ and $\bar{\vartheta}(\bar{t}_{i+1})$, $(\forall t \in [0, \bar{t}_{i+1}])$ have to be found. Nevertheless, this is not possible because the functions (3.15) depend on $\bar{t}_{i+1}$ over the equations (2.18), (3.6), (3.10), (3.11) – (3.16).

However, a physical consideration leads to the conclusion that the constraints must be convex. For a movement of a quadrotor between a start point and an end point in a transition time interval $t_{\mathrm{t}}^i = \bar{t}_{i+1}$ the feedforward controller has to generate a fixed actuation energy. If the movement should be performed in a smaller transition time interval, the actuation energy must be provided in less time. This leads to a higher amplitude of the actuation energy and higher amplitudes of the angles $\phi(t)$ and $\vartheta(t)$, because the quadrotor accelerates faster. As described in Chapter 10, an acceleration is performed by the quadrotor by a rotation around its axes causing a deflection of the angles $\phi(t)$ and $\vartheta(t)$. Furthermore, the speed $v(t)$ increases, because the same way has to be covered in less time. This results in monotonically falling courses of the functions $\bar{u}_j(\bar{t}_{i+1})$, $\bar{\phi}(\bar{t}_{i+1})$, $\bar{\vartheta}(\bar{t}_{i+1})$. For the functions $\underline{u}_j(\bar{t}_{i+1})$, $\underline{\phi}(\bar{t}_{i+1})$, $\underline{\vartheta}(\bar{t}_{i+1})$ monotonically increasing courses result. The courses of the functions in Fig. 3.1 and Fig. 3.2 approve the physical considerations.

Hence, it is supposed that the constraints of (3.18) are convex. With this assumption the constraints define a lower bound of the domain of the cost function J. If the optimisation problem is convex, a solution of (3.18) corresponds to a global optimum. Hence, the lower bound, shown in Fig. 3.3, states the minimum transition time interval $t_{\mathrm{t,min}}$.

Solution of the optimisation problem. An utilisation of the gradient descent method introduced in Section 2.4 to solve optimisation problem (3.18) requires an extension of Algorithm 1 on p. 39 to evaluate the constraints of the problem. Algorithm 2 summarises the method. The algorithm starts with a sufficiently large transition time interval $\bar{t}_{i+1,0}$. It plans the trajectory $w^i(t)$ and checks the fulfilment of the constraints. It reduces the transition time interval by the step size s until one constraint is violated.

Algorithm 2 Determination of the minimum transition time interval $t_{\mathrm{t,min}}$

Given: Start time $\bar{t}_{i+1,0} = t_{\mathrm{t,d}} \in D_f$.
Conditions on the start point $\boldsymbol{W}^i(\bar{t}_i)$ and the end point $\boldsymbol{W}^i(\bar{t}_{i+1})$.
Constraints given by (3.17).
Step size s.

1: Set $\Delta\bar{t}_{i+1,k} = -\nabla J(\bar{t}_{i+1,k}) = -1$.
2: Set $\bar{t}_{i+1,k} = \bar{t}_{i+1,0}$.
3: Set $\bar{t}_{i+1,k+1} = \bar{t}_{i+1,k} + s\,\Delta\bar{t}_{i+1,k}$.
4: Determine trajectory (3.10) with $\boldsymbol{W}^i(\bar{t}_i)$, $\boldsymbol{W}^i(\bar{t}_{i+1})$, $\bar{t}_{i+1,k}$ and (2.10), (3.6).
5: Transform trajectory (3.10) in the monomial base (2.15) with (2.18).
6: Determine $\boldsymbol{u}_{\mathrm{ff,1}}(t)$ with (3.11).
7: Determine $\phi(t)$, $\vartheta(t)$, $v(t)$ with (3.12).
8: Determine extrema $\underline{u}_j(\bar{t}_{i+1,k})$, $\bar{u}_j(\bar{t}_{i+1,k})$, $\underline{\phi}(\bar{t}_{i+1,k})$, $\bar{\phi}(\bar{t}_{i+1,k})$, $\underline{\vartheta}(\bar{t}_{i+1,k})$, $\bar{\vartheta}(\bar{t}_{i+1,k})$, $\underline{v}(\bar{t}_{i+1,k})$, $\bar{v}(\bar{t}_{i+1,k})$ with (3.15).
9: Summarise extrema in (3.16).
10: If

$$\begin{aligned}
&\underline{u} + \rho\,\bar{u} \leq \hat{\boldsymbol{u}}_j(\bar{t}_{i+1,k}) \leq \bar{u} - \rho\,\bar{u}\\
\wedge\;& \kappa\,\underline{\phi} \leq \hat{\boldsymbol{\phi}}(\bar{t}_{i+1,k}) \leq \kappa\,\bar{\phi}\\
\wedge\;& \kappa\,\underline{\vartheta} \leq \hat{\boldsymbol{\vartheta}}(\bar{t}_{i+1,k}) \leq \kappa\,\bar{\vartheta}\\
\wedge\;& -v_{\max} \leq \hat{\boldsymbol{v}}(\bar{t}_{i+1,k}) \leq v_{\max}
\end{aligned}$$

Set $\bar{t}_{i+1,k} = \bar{t}_{i+1,k+1}$, go to step 4.
Otherwise set $t_{\mathrm{t,min}} = \bar{t}_{i+1,k}$.

Result: Minimum transition time interval $t_{\mathrm{t,min}}$.

Remark. The optimisation problem is characterised by the fact that the cost function $J = \bar{t}_{i+1}$ has a slope of $\frac{\partial}{\partial \bar{t}_{i+1}} J = 1$ over the entire domain D_f due to its linearity. The step size $s = 0.1$ is chosen, because the computation time of the algorithm equals the sampling time $T_{\mathrm{S}} = 8\,\mathrm{ms}$ of the system. With this step size an acceptable accuracy of $\Delta\bar{t}_{i+1} = 0.1\,\mathrm{s}$ results. A smaller step size leads to a significantly increased computation time.

The default transition time interval (3.9) is chosen as the start time of the algorithm and corresponds to an initial solution of (3.18), because with this time it is ensured that the constraints of (3.18) are fulfilled. The algorithm performs a descent in the direction of the negative gradient as shown in Fig. 3.3. The constraints are evaluated in each iteration step and the algorithm terminates if one constraint is violated.

3.2.4 *Planning task and planning algorithm*

Planning of a partial trajectory $\boldsymbol{w}^i(t)$. The task for planning the partial trajectories $\boldsymbol{w}^i(t)$ of the agents for a general movement in the 3D airspace based on piecewise Bézier curves is as follows:

Given:
- Conditions on the start point $\boldsymbol{W}^i(\bar{t}_i)$ and the end point $\boldsymbol{W}^i(\bar{t}_{i+1})$ of the trajectory given by (3.2).
- Type of determination of the transition time interval t_{t}^i.

Find: Trajectory $\boldsymbol{w}^i(t)$ in the form (3.19) that satisfies the conditions $\boldsymbol{W}^i(\bar{t}_i)$, $\boldsymbol{W}^i(\bar{t}_{i+1})$.

Boundary conditions: System and actuator limitations (1.8) – (1.10) on p. 17.

For the determination of one part of the entire trajectory, the control points calculated by (3.6), the transition time interval given by (3.9) or by the solution of optimisation problem (3.18) and the Bernstein polynomials given by (2.10) are required. For the quadrotors the resulting trajectory $\boldsymbol{w}^i(t)$ is given by

$$\boldsymbol{w}^i(t) = \begin{pmatrix} w_{\mathrm{x}}^i(t) \\ w_{\mathrm{y}}^i(t) \\ w_{\mathrm{z}}^i(t) \end{pmatrix} = \begin{pmatrix} \sum_{k=0}^{9} b_{k,\mathrm{x}}^i B_k^9(t) \\ \sum_{k=0}^{9} b_{k,\mathrm{y}}^i B_k^9(t) \\ \sum_{k=0}^{9} b_{k,\mathrm{z}}^i B_k^9(t) \end{pmatrix}, \quad t \in [\bar{t}_i, \bar{t}_{i+1}] \tag{3.19}$$

and satisfies the requirement of the C^4 – continuity. Although a trajectory of order $m = 5$ is sufficient for the z-component, as it can be seen in (2.21), a trajectory of order $m = 9$ is planned for consistency with the x-component and the y-component.

The trajectory planning method for the partial trajectories $\boldsymbol{w}^i(t)$ is summarised in Algorithm 3, which is used by the event-based control unit to solve Problem 1.4.

Algorithm 3 Planning of a partial trajectory $\boldsymbol{w}^i(t)$

Given: Conditions $\boldsymbol{W}^i(\bar{t}_i)$, $\boldsymbol{W}^i(\bar{t}_{i+1})$ (3.2).
Constraints given by (3.17).
Start time instant $\bar{t}_i$.
Default transition time interval $t_{\mathrm{t,d}}$.

1: Choose method of determining the transition time interval t_{t}^i.
2: If $t_{\mathrm{t}} = t_{\mathrm{t,d}}$:
Plan trajectory (3.19) with $\boldsymbol{W}^i(\bar{t}_i)$, $\boldsymbol{W}^i(\bar{t}_{i+1})$, $t_{\mathrm{t,d}}$ and (2.10), (3.6).
Transform trajectory (3.19) in the monomial base (2.15) with (2.18).
If $t_{\mathrm{t}} = t_{\mathrm{t,min}}$:
Determine $t_{\mathrm{t,min}}$ by executing Algorithm 2 on p. 56.
Plan trajectory (3.19) with $\boldsymbol{W}^i(\bar{t}_i)$, $\boldsymbol{W}^i(\bar{t}_{i+1})$, $t_{\mathrm{t,min}}$ and (2.10), (3.6).
Transform trajectory (3.19) in the monomial base (2.15) with (2.18).

Result: Trajectory $\boldsymbol{w}^i(t)$ in the form (2.15) that fulfils the limitations (1.8) – (1.10).

Planning of the entire trajectory $\boldsymbol{w}(t)$. By combining the separately planned partial trajectories $\boldsymbol{w}^i(t)$, which have to satisfy eqn. (2.14) the entire trajectory $\boldsymbol{w}(t)$ is obtained.

The combination of the time instant and position at a transition between two partial trajectories is called an *intermediate point* $P_i(t)$, $(i = 1, \ldots, N)$. For the quadrotor example the matrix

$$\boldsymbol{T} = \begin{pmatrix} t_{\text{start}} & t_{\text{t}}^{i} & t_{\text{t}}^{i+1} & t_{\text{t}}^{i+2} & \ldots & t_{\text{t}}^{i+N} \\ \boldsymbol{b}_{\text{x}}^{1} & \boldsymbol{b}_{\text{x}}^{i} & \boldsymbol{b}_{\text{x}}^{i+1} & \boldsymbol{b}_{\text{x}}^{i+2} & \ldots & \boldsymbol{b}_{\text{x}}^{i+N} \\ \boldsymbol{b}_{\text{y}}^{1} & \boldsymbol{b}_{\text{y}}^{i} & \boldsymbol{b}_{\text{y}}^{i+1} & \boldsymbol{b}_{\text{y}}^{i+2} & \ldots & \boldsymbol{b}_{\text{y}}^{i+N} \\ \boldsymbol{b}_{\text{z}}^{1} & \boldsymbol{b}_{\text{z}}^{i} & \boldsymbol{b}_{\text{z}}^{i+1} & \boldsymbol{b}_{\text{z}}^{i+2} & \ldots & \boldsymbol{b}_{\text{z}}^{i+N} \end{pmatrix}, \quad i = 1, \ldots, N-1 \tag{3.20}$$

is introduced, where in the first column the control points

$$\boldsymbol{b}_j^1 = \begin{pmatrix} b_{0,j}^1 & b_{1,j}^1 & b_{2,j}^1 & b_{3,j}^1 & b_{4,j}^1 \end{pmatrix}^{\text{T}}, \quad j \in \{\text{x}, \text{y}, \text{z}\} \tag{3.21}$$

at the start point are given and in the remaining columns the control points

$$\boldsymbol{b}_j^i = \begin{pmatrix} b_{5,j}^i & b_{6,j}^i & b_{7,j}^i & b_{8,j}^i & b_{9,j}^i \end{pmatrix}^{\text{T}}, \quad i = 1, \ldots, N, \ j \in \{\text{x}, \text{y}, \text{z}\} \tag{3.22}$$

at the end point are stated. The matrix represents arbitrary trajectories $\boldsymbol{w}(t)$ consisting of N partial trajectories $\boldsymbol{w}^i(t)$ only implicitly, but in a compact way. The first row contains the start time instant t_{start}, at which the first partial trajectory $\boldsymbol{w}^1(t)$ begins. Furthermore, it contains the transition time intervals t_{t}^i of all following partial trajectories $\boldsymbol{w}^i(t)$, $(i = 2, \ldots, N)$, whereby their start time instants and end time instants are denoted by $\bar{t}_i$ and $\bar{t}_{i+1}$ according to (2.13) on p. 35 , respectively and are given implicitly by:

$$\bar{t}_{i+1} = \bar{t}_i + t_{\text{t}}^i, \quad i = 1, \ldots, N.$$

The remaining rows contain the control points of a partial trajectory of the x, y and z components at an intermediate point. Thus, each partial trajectory $\boldsymbol{w}^i(t)$ is planned using the control points of column i for the start point and column $i+1$ for the end point. Using (2.10) and inserting these values of $\boldsymbol{T}$ in (3.19) gives the partial trajectories $\boldsymbol{w}^i(t)$ and as result the entire trajectory $\boldsymbol{w}(t)$.

3.3 Planning of a circular trajectory

3.3.1 *Circular segments*

In order to enable the agents to move on circular trajectories with a constant speed and to change the speed and the height of the movement, the planning method has to be modified. In order to ease the method, the circular trajectory $\boldsymbol{w}_{\text{c}}(t)$ is planned using cylindrical coordinates.

Fragmentation of the circle into four quarter circle segments. For planning, the circular trajectory is split into four quarter circle segments $w^i_{cs}(t)$. This approach allows more flexibility in planning the circular trajectory than planning the trajectory as one Bézier curve. For example, the speed or the height of the movement can be changed within one quarter circle segment without changing the other quarter circle segments. For the quadrotor example, at the transitions between the circular segments, still C^4 – continuity is required to ensure a smooth transition.

In order to plan the circular trajectory, four intermediate points $P_i(t)$, $(i = 1, \ldots, 4)$ are defined between which the quarter circle segments are generated. The points are placed on the coordinate axes around the origin of the coordinate system depending on the radius of the circle, as shown in Fig. 3.4, to simplify the further calculations. For the planning the radius of the circle, the start position of the agent and the direction of rotation must be given. For simplicity, the start position is chosen to be one of the intermediate points $P_i(t)$, $(i = 1, \ldots, 4)$. As the agents should move on a circle at a constant height $z(t) = z_c$, the circle segments are planned in the xy-plane at this height with no speed component in z-direction ($\dot{z}(t) = 0$). The planning of a trajectory for a height change is described in Section 3.4.2.

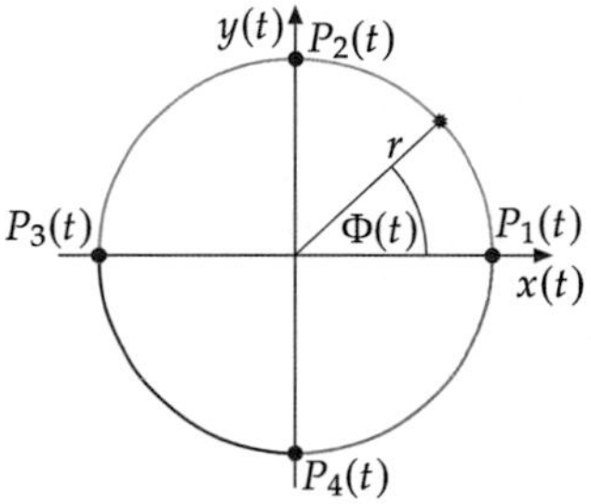

Fig. 3.4: Planning of a circular trajectory.

First, a circular segment is planned between the points $P_1(t)$ and $P_2(t)$, which forces the agent to rotate in mathematically positive direction. After that any further circular segment can be planned by using this segment and the properties of the Bézier curves, as described in Section 2.3.1.

Determination of the control points. The position of the control points $\boldsymbol{b}^i_k \in \mathbb{R}^3$ are determined by using cylindrical coordinates. A circle at a constant height $z(t) = z_c$ in the three-dimensional space in cylindrical coordinates is given by

$$\boldsymbol{w}^i_{cs}(t) = \begin{pmatrix} w^i_{cs,x}(t) \\ w^i_{cs,y}(t) \\ w^i_{cs,z}(t) \end{pmatrix} = \begin{pmatrix} r\cos(\omega t) \\ r\sin(\omega t) \\ z_c \end{pmatrix} \tag{3.23}$$

with the radius r of the circle and the angular speed

$$\omega = \frac{v}{r} = \frac{2\pi}{t_\mathrm{u}} = \frac{\pi}{2\,t^i_\mathrm{tc}} \tag{3.24}$$

as shown in Fig. 3.4. The time interval $t_\mathrm{u} = 4\,t^i_\mathrm{tc}$ is the period of an agent on the circle, which is determined by the addition of the transition time intervals

$$t^1_\mathrm{tc} = t^2_\mathrm{tc} = t^3_\mathrm{tc} = t^4_\mathrm{tc} = t^i_\mathrm{tc} = \frac{1}{4}\frac{2\pi r}{v} = \frac{\pi r}{2v} \tag{3.25}$$

of the four quarter circle segments. The transition time interval t^i_tc can be chosen arbitrarily large, which results in a small angular speed ω. The lower bound of t^i_tc is determined with Algorithm 2 on p. 56, which results in a large angular speed, whereby the limitations (1.8), (1.9) and (1.11) are satisfied. If the planning method is used in combination with the prediction method stated in Section 4.2 to include the movement of nearby agents in a set, the limits of t_tc are given by Theorem 3.1 on p. 64. The angle over time called *phase* is given by

$$\Phi^i(t) = \omega\, t. \tag{3.26}$$

In order to satisfy the required C^4 – continuity between two segments at the intermediate points $P_i(t)$, $(i = 1, \ldots, 4)$, the position and the first four derivatives of each segment (3.23) have to coincide at the start point and at the end point of two segments $\boldsymbol{w}^i_\mathrm{cs}(t)$ and $\boldsymbol{w}^{i+1}_\mathrm{cs}(t)$. In order to calculate the boundary conditions $\boldsymbol{W}^i_\mathrm{cs}(t)$ of segment i at an intermediate point $P_i(t)$ the position of this point together with the first four derivatives at this point are required as stated in Corollary 3.1.

Lemma 3.1 (Boundary conditions on a circular segment). *For a given radius r, a constant speed v and a constant height z_c of a circular trajectory, the boundary conditions in the form* (3.2) *on a circular segment at an arbitrary phase $\Phi^i(t)$ are given with* (3.24), (3.26) *by*

$$\boldsymbol{W}^i_\mathrm{cs}(t) = \begin{pmatrix} \boldsymbol{w}^{i\mathrm{T}}_\mathrm{cs}(t) \\ \dot{\boldsymbol{w}}^{i\mathrm{T}}_\mathrm{cs}(t) \\ \ddot{\boldsymbol{w}}^{i\mathrm{T}}_\mathrm{cs}(t) \\ \boldsymbol{w}^{(3)i\mathrm{T}}_\mathrm{cs}(t) \\ \boldsymbol{w}^{(4)i\mathrm{T}}_\mathrm{cs}(t) \end{pmatrix} = \begin{pmatrix} r\cos(\Phi^i(t)) & r\sin(\Phi^i(t)) & z_\mathrm{c} \\ -\,\omega\, r\sin(\Phi^i(t)) & \omega\, r\cos(\Phi^i(t)) & 0 \\ -\,\omega^2 r\cos(\Phi^i(t)) & -\,\omega^2 r\sin(\Phi^i(t)) & 0 \\ \omega^3 r\sin(\Phi^i(t)) & -\,\omega^3 r\cos(\Phi^i(t)) & 0 \\ \omega^4 r\cos(\Phi^i(t)) & \omega^4 r\sin(\Phi^i(t)) & 0 \end{pmatrix}, \quad i = 1, \ldots, 4. \tag{3.27}$$

For the quarter circle segment between the end points $P_1(t)$ and $P_2(t)$ the start time instant is chosen to be $\bar{t}_i = 0$ so that for the end time instant $\bar{t}_{i+1} = t^i_\mathrm{tc}$ holds. With

Corollary 3.1 the boundary conditions on the start point and the end point for this segment are given by

$$\boldsymbol{W}_{\text{cs}}^{i}(\bar{t}_i) = \begin{pmatrix} r & 0 & z_{\text{c}} \\ 0 & v & 0 \\ -v^2\, r^{-1} & 0 & 0 \\ 0 & -v^3\, r^{-2} & 0 \\ v^4\, r^{-3} & 0 & 0 \end{pmatrix}, \quad \boldsymbol{W}_{\text{cs}}^{i}(\bar{t}_{i+1}) = \begin{pmatrix} 0 & r & z_{\text{c}} \\ -v & 0 & 0 \\ 0 & -v^2\, r^{-1} & 0 \\ v^3\, r^{-2} & 0 & 0 \\ 0 & v^4\, r^{-3} & 0 \end{pmatrix}. \tag{3.28}$$

It can be seen that the start point and the end point of the segment correspond to the points $P_1(t)$ and $P_2(t)$. The values of the first derivative indicate the movement in the mathematically positive direction of rotation. The control points can be determined by inserting (3.28) into (3.6) and using (3.24) as

$$\begin{aligned}
\boldsymbol{b}_{\text{x}}^{i} &= \begin{pmatrix} b_{0,\text{x}}^{i} \\ b_{1,\text{x}}^{i} \\ b_{2,\text{x}}^{i} \\ b_{3,\text{x}}^{i} \\ b_{4,\text{x}}^{i} \end{pmatrix} = \begin{pmatrix} r \\ r \\ r(1-\frac{\pi^2}{288}) \\ r(1-\frac{\pi^2}{96}) \\ r(1-\frac{\pi^2}{48}+\frac{\pi^2}{48384}) \end{pmatrix}, \quad
\boldsymbol{b}_{\text{x}}^{i} = \begin{pmatrix} b_{5,\text{x}}^{i} \\ b_{6,\text{x}}^{i} \\ b_{7,\text{x}}^{i} \\ b_{8,\text{x}}^{i} \\ b_{9,\text{x}}^{i} \end{pmatrix} = \begin{pmatrix} r(\frac{2\pi}{9}-\frac{\pi^3}{1008}) \\ r(\frac{\pi}{6}-\frac{\pi^3}{4032}) \\ r\frac{\pi}{9} \\ r\frac{\pi}{18} \\ 0 \end{pmatrix} \\
\boldsymbol{b}_{\text{y}}^{i} &= \begin{pmatrix} b_{0,\text{y}}^{i} \\ b_{1,\text{y}}^{i} \\ b_{2,\text{y}}^{i} \\ b_{3,\text{y}}^{i} \\ b_{4,\text{y}}^{i} \end{pmatrix} = \begin{pmatrix} 0 \\ r\frac{\pi}{18} \\ r\frac{\pi}{9} \\ r(\frac{\pi}{6}-\frac{\pi^3}{4032}) \\ r(\frac{2\pi}{9}-\frac{\pi^3}{1008}) \end{pmatrix}, \quad
\boldsymbol{b}_{\text{y}}^{i} = \begin{pmatrix} b_{5,\text{y}}^{i} \\ b_{6,\text{y}}^{i} \\ b_{7,\text{y}}^{i} \\ b_{8,\text{y}}^{i} \\ b_{9,\text{y}}^{i} \end{pmatrix} = \begin{pmatrix} r(1-\frac{\pi^2}{48}+\frac{\pi^2}{48384}) \\ r(1-\frac{\pi^2}{96}) \\ r(1-\frac{\pi^2}{288}) \\ r \\ r \end{pmatrix} \\
\boldsymbol{b}_{\text{z}}^{i} &- \begin{pmatrix} b_{0,\text{z}}^{i} \\ b_{1,\text{z}}^{i} \\ b_{2,\text{z}}^{i} \\ b_{3,\text{z}}^{i} \\ b_{4,\text{z}}^{i} \end{pmatrix} - \begin{pmatrix} z_{\text{c}} \\ 0 \\ 0 \\ 0 \\ 0 \end{pmatrix}, \quad
\boldsymbol{b}_{\text{z}}^{i} = \begin{pmatrix} b_{5,\text{z}}^{i} \\ b_{6,\text{z}}^{i} \\ b_{7,\text{z}}^{i} \\ b_{8,\text{z}}^{i} \\ b_{9,\text{z}}^{i} \end{pmatrix} = \begin{pmatrix} 0 \\ 0 \\ 0 \\ 0 \\ z_{\text{c}} \end{pmatrix}.
\end{aligned} \tag{3.29}$$

As in the previous section the superscript i denotes the control points for a part of the piecewise Bézier curve in the interval $[\bar{t}_i, \bar{t}_{i+1}]$. At the start point the control points in the form (3.21) with indices 0 – 4 apply and at the end point the control points in the form (3.22) with indices 5 – 9 hold. The Bézier curve $\boldsymbol{w}_{\text{cs}}^{i}(t)$ of the circular segment, shown in Fig. 3.5, can be planned by using (2.10), (2.11) and (3.29) and the transition time interval t_{tc}^{i} of the segment.

Remark. As it can be seen in (3.29) the control points depend only on the radius of the circle, but they do not depend on the transition time interval t_{tc}^{i} and the angular speed ω.

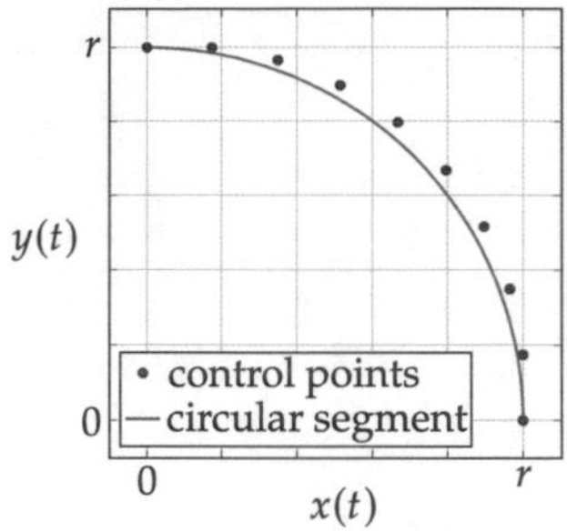

Fig. 3.5: Bézier curve of a circular segment.

The radius scales the position of the control points and does not change the shape of the segment. Hence, the circular segment has the same shape for different time intervals t^i_{tc}.

Verification of the circular form. The calculation of the speed $v(t)$ of the agent on its trajectory as a function based on Bézier curves is complex. For a simpler approximation of the speed and hence to save computation time, it is verified whether the quarter circle segment, shown in Fig. 3.5, has an exact circular shape. As a result, the entire circular trajectory coincides with an ideal circle. As a measure of the quality of the shape of the circular segment the distance of a point on the segment to the origin of the coordinate system is compared with the radius of an ideal quarter circle, which is constant over time using the error function

$$f(t) = \frac{||\boldsymbol{w}^i_{\mathrm{cs}}(t)|| - r}{r}.$$

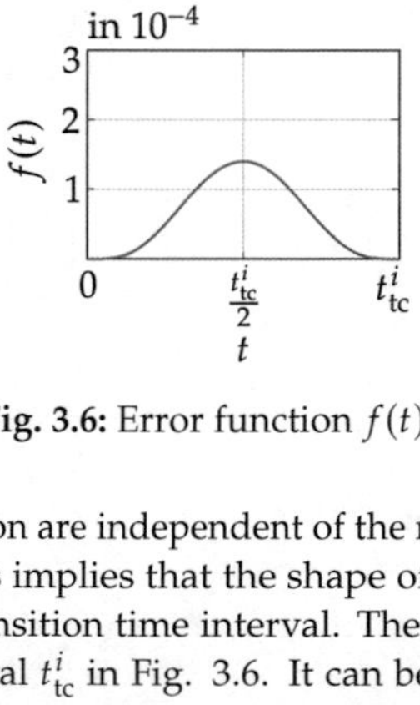

Fig. 3.6: Error function $f(t)$.

The values of the error function are independent of the radius r due to the invariance of the shape of the segment. This implies that the shape of the error function only varies in width depending on the transition time interval. The error function is shown for an arbitrary transition time interval t^i_{tc} in Fig. 3.6. It can be seen that the quarter circle is closely approximated by the circular segment, which states that the entire circle is closely approximated by a circular trajectory determined with the method given in this section.

Speed on the circular segment. The determination of the constant speed on the circular segment is necessary to check whether the limitation (1.11) is satisfied all the time. As the Bézier curve approximates the circular path sufficiently, the speed can be determined by using (3.23) as

$$\begin{aligned} v(t_{\mathrm{tc}}^i) &= \left\| \frac{\mathrm{d}}{\mathrm{d}t} \boldsymbol{w}_{\mathrm{cs}}^i(t) \right\| = \left\| \begin{pmatrix} -\omega\, r \sin(\omega t) \\ \omega\, r \cos(\omega t) \\ \dot{z}_{\mathrm{c}} \end{pmatrix} \right\| \\ &= \omega\, r \sqrt{\sin^2(\omega t) + \cos^2(\omega t)} = \omega\, r = r \frac{\pi}{2\, t_{\mathrm{tc}}^i}, \end{aligned} \tag{3.30}$$

in which the addition theorem $\sin^2(x) + \cos^2(x) = 1$ is applied and the fact is used that $\dot{z}_{\mathrm{c}} = 0$ holds due to the movement at a constant height. The speed on the curve is only depending on the transition time interval t_{tc}^i.

Validation of conditions on the trajectory planning for the prediction method. The prediction method stated in Section 4.2 generates a set $\mathcal{P}(t, t_k)$ (4.4) on p. 87 to include all possible future positions of nearby agents. For the prediction, knowledge about the trajectory planning method is used. In order to simplify the calculations the boundary conditions stated in Assumption 1.4 on p. 17 are considered. As it can be seen in (3.28) for the determination of the circular trajectory the assumption is violated. Hence, this paragraph analyses under which condition the prediction method still generates a feasible inclusion $\mathcal{P}(t, t_k)$ despite the violation of Assumption 1.4.

The planning of a quarter circle segment in compliance with the boundary conditions of Assumption 1.4 does not make use of all possible degrees of freedom. This fact leads to control points that have a large distance to the curve and thus form an oversize convex hull compared to the space required by the curve. Hence, if the quarter circle segment planned with the boundary conditions (3.28) is inside the convex hull formed by the control points that are placed with respect to Assumption 1.4, the prediction method stated in Theorem 4.1 on p. 87 is still applicable. Theorem 3.1 provides a condition with which the segment $\boldsymbol{w}_{\mathrm{cs}}^i(t)$ is inside this convex hull.

Theorem 3.1 (Locations of control points in the cylindrical coordinate system). *The control points of a Bézier curve fulfilling the boundary conditions* (3.28) *lie inside the convex hull formed by the control points of a Bézier curve fulfilling the boundary conditions of Assumption* 1.4 *if the transition time interval* t^i_{tc} *is in the interval*

$$0 \leq t^i_{\text{tc}} \leq \frac{9}{4\,\omega} \tag{3.31}$$

with the angular speed ω *of the trajectory.*

Proof. See Appendix B.1. ■

Remark. For the scenarios considered in this thesis, the angular speeds ω of the agents on their circular segments $\boldsymbol{w}^i_{\text{cs}}(t)$ are chosen so that the transition time interval fulfils condition (3.31). Hence, the prediction of the movement of nearby agents is possible despite the violation of Assumption 1.4 by the planning method for the circular trajectory.

Example 3.1. The result of Theorem 3.1 is shown in Fig. 3.7. A quarter circle segment shown as the blue line is planned between the points $P_1(t)$ and $P_2(t)$ with the radius $r = 1\,\text{m}$ and the angular speed $\omega = 0.7854\frac{\text{m}}{\text{s}}$ with boundary conditions fulfilling Assumption 1.4. The convex hull formed by the corresponding control points is illustrated by the red frame. A quarter circle segment with the boundary conditions (3.28) is shown as the orange line. The transition time interval is given by $t^i_{\text{tc}} = 2\,\text{s}$ and fulfils condition (3.31) as

$$0 \leq t^i_{\text{tc}} = 2\,\text{s} \leq \frac{9}{4\,\omega} = 2.86\,\text{s}.$$

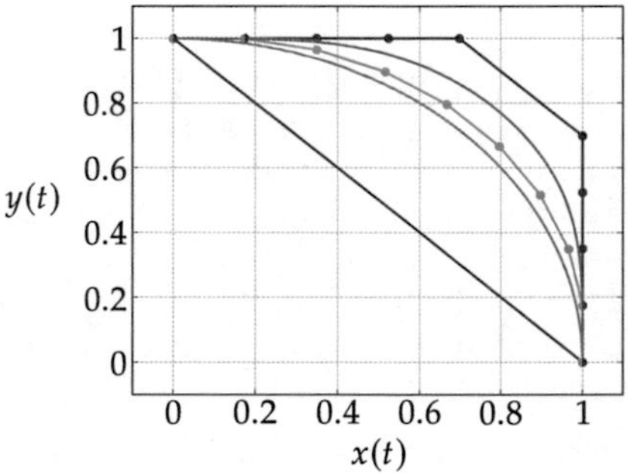

Fig. 3.7: Convex hulls of quarter circle segments fulfilling the boundary conditions (1.12) and (3.28).

It can be seen that the curve together with its green convex hull is inside the red convex hull. Hence, the prediction method overestimates the movement of the agent, but the prediction result is still valid. Furthermore, it can be seen that using all degrees of freedom for placing the control points results in a circular curve, which corresponds much more to an ideal circle. □

3.3.2 *Planning of a circular trajectory using quarter circle segments*

Combination of four segments to a circular trajectory. The entire circular trajectory $w_c(t)$ is generated by combining four quarter circle segments. In order to ensure C^4 – continuity at the intermediate points $P_i(t)$, $(i = 1, \dots, 4)$ of the trajectories, the boundary conditions on the quarter circle segments are determined by inserting the phase $\Phi^i(t)$ at which the intermediate points are located in (3.27). The phases $\Phi^i(t)$ and the transition time intervals t^i_{tc} associated with the intermediate points are given in Tab. 3.3. The boundary conditions of the intermediate points $P_i(t)$ shown in Fig. 3.4 are given by

$$\boldsymbol{W}_{\text{P1}} = \begin{pmatrix} r & 0 & z_c \\ 0 & v & 0 \\ -v^2\,r^{-1} & 0 & 0 \\ 0 & -v^3\,r^{-2} & 0 \\ v^4\,r^{-3} & 0 & 0 \end{pmatrix}, \quad \boldsymbol{W}_{\text{P2}} = \begin{pmatrix} 0 & r & z_c \\ -v & 0 & 0 \\ 0 & -v^2\,r^{-1} & 0 \\ v^3\,r^{-2} & 0 & 0 \\ 0 & v^4\,r^{-3} & 0 \end{pmatrix}$$

$$\boldsymbol{W}_{\text{P3}} = \begin{pmatrix} -r & 0 & z_c \\ 0 & -v & 0 \\ v^2\,r^{-1} & 0 & 0 \\ 0 & v^3\,r^{-2} & 0 \\ -v^4\,r^{-3} & 0 & 0 \end{pmatrix}, \quad \boldsymbol{W}_{\text{P4}} = \begin{pmatrix} 0 & -r & z_c \\ v & 0 & 0 \\ 0 & v^2\,r^{-1} & 0 \\ -v^3\,r^{-2} & 0 & 0 \\ 0 & -v^4\,r^{-3} & 0 \end{pmatrix}.$$

The boundary conditions $\boldsymbol{W}_{\text{P1}}$ and $\boldsymbol{W}_{\text{P2}}$ were already stated in (3.28) and are given here for the sake of completeness.

Due to the properties of the Bézier curves, the generation of one quarter circle segment is sufficient for the planning of the entire circular trajectory. The other quarter circle segments are derived by mirroring the already planned segment on the x-axis and the y-axis and by point reflection around the origin of the coordinate system to save computation time.

Direction of rotation on the circle. In order to enable a circular movement both in mathematically positive direction of rotation and in mathematically negative direction of rotation, (3.23) is extended by the factor d_r as

$$w^i_{\text{cs}}(t) = \begin{pmatrix} r\cos(\Phi(t)) \\ d_r\, r\sin(\Phi(t)) \\ z_c \end{pmatrix} \text{ with } d_r = \begin{cases} 1, & \text{mathematically positive direction of rotation} \\ -1, & \text{mathematically negative direction of rotation.} \end{cases}$$

The factor d_r must be taken into account in the determination of the derivatives. When determining the boundary conditions on the trajectory, the direction of rotation has to be considered by the phases $\Phi^i(t)$ and the transition time intervals t^i_{tc}. Table 3.3 states the time instants at which the agent passes the intermediate points $P_i(t)$, $(i = 1, \dots, 4)$ with their corresponding phases with respect to the direction of rotation.

Table 3.3: Transition time intervals and phases of the trajectory.

Position	math. pos. direction of rotation		math. neg. direction of rotation	
	Transition time t^i_{tc}	Phase $\Phi^i(t)$	Transition time t^i_{tc}	Phase $\Phi(t)^i$
P_1	0	0°	0	0°
P_2	t^i_{tc}	90°	$3\,t^i_{\text{tc}}$	270°
P_3	$2\,t^i_{\text{tc}}$	180°	$2\,t^i_{\text{tc}}$	180°
P_4	$3\,t^i_{\text{tc}}$	270°	t^i_{tc}	90°

The determination of the control points and the planning of the trajectory in the mathematically negative direction of rotation is the same as for the mathematically positive direction of rotation.

Start positions of the agents on the circular path. Basically, the agents can start their movement from any point on the circle. In order to simplify the implementation of the planning method, the method is designed in a way so that the agents are able to start from the four positions shown in Fig. 3.4. After a start position and an end position has been selected, the corresponding start time instant t_{start} and the end time instant t_{end} for the entire circular trajectory $w_{\text{c}}(t)$ can be taken from Tab. 3.3 for the desired direction of rotation.

3.3.3 *Planning task and planning algorithm*

Planning of a circular trajectory. The task for planning circular trajectories of the agents based on Bézier curves using cylindrical coordinates is given by:

Given:
- Intermediate points $P_i(t)$, $(i = 1, \dots, 4)$ with corresponding phases $\Phi^i(t)$.
- Start time instant t_{start}, start phase Φ_{start}.
- Transition time interval t^i_{tc}.
- Direction of rotation d_{r}.
- Radius r of the circle.

Find: Circular trajectory $w_{\text{c}}(t)$ that satisfies start and end conditions $\boldsymbol{W}^i_{\text{cs}}(t)$ derived with (3.27).

Boundary conditions: System and actuator limitations (1.8), (1.9), (1.11) on p. 17.

The planning method for a circular trajectory is summarised in the following algorithm that is used by the event-based control unit to solve Problem 1.4 for a circular movement.

Algorithm 4 Planning of a circular trajectory with a constant speed

---

Given: Intermediate points $P_i(t)$, $(i = 1, \ldots, 4)$ with corresponding phases $\Phi^i(t)$.
Start time instant t_{start}, start phase Φ_{start}.
Constant speed v with corresponding transition time interval t^i_{tc}.
Direction of rotation d_{r}.
Radius r of the circle.

1: Determine boundary requirement $\boldsymbol{W}_{\text{P1}}$ with (3.27) and $\Phi^i(t) = \Phi_{\text{start}}$.
Determine boundary requirement $\boldsymbol{W}_{\text{P2}}$ with (3.27) and $\Phi^i(t) = \text{mod}(\Phi_{\text{start}} + 90°, 360°)$.
Determine boundary requirement $\boldsymbol{W}_{\text{P3}}$ with (3.27) and $\Phi^i(t) = \text{mod}(\Phi_{\text{start}} + 180°, 360°)$.
Determine boundary requirement $\boldsymbol{W}_{\text{P4}}$ with (3.27) and $\Phi^i(t) = \text{mod}(\Phi_{\text{start}} + 270°, 360°)$.

2: Set $i = 1$, $j = 2$.

3: If $v = 0\frac{\text{m}}{\text{s}}$:
Determine control points $\boldsymbol{b}^i_{\text{x}}, \boldsymbol{b}^i_{\text{y}}, \boldsymbol{b}^i_{\text{z}}$ of $\boldsymbol{w}^i_{\text{cs}}(t)$ and plan the circular segment $\boldsymbol{w}^i_{\text{cs}}(t)$ by executing Algorithm 3 on p. 57 with $\boldsymbol{W}^i_{\text{cs}}(\bar{t}_i) = \boldsymbol{W}_{\text{P}i}$, $\boldsymbol{W}^i_{\text{cs}}(\bar{t}_{i+1}) = \boldsymbol{W}_{\text{P}i}$, $\bar{t}_i$, $t^i_{\text{tc}} = t_{\text{t,stood}}$ given by (3.32).
If $v > 0\frac{\text{m}}{\text{s}}$:
Determine control points $\boldsymbol{b}^i_{\text{x}}, \boldsymbol{b}^i_{\text{y}}, \boldsymbol{b}^i_{\text{z}}$ of $\boldsymbol{w}^i_{\text{cs}}(t)$ and plan the circular segment $\boldsymbol{w}^i_{\text{cs}}(t)$ by executing Algorithm 3 with $\boldsymbol{W}^i_{\text{cs}}(\bar{t}_i) = \boldsymbol{W}_{\text{P}i}$, $\boldsymbol{W}^i_{\text{cs}}(\bar{t}_{i+1}) = \boldsymbol{W}_{\text{P}j}$, $\bar{t}_i$, t^i_{tc} given by (3.25).

4: Set $i = i + 1$, $j = \text{mod}(j + 1, 4)$.

5: If $i \leq 4$:
go to step 2.
Else:
Combine $\boldsymbol{b}^i_{\text{x}}, \boldsymbol{b}^i_{\text{y}}, \boldsymbol{b}^i_{\text{z}}$ $t_{\text{t}i}$ for all i and t_{start} in the matrix $\boldsymbol{T}$ of the form (3.20).

Result: Circular trajectory $\boldsymbol{w}_{\text{c}}(t)$ in $\mathbb{R}^3$ in the form (2.15) that fulfils the limitations (1.8), (1.9), (1.11).
Circular trajectory $\boldsymbol{w}_{\text{c}}(t)$ in the form (2.11) implicitly given by $\boldsymbol{T}$.

Remark. In order to plan a trajectory with the speed $v = 0\frac{\text{m}}{\text{s}}$ with Algorithm 4, it might be obvious to set the transition time to $t^i_{\text{tc}} = \infty$. The standstill of an agent would be achieved by an infinite time passing before the next intermediate point is reached. However, the implemented algorithms require finite transition time intervals for the trajectory planning. Hence, the implementation of the method requires for the planning of a standstill that a trajectory is determined from a start point with speed $v = 0\frac{\text{m}}{\text{s}}$ to the same point within a finite transition time interval. This results in a trajectory that forces an agent to remain at the start point for the entire transition time interval. To ensure that the algorithms introduced in the remainder of this thesis operate correctly, a transition time interval of

$$t_{\mathrm{t,stood}} \geq \frac{2\pi r}{v_{\min}} \tag{3.32}$$

has to be chosen.

Remark. For the robot model (2.22) the circular trajectory is given by

$$\boldsymbol{w}_{\mathrm{c}}(t) = \begin{pmatrix} r\cos(\omega t) & r\sin(\omega t) \end{pmatrix}^{\mathrm{T}}$$

using polar coordinates. As the robots move only in the xy-plane, the height z is not necessary.

Representation of periodic trajectories. For periodic trajectories $\bar{\boldsymbol{T}}$ (e.g. circular trajectories with constant speed and height) consisting of N different partial trajectories, a matrix $\boldsymbol{T}$ of the form (3.20) on p. 58 with $N+1$ columns is sufficient to specify the trajectory over the entire time course. Each column i of a periodic trajectory exceeding the N-th column is equivalent to the column $\mathrm{mod}(i,\, N)+1$.

All circular trajectories $\boldsymbol{T} = \begin{pmatrix} \tilde{\boldsymbol{T}} & \bar{\boldsymbol{T}} \end{pmatrix}$ can be split into a periodic part $\bar{\boldsymbol{T}}$ and a non-periodic part $\tilde{\boldsymbol{T}}$. The non-periodic part contains a trajectory with speed changes and height changes, determined in Sections 3.4.1 and 3.4.2. The periodic part contains a subsequent circular movement with constant speed and height.

Periodic continuation of a circular trajectory. With Algorithm 4 one circular trajectory for an agent is planned. If the agent should move on the circle several times, a periodic continuation of the circular trajectory is planned with Algorithm 5.

Algorithm 5 Periodic continuation of an arbitrary circular trajectory

---

Given: Matrix $\tilde{\boldsymbol{T}}$ of the non-periodic trajectory $\tilde{\boldsymbol{w}}_{\mathrm{c}}(t)$.
End time instant t_{end} of $\tilde{\boldsymbol{w}}_{\mathrm{c}}(t)$.
Phase $\Phi(t_{\mathrm{end}})$.

1: Determine the radius $r = \sqrt{b_{0,\mathrm{x}}^{N2} + b_{0,\mathrm{y}}^{N2}}$ in the xy-plane using the last control points $\boldsymbol{b}_{\mathrm{x}}^{N}$, $\boldsymbol{b}_{\mathrm{y}}^{N}$ of $\tilde{\boldsymbol{w}}_{\mathrm{c}}(t)$.
Determine the speed $v = \sqrt{\dot{\tilde{w}}_{\mathrm{c,x}}^{2}(t_{\mathrm{end}}) + \dot{\tilde{w}}_{\mathrm{c,y}}^{2}(t_{\mathrm{end}}) + \dot{\tilde{w}}_{\mathrm{c,z}}^{2}(t_{\mathrm{end}})}$ at the end point of $\tilde{\boldsymbol{w}}_{\mathrm{c}}(t)$.

2: Determine $\bar{\boldsymbol{T}}$ as the periodic continuation of $\tilde{\boldsymbol{w}}_{\mathrm{c}}(t)$ by Algorithm 4 with the start time instant $\bar{t}_{i+1}$ and the speed v determined in step 1.

3: Generate matrix $\boldsymbol{T} = \begin{pmatrix} \tilde{\boldsymbol{T}} & \bar{\boldsymbol{T}} \end{pmatrix}$.

Result: Matrix $\boldsymbol{T}$ of the trajectory $\boldsymbol{w}_{\mathrm{c}}(t)$, that periodically continues $\tilde{\boldsymbol{T}}$ with the speed of the last intermediate point.

3.4 Two planning scenarios

3.4.1 *Trajectory for a change of the speed*

Use of a transition segment. In this section a quarter circle segment is planned, which leads to a change of the speed of an agent. This *transition segment* $\boldsymbol{w}_{\mathrm{t}}^{i}(t)$ is used to change the circular movement with the constant angular speed ω_{s} of an agent at the start point to a different speed ω_{e} at the end point. Furthermore, it is used to force the agent from the hover flight into a circular movement within a quarter circle. In this section a transition segment between the points $P_1(t)$ and $P_2(t)$, as shown in Fig. 3.4, with a mathematically positive direction of rotation is planned. Segments between the other points can be generated in the same way.

Determination of the control points. The boundary conditions are determined with (3.23) and (3.27) as

$$\boldsymbol{W}_{\mathrm{t}}^{i}(\bar{t}_i) = \begin{pmatrix} r & 0 & z_{\mathrm{c}} \\ 0 & v_{\mathrm{s}} & 0 \\ -v_{\mathrm{s}}^2 r^{-1} & 0 & 0 \\ 0 & -v_{\mathrm{s}}^3 r^{-2} & 0 \\ v_{\mathrm{s}}^4 r^{-3} & 0 & 0 \end{pmatrix}, \quad \boldsymbol{W}_{\mathrm{t}}^{i}(\bar{t}_{i+1}) = \begin{pmatrix} 0 & r & z_{\mathrm{c}} \\ -v_{\mathrm{e}} & 0 & 0 \\ 0 & -v_{\mathrm{e}}^2 r^{-1} & 0 \\ v_{\mathrm{e}}^3 r^{-2} & 0 & 0 \\ 0 & v_{\mathrm{e}}^4 r^{-3} & 0 \end{pmatrix} \tag{3.33}$$

where at the start time instant $\bar{t}_i$ the speed v_{s} applies and at the end time instant $\bar{t}_{i+1}$ the speed v_{e} holds. With a time shift without loss of generality $\bar{t}_i = 0$ and $\bar{t}_{i+1} = t_{\mathrm{t}}^{i}$ applies. The control points in $\mathbb{R}^3$ can be determined by inserting (3.33) in (3.6) as

$$
\boldsymbol{b}_{\mathrm{x}}^{i} = \begin{pmatrix} r \\ r \\ r - \frac{t_{\mathrm{t}}^{i2}}{72} v_{\mathrm{s}}^{2} r^{-1} \\ r - \frac{t_{\mathrm{t}}^{i2}}{24} v_{\mathrm{s}}^{2} r^{-1} \\ r - \frac{t_{\mathrm{t}}^{i2}}{12} v_{\mathrm{s}}^{2} r^{-1} + \frac{t_{\mathrm{t}}^{i4}}{3024} v_{\mathrm{s}}^{4} r^{-3} \end{pmatrix}, \quad \boldsymbol{b}_{\mathrm{x}}^{i} = \begin{pmatrix} \frac{4\, t_{\mathrm{t}}^{i}}{9} v_{\mathrm{e}} - \frac{t_{\mathrm{t}}^{i3}}{126} v_{\mathrm{e}}^{3} r^{-2} \\ \frac{t_{\mathrm{t}}^{i}}{3} v_{\mathrm{e}} - \frac{t_{\mathrm{t}}^{i3}}{504} v_{\mathrm{e}}^{3} r^{-2} \\ \frac{2\, t_{\mathrm{t}}^{i}}{9} v_{\mathrm{e}} \\ \frac{t_{\mathrm{t}}^{i}}{9} v_{\mathrm{e}} \\ 0 \end{pmatrix}
$$

$$
\boldsymbol{b}_{\mathrm{y}}^{i} = \begin{pmatrix} 0 \\ \frac{t_{\mathrm{t}}^{i}}{9} v_{\mathrm{s}} \\ \frac{2\, t_{\mathrm{t}}^{i}}{9} v_{\mathrm{s}} \\ \frac{t_{\mathrm{t}}^{i}}{3} v_{\mathrm{s}} - \frac{t_{\mathrm{t}}^{i3}}{504} v_{\mathrm{s}}^{3} r^{-2} \\ \frac{4\, t_{\mathrm{t}}^{i}}{9} v_{\mathrm{s}} - \frac{t_{\mathrm{t}}^{i3}}{126} v_{\mathrm{s}}^{3} r^{-2} \end{pmatrix}, \quad \boldsymbol{b}_{\mathrm{y}}^{i} = \begin{pmatrix} r - \frac{t_{\mathrm{t}}^{i2}}{12} v_{\mathrm{e}}^{2} r^{-1} + \frac{t_{\mathrm{t}}^{i4}}{3024} v_{\mathrm{e}}^{4} r^{-3} \\ r - \frac{t_{\mathrm{t}}^{i2}}{24} v_{\mathrm{e}}^{2} r^{-1} \\ r - \frac{t_{\mathrm{t}}^{i2}}{72} v_{\mathrm{e}}^{2} r^{-1} \\ r \\ r \end{pmatrix} \tag{3.34}
$$

$$
\boldsymbol{b}_{\mathrm{z}}^{i} = \begin{pmatrix} z_{\mathrm{c}} \\ 0 \\ 0 \\ 0 \\ 0 \end{pmatrix}, \quad \boldsymbol{b}_{\mathrm{z}}^{i} = \begin{pmatrix} 0 \\ 0 \\ 0 \\ 0 \\ z_{\mathrm{c}} \end{pmatrix}.
$$

Remark. As it can be seen in (3.34) the control points depend on the speeds v_{s} and v_{e} at the start point and the end point of the trajectory $\boldsymbol{w}_{\mathrm{t}}^{i}(t)$ and the transition time interval t_{t}^{i}. Hence, in contrast to the circular segment $\boldsymbol{w}_{\mathrm{cs}}^{i}(t)$, a change of t_{t}^{i} affects the shape of the segment. Due to the convex hull property of the Bézier curves the impact of a change of t_{t}^{i} on the shape of the segment can be evaluated in advance.

Determination of the transition time interval t_{t}^{i}. As it is stated in Sections 3.2.2 and 3.2.3, the transition time interval can be determined in two ways for two scenarios. First, for the reactive trajectory the transition time interval is determined by solving optimisation problem (3.18). As the circular shape of the segment is affected by t_{t}^{i} with this transition time interval, the circular shape of the trajectory can no longer be maintained. This short transition time interval influences the locations of the control points in a way so that the segment is less circular between two points $P_i(t)$, $(i = 1, \ldots, 4)$, as shown in Fig. 3.8. In the figure an ideal circular segment is illustrated in blue and the transition segment with the minimised transition time interval is coloured in red.

Second, the transition time interval is determined in a way so that the following two conditions are fulfilled:

- In order to obtain a shape of the transition segment, which is as similar as possible to an ideal circle, the positions of the control points are limited to the squares shown

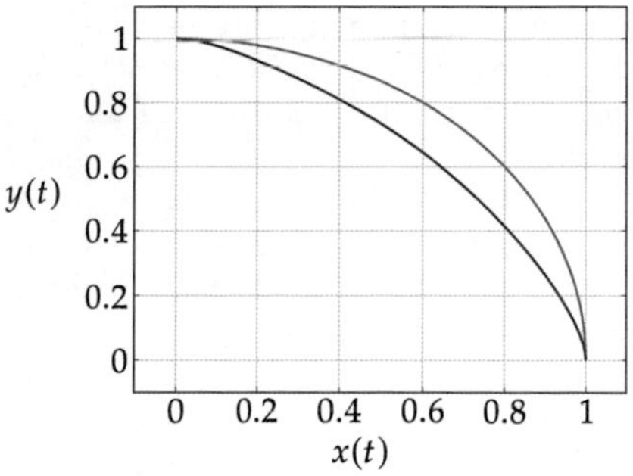

Fig. 3.8: Shape of a circular segment with transition time intervals determined by (3.9) (blue) and (3.18) (red).

in Fig. 3.9. The components of the control points affected by t_{t}^{i} may be a maximum of 10% smaller or larger than the radius r of the circle:

$$0.9\,r \le |\boldsymbol{b}_{\mathrm{x}}^{i}| \le 1.1\,r \quad \wedge \quad 0.9\,r \le |\boldsymbol{b}_{\mathrm{y}}^{i}| \le 1.1\,r, \quad i = 0, \ldots, N, \tag{3.35}$$

where the conditions have to be satisfied element-wise.

- The speed along the trajectory should remain in the intervals

$$\begin{aligned} v_{\mathrm{s}} &\le v(t, t_{\mathrm{t}}^{i}) = \left\|\tfrac{\mathrm{d}}{\mathrm{d}t}\boldsymbol{w}_{\mathrm{t}}^{i}(t)\right\| \le v_{\mathrm{e}}, \quad t \in [\bar{t}_i, \bar{t}_{i+1}], \quad (I) \\ v_{\mathrm{e}} &\le v(t, t_{\mathrm{t}}^{i}) = \left\|\tfrac{\mathrm{d}}{\mathrm{d}t}\boldsymbol{w}_{\mathrm{t}}^{i}(t)\right\| \le v_{\mathrm{s}}, \quad t \in [\bar{t}_i, \bar{t}_{i+1}], \quad (II) \end{aligned} \tag{3.36}$$

in which v_{s} is the speed at the start point and v_{e} corresponds to the speed at the end point of the transition segment.

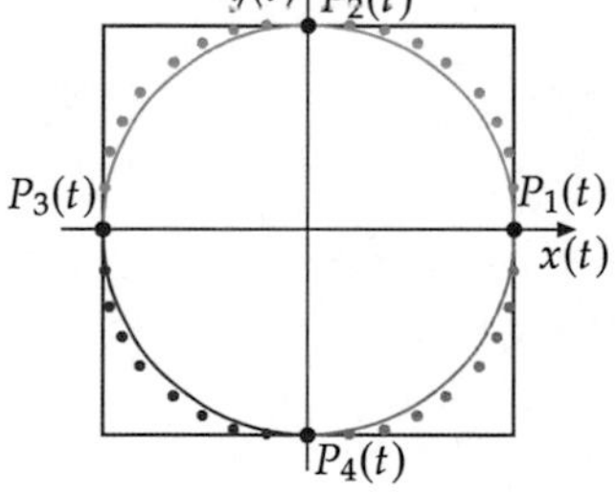

Fig. 3.9: Limitation of the positions of the control points.

Remark. It must be noted that an exact circular shape is not possible since the control points at the start point and at the end point of the segment are not located symmetrically. This is caused by the different boundary conditions on the trajectory due to the different speeds at the start point and the end point.

Inequality (I) of (3.36) describes an accelerating movement, while inequality (II) indicates a decelerating movement. Figure 3.10 shows the speed $v(t, t_\mathrm{t}^i)$ in the interval $[0, t_\mathrm{t}^i]$ for different transition time intervals t_t^i on a circular trajectory with radius $r = 1\,\mathrm{m}$ and the speed $v_\mathrm{s} = 0.5\,\frac{\mathrm{m}}{\mathrm{s}}$ at the start point and the speed $v_\mathrm{e} = 1\,\frac{\mathrm{m}}{\mathrm{s}}$ at the end point. It can be seen that the function has a local extremum depending on t_t^i. If t_t^i is too small, the upper bound of (3.36) (I) is violated (blue line in Fig. 3.10), if t_t^i is too large, the lower bound of (3.36) (I) cannot be fulfilled (red line in Fig. 3.10). With an appropriate transition time interval the conditions are fulfilled (green line in Fig. 3.10).

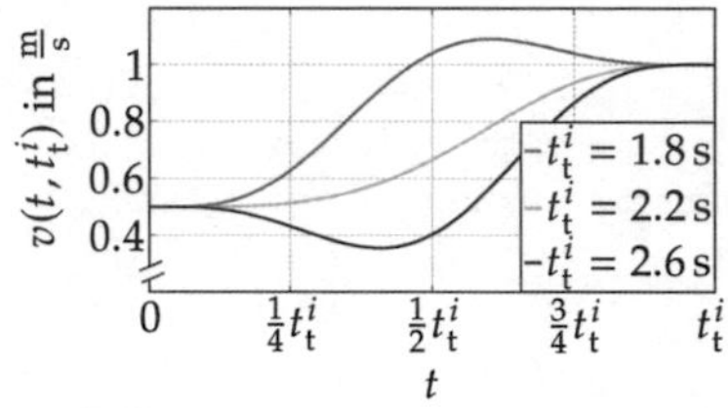

Fig. 3.10: Function of the speed $v(t, t_\mathrm{t}^i)$ for different times t_t^i and $\Delta v = 0.5\,\frac{\mathrm{m}}{\mathrm{s}}$.

The transition time interval is determined with Algorithm 6 in a way so that the transition segment fulfils conditions (3.35) and (3.36). The planning method is summarised in Algorithm 10 on p. 80.

Algorithm 6 Determination of a feasible transition time interval for a change of the speed

Given: Boundary conditions determined with (3.27) according to (3.33).
Radius r of the circle.
Start speed v_s.
End speed v_e.

1: If $v_\mathrm{s} > v_\mathrm{e}$: Set $t_\mathrm{min} = \frac{\pi r}{2 v_\mathrm{s}}$, $t_\mathrm{max} = \frac{\pi r}{2 v_\mathrm{e}}$.
If $v_\mathrm{s} < v_\mathrm{e}$: Set $t_\mathrm{min} = \frac{\pi r}{2 v_\mathrm{e}}$, $t_\mathrm{max} = \frac{\pi r}{2 v_\mathrm{s}}$.
2: Set $k = 1, n = 0$.
3: Plan transition segment $\boldsymbol{w}_\mathrm{t}^i(t)$ using Algorithm 3 on p. 57 with $t_\mathrm{t}^i = t_\mathrm{min} + n \cdot 10^{-k}$.
4: Determine $v(t, t_\mathrm{t}^i)$ of $\boldsymbol{w}_\mathrm{t}^i(t)$ with (3.30).
Determine $v_\mathrm{min} = \min_{t \in [\bar{t}_i, \bar{t}_{i+1}]} v(t, t_\mathrm{t}^i)$.
Determine $v_\mathrm{max} = \max_{t \in [\bar{t}_i, \bar{t}_{i+1}]} v(t, t_\mathrm{t}^i)$.
5: If $v_\mathrm{min} > v_\mathrm{e} \wedge v_\mathrm{max} < v_\mathrm{s}$:
  If $t_\mathrm{t}^i < t_\mathrm{max}$:
    Set $n = n + 1$; go to step 3.

If $t_t^i \geq t_{max}$:
Set $k = k + 1$, $n = 0$; go to step 3.
Else:
Terminate the algorithm.

Result: Transition time interval t_t^i for planning the transition segment for a change of the speed fulfilling conditions (3.35) and (3.36).

Remark. For the calculation of t_t^i, the maximum speed is not important but only the magnitude of the change of the speed. First, in the algorithm the upper and lower limits of the speed are defined depending on the direction of change of the speed (accelerating movement or decelerating movement). Due to the circular shape of the segment ensured by satisfying condition (3.35), the value of $v(t, t_{min})$ is greater than the upper bound. In the same way, the value of $v(t, t_{max})$ is smaller than the lower bound. Hence, the time interval t_t^i is increased from t_{min} to t_{max} and the extrema are evaluated recursively. If the values of the extrema are between the bounds, the algorithm terminates. If a value of an extremum is outside the bounds, the determination of t_t^i restarts and t_t^i is increased with a smaller step size k.

Verification of a constant direction of movement. The implementation of the planning algorithms requires the following considerations. According to Assumption 1.3 on p. 17 the agents are required to move constantly in one direction on their circular path (mathematically positive direction or negative direction). For an accelerating movement, as shown in Fig. 3.8, the assumption is always fulfilled. For decreasing transition time intervals resulting in faster accelerations the deviation of the transition segment to an ideal quarter circle increases, but the monotonous direction of movement is ensured.

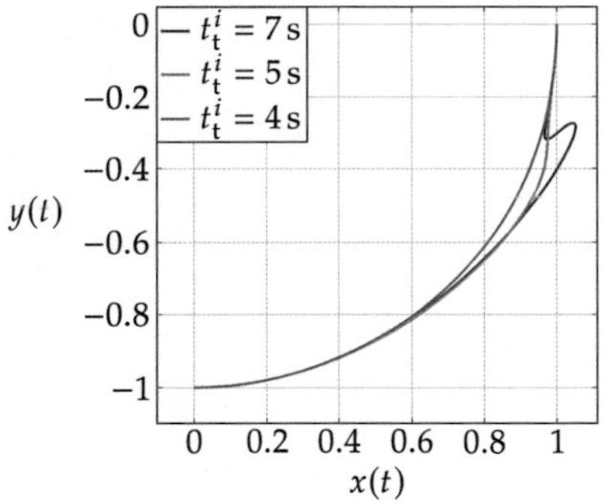

Fig. 3.11: Shape of a circular segment for a decelerating movement.

In contrast, for a decelerating movement the transition time intervals increase to elapse more time until the next intermediate point. In Fig. 3.11 the transition segments for a decelerating movement for different transition time intervals are illustrated. During the quarter circle segment with radius $r = 1\,\mathrm{m}$ the agent reduces its speed from $v_\mathrm{s} = 0.6\frac{\mathrm{m}}{\mathrm{s}}$ to $v_\mathrm{e} = 0.2\frac{\mathrm{m}}{\mathrm{s}}$. With a transition time interval $t_\mathrm{t}^i = 4\,\mathrm{s}$ nearly an ideal quarter circle is planned. The trajectory determined with $t_\mathrm{t}^i = 5\,\mathrm{s}$ obviously deviates from the ideal quarter circle, nevertheless a monotonous movement results. Increasing the transition time interval to $t_\mathrm{t}^i = 7\,\mathrm{s}$ results in a transition segment with which the agent reduces its speed to the end of the segment massively and moves a short distance in the opposite direction to the desired direction of movement.

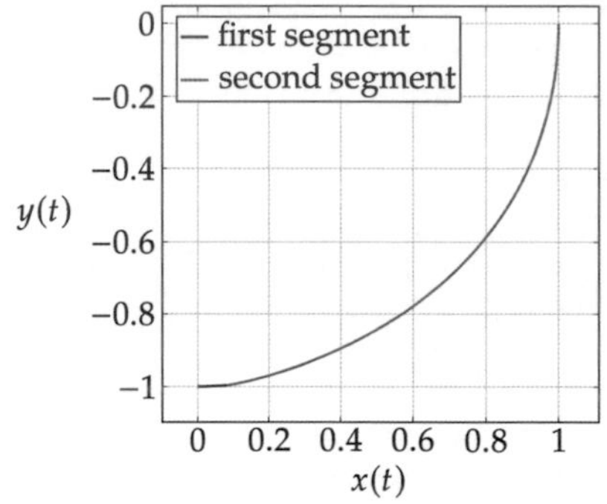

Fig. 3.12: Shape of a circular segment for a decelerating movement with an additional intermediate point.

This effect can be prevented by inserting another intermediate point with the boundary condition $v = 0\frac{\mathrm{m}}{\mathrm{s}}$ into the quarter circle segment. The distance $\Delta\Phi$ to this intermediate point and the corresponding transition time interval t_t1^i depend on the dynamics of the agent and are derived with Algorithm 8 on p. 75. The boundary conditions at the intermediate point are determined with (3.27). The transition time interval to the end point of the segment is given by $t_\mathrm{t2}^i = t_\mathrm{t}^i - t_\mathrm{t1}^i$. In Fig. 3.12 the trajectory with $t_\mathrm{t}^i = 7\,\mathrm{s}$ is shown after applying the approach. In the first part of the segment the agent slows down to a standstill within $t_\mathrm{t1}^i = 0.1\,\mathrm{s}$ (red line in Fig. 3.12). In the second part the agent covers within $t_\mathrm{t2}^i = 6.9\,\mathrm{s}$ the remaining distance to the end of the quarter circle segment (blue line in Fig. 3.12). The verification approach is summarised in Algorithm 7.

Algorithm 7 Verification of constant direction of movement

Given: Transition segment $w_\mathrm{t}^i(t)$.
Transition time interval t_t^i.
Number n of iterations.

1: If $v_\mathrm{s} > v_\mathrm{e}$: Set $t_\mathrm{min} = \frac{\pi r}{2 v_\mathrm{s}}$, $t_\mathrm{max} = \frac{\pi r}{2 v_\mathrm{e}}$.
If $v_\mathrm{s} < v_\mathrm{e}$: Set $t_\mathrm{min} = \frac{\pi r}{2 v_\mathrm{e}}$, $t_\mathrm{max} = \frac{\pi r}{2 v_\mathrm{s}}$.

2: Set $t_1 = 0$.
3: Set $t_2 = t_1 + \frac{t_t^i}{n}$.
4: Set $\Phi_1 = \Phi(w_x(t_1), w_y(t_1))$.
Set $\Phi_2 = \Phi(w_x(t_2), w_y(t_2))$.
Set $\dot{\Phi}(t) = \frac{\Phi_2 - \Phi_1}{t_2 - t_1}$.
5: If $\dot{\Phi}(t) < 0$:
Set $u_{\text{phase}} = 0$.
Terminate the algorithm.
If $t_2 < t_t$:
Set $\Phi_1 = \Phi_2$.
Set $t_1 = t_2$.
Go to step 2.
Else:
Set $u_{\text{phase}} = 1$
Terminate the algorithm.

Result: Boolean variable $u_{\text{phase}} \in [0, 1]$ that states if the phase $\Phi^i(t)$ of $w_t^i(t)$ is monotonous.

Remark. For both movements of the agents, in mathematically positive or negative direction, the phase $\Phi^i(t)$ is defined to increase over time ($\dot{\Phi}^i(t) \geq 0$). In order to verify if the phase $\Phi^i(t)$ is monotonous, eqn. (2.3) can be used by inserting $w_{t,x}^i(t)$ and $w_{t,y}^i(t)$ for x and y. However, the determination of the derivative $\dot{\Phi}^i(t)$ using analytical methods is quite complex. Hence, a numerical method based on difference quotients is used instead. The course of the phase $\Phi^i(t)$ is divided into n segments and their secants are determined successively. If a secant with a negative gradient ($\dot{\Phi}^i(t) < 0$) is found, the course of the phase is not monotonous and Assumption 1.3 is violated. If no secant with a negative gradient is found, Assumption 1.3 on p. 17 is satisfied.

Algorithm 8 Determination of an intermediate point

Given: Boundary condition at the start point determined with (3.27).
Radius r of the circle.
Start speed v_s.
End speed v_e.
Constraints given by (3.17) on p. 54.

1: Set $i = 1$, $s = 10^{-3}$ m.
2: Set $d = s \cdot i$, $\Delta\Phi = \frac{d}{r} \cdot \frac{360°}{2\pi}$.
3: Determine the boundary condition at the intermediate point with (3.27) using $\Phi^i(t) = \Phi^i(\bar{t}_i) + \Delta\Phi$ and $v = v_s$.
4: Determine transition time interval t_{t1}^i with Algorithm 2 on p. 56.
Verify the constant direction of movement with Algorithm 7.

5: If $u_{\text{phase}} = 1$:
Terminate the algorithm.
If $u_{\text{phase}} = 0$:
Set $i = i + 1$.
Go to step 2.

Result: Distance $\Delta\Phi$ to the intermediate point and transition time interval t^i_{t1} to this point.

3.4.2 *Trajectory for a change of the height*

Use of a transition segment. In addition to the change of the speed on a circular trajectory, a transition segment can also be used to change the height of the circular movement. At the start point and at the end point of the segment the same speed v in the xy-plane and no speed in z-direction should apply. In this section a transition segment between the points $P_1(t)$ and $P_2(t)$, as shown in Fig. 3.4, with a mathematically positive direction of rotation is planned. Segments between the other points can be determined in the same way.

Determination of the control points. The boundary conditions in the xy-plane are identical to them in (3.28). The boundary conditions for the z-direction are stated in the last columns of $\boldsymbol{W}^i_{\text{t}}(\bar{t}_i)$ and $\boldsymbol{W}^i_{\text{t}}(\bar{t}_{i+1})$:

$$\boldsymbol{W}^i_{\text{t}}(\bar{t}_i) = \begin{pmatrix} r & 0 & z_{\text{s}} \\ 0 & v & 0 \\ -v^2\,r^{-1} & 0 & 0 \\ 0 & -v^3\,r^{-2} & 0 \\ v^4\,r^{-3} & 0 & 0 \end{pmatrix}, \quad \boldsymbol{W}^i_{\text{t}}(\bar{t}_{i+1}) = \begin{pmatrix} 0 & r & z_{\text{s}} \pm \bar{z} \\ -v & 0 & 0 \\ 0 & -v^2\,r^{-1} & 0 \\ v^3\,r^{-2} & 0 & 0 \\ 0 & v^4\,r^{-3} & 0 \end{pmatrix}. \tag{3.37}$$

The parameter z_{s} denotes the height at the start point and $\bar{z}$ indicates the change of the height. With (3.37) the control points can be determined with (3.6) as

$$
\boldsymbol{b}_{\mathrm{x}}^{i} = \begin{pmatrix} r \\ r \\ r - \frac{t_{\mathrm{t}}^{i2}}{72} v^2 r^{-1} \\ r - \frac{t_{\mathrm{t}}^{i2}}{24} v^2 r^{-1} \\ r - \frac{t_{\mathrm{t}}^{i2}}{12} v^2 r^{-1} + \frac{t_{\mathrm{t}}^{i4}}{3024} v^4 r^{-3} \end{pmatrix}, \quad \boldsymbol{b}_{\mathrm{x}}^{i} = \begin{pmatrix} \frac{4\,t_{\mathrm{t}}^{i}}{9} v - \frac{t_{\mathrm{t}}^{i3}}{126} v^3 r^{-2} \\ \frac{t_{\mathrm{t}}^{i}}{3} v - \frac{t_{\mathrm{t}}^{i3}}{504} v^3 r^{-2} \\ \frac{2\,t_{\mathrm{t}}^{i}}{9} v \\ \frac{t_{\mathrm{t}}^{i}}{9} v \\ 0 \end{pmatrix}
$$

$$
\boldsymbol{b}_{\mathrm{y}}^{i} = \begin{pmatrix} 0 \\ \frac{t_{\mathrm{t}}^{i}}{9} v \\ \frac{2\,t_{\mathrm{t}}^{i}}{9} v \\ \frac{t_{\mathrm{t}}^{i}}{3} v - \frac{t_{\mathrm{t}}^{i3}}{504} v^3 r^{-2} \\ \frac{4\,t_{\mathrm{t}}^{i}}{9} v - \frac{t_{\mathrm{t}}^{i3}}{126} v^3 r^{-2} \end{pmatrix}, \quad \boldsymbol{b}_{\mathrm{y}}^{i} = \begin{pmatrix} r - \frac{t_{\mathrm{t}}^{i2}}{12} v^2 r^{-1} + \frac{t_{\mathrm{t}}^{i4}}{3024} v^4 r^{-3} \\ r - \frac{t_{\mathrm{t}}^{i2}}{24} v^2 r^{-1} \\ r - \frac{t_{\mathrm{t}}^{i2}}{72} v^2 r^{-1} \\ r \\ r \end{pmatrix}
$$

$$
\boldsymbol{b}_{\mathrm{z}}^{i} = \begin{pmatrix} z_{\mathrm{s}} \\ z_{\mathrm{s}} \\ z_{\mathrm{s}} \\ z_{\mathrm{s}} \\ z_{\mathrm{s}} \end{pmatrix}, \quad \boldsymbol{b}_{\mathrm{z}}^{i} = \begin{pmatrix} z_{\mathrm{s}} \pm \bar{z} \\ z_{\mathrm{s}} \pm \bar{z} \\ z_{\mathrm{s}} \pm \bar{z} \\ z_{\mathrm{s}} \pm \bar{z} \\ z_{\mathrm{s}} \pm \bar{z} \end{pmatrix}.
$$

As for the transition segment for the change of the speed, the control points for the x-direction and the y-direction depend on the transition time interval t_{t}^{i} and the speed v. At the start point and at the end point the control points are each located in one plane.

Determination of the transition time interval t_{t}^{i}. As for the transition segment for a change of the speed the transition time interval can be calculated by solving the optimisation problem (3.18) to derive the minimum transition time interval.

Otherwise the transition time interval t_{t}^{i} can be determined in two ways for two different scenarios: On the one hand the speed of the circular movement in the xy-plane should remain constant. This requirement increases the total speed $v(t, t_{\mathrm{t}}^{i})$ of the movement due to the additional change of the height. In this case the transition time interval is given by the time interval t_{tc}^{i} of the circular segment as

$$
t_{\mathrm{t}}^{i} = t_{\mathrm{tc}}^{i}.
$$

On the other hand the total speed $v(t, t_{\mathrm{t}}^{i})$ in the 3D space should remain constant. In order to achieve this aim, t_{t}^{i} is derived recursively with Algorithm 9 on p. 78. In Fig. 3.13 the speed $v(t, t_{\mathrm{t}}^{i})$ derived with (3.30) is shown in the interval $[0, t_{\mathrm{t}}^{i}]$ for different time intervals t_{t}^{i} for a trajectory that leads to a change of the height of $\bar{z} = 0.5\,\mathrm{m}$ with a constant speed $v = 1\,\frac{\mathrm{m}}{\mathrm{s}}$ at the start point and at the end point. As it can be seen, t_{t}^{i} influences the extrema of the function. As a constant speed over the entire interval $[0, t_{\mathrm{t}}^{i}]$ is desired, the deviations of the extrema from the constant speed v should be as small as possible. The function $v(t, t_t^i)$ has n extrema at the time instants t_l. The variance of the extrema as a measure of the deviations around v is computed with

$$\sigma^2(t_t^i) = \frac{1}{n} \sum_{l=0}^{n} (v(t_l, t_t^i) - v)^2. \tag{3.38}$$

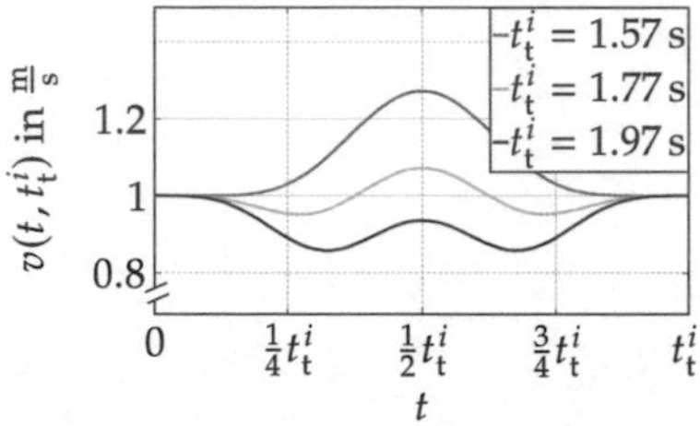

Fig. 3.13: Function of the speed $v(t, t_t^i)$ for different transition time intervals t_t^i and $\bar{z} = 0.5\,\mathrm{m}$.

For an iterative calculation of the transition time interval t_t^i with Algorithm 9 a minimum transition time interval $t_{\min} = t_{tc}^i$ is set. As the speed on the circular trajectory in the 2D plane is constant, with the choice of this time interval as the minimum transition time interval, it is intuitively clear that t_t^i has to be larger than $t_{\min}$ to obtain a constant total speed in the 3D space, because a longer distance needs to be covered due to the additional change of the height. As the distance between the locations of the control points increases with an increasing time t_t^i the condition (3.35) has to be satisfied to ensure the control points to be in the square shown in Fig. 3.9. By using Algorithm 9 the transition time interval t_t^i is feasible for a change of the height and it satisfies condition (3.35). The planning method is summarised in Algorithm 10 in the next section.

Algorithm 9 Determination of a feasible transition time interval for a change of the height

Given: Boundary conditions determined with (3.27).
Radius r of the circle.
Change of the height $\bar{z}$.
Constant speed v.

1: Set $t_{\min} = \frac{\pi r}{2 \bar{v}}$, $n = 0\,\mathrm{s}$.
2: Set $t_t^i = 0.01\,n + t_{\min}$.
3: Plan transition segment $\boldsymbol{w}_t^i(t)$ using Algorithm 3 on p. 57 with $t_t^i = 0.01\,n + t_{\min}$. If condition (3.35) is violated, go to step 7.
4: Determine $v(t, t_t^i)$ of $\boldsymbol{w}_t^i(t)$ with (3.30) and its extrema.
5: Determine $\sigma^2(t_t^i)$ with (3.38).
6: If $\sigma^2(t_t^i) > \bar{\sigma}$:

Set $n = n + 1$, go to step 2.
If $\sigma^2(t_t^i) \leq \bar{\sigma}$:
go to step 7.
7: Terminate the algorithm.

Result: Transition time interval t_t^i for planning the transition segment for a change of the height with a small speed variation around the constant speed v.

Remark. Algorithm 9 computes a feasible transition time interval t_t^i iteratively with a step size of 0.01 s. The step size is sufficient to find a trajectory with a nearly constant speed. A smaller step size would increase the computation effort significantly and would improve the result only marginally. A feasible transition time interval is found if

$$\sigma^2(t_t^i) \leq \bar{\sigma} \tag{3.39}$$

applies, where $\bar{\sigma}$ is an appropriately chosen tolerable deviation.

The algorithm is used because the function $\sigma^2(t_t^i)$ is algebraically complex to analyse. The function includes the square root of a polynomial of 18-th order, which would have to be calculated for each t_t^i. The complexity of the curve of $\sigma^2(t_t^i)$ can be seen in Fig. 3.14. The jagged course of $\sigma^2(t_t^i)$ results, because with an increasing time t_t^i the position of the control points changes slightly, leading to minor changes of the shape of the trajectory. This influences the speed on the trajectory and leads to slightly larger and smaller deviations of the extrema of $v(t, t_t^i)$ from the constant speed v.

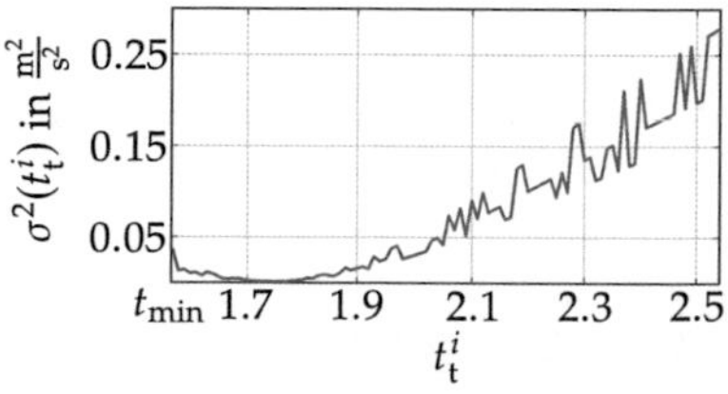

Fig. 3.14: Function $v(t, t_t^i)$ for $r = 1$ m, $v = 1\frac{m}{s}$ and $\bar{z} = 0.5$ m.

3.4.3 *Planning task and planning algorithm*

The task for planning trajectories for a change of the height or a change of the speed is stated as:

Given:
- Intermediate points $P_i(t)$, $(i = 1, \dots, 4)$ with corresponding phases $\Phi^i(t)$.
- Start time instant $\bar{t}_i$, start phase $\Phi(\bar{t}_i)$.
- Start speed v_s, end speed v_e or constant speed v, change of the height $\bar{z}$.
- Type of determination of the transition time interval t_t^i.
- Direction of rotation d_r.
- Radius r of the circle.

Find: Transition segment $\boldsymbol{w}_t^i(t)$ that satisfies the requirements $\boldsymbol{W}_t^i(\bar{t}_i)$, $\boldsymbol{W}_t^i(\bar{t}_{i+1})$ derived with (3.27) and the conditions (3.35), (3.36) or (3.39).

Boundary conditions: System and actuator limitations (1.8), (1.9), (1.11) on p. 17.

The trajectory planning method is summarised in Algorithm 10.

Algorithm 10 Planning of a transition segment

Given: Start position $P_i(t)$, $(i = 1, \dots, 4)$.
Intermediate points $P_i(t)$, $(i = 1, \dots, 4)$ with corresponding phases $\Phi^i(t)$.
Start time instant $\bar{t}_i$, start phase $\Phi(\bar{t}_i)$.
Start speed v_s, end speed v_e or constant speed v, change of the height $\bar{z}$.
Type of determination of the transition time interval t_t^i.
Direction of rotation d_r.
Radius r of the circle.
Type of transition segment to be planned.

1: Transition segment for a change of the speed: go to step 2.
Transition segment for a change of the height: go to step 7.

2: Determine the boundary conditions at the start point with (3.27) using $\Phi^i(t) = \Phi(\bar{t}_i)$ and $v = v_s$.
Determine the boundary conditions at the end point with (3.27) using $\Phi^i(t) = \mathrm{mod}\,(\Phi(\bar{t}_i) + 90°, 360°)$ and $v = v_e$.

3: If $t_t^i = t_{t,\min}$:
  Determine transition time interval $t_{t,\min}$ by executing Algorithm 2 on p. 56.
Otherwise:
  Determine transition time interval t_t^i by executing Algorithm 6 on p. 72.

4: Plan transition segment $\boldsymbol{w}_t^i(t)$ for a change of the speed by executing Algorithm 3 on p. 57.

5: Verify the constant direction of movement with Algorithm 7 on p. 74.
6: If $u_{\text{phase}} = 1$:
  Terminate the algorithm.
 If $u_{\text{phase}} = 0$:
  Insert intermediate point at position $\Phi^i(t) + \Delta\Phi$ with Algorithm 8 on p. 75.
  Go to step 2 using $\Phi^i(t) + \Delta\Phi$ for the end point of the first part of the segment and for the start point of the second part of the segment.
7: Determine the boundary conditions at the start point with (3.27) using $\Phi^i(t) = \Phi(\bar{t}_i)$.
 Determine the boundary conditions at the end point with (3.27) using $\Phi^i(t) = \text{mod}\,(\Phi(\bar{t}_i) + 90°, 360°)$ and $\bar{z}$.
8: If $t_t^i = t_{t,\min}$:
  Determine transition time interval $t_{t,\min}$ by executing Algorithm 2.
 Otherwise:
  Determine transition time interval t_t^i by executing Algorithm 9 on p. 78.
9: Plan transition segment $\boldsymbol{w}_t^i(t)$ for a change of the height by executing Algorithm 3.

Result: Transition segment $\boldsymbol{w}_t^i(t)$ for a change of the speed or a change of the height in the form (2.15) that fulfil the limitations (1.8), (1.9), (1.11).

Remark. The transition segments are used by the event-based control unit to change the trajectory of an agent. A change of the trajectory is invoked after the generation of an event at an arbitrary time instant. Hence, with the trajectory planning method, it is possible to start the transition segments not only at the intermediate points $P_i(t)$, $(i = 1, \ldots, 4)$, but at any points on the circular path.

Prediction and event generation 4

This chapter presents two basic methods, which are necessary for the event-based control of the agents to satisfy their control aims. First, a prediction method is given, which includes all possible future positions of a nearby agent with an uncertain movement in a set (Section 4.2). Methods for both, a general movement and a circular movement of agents are given. Second, a method for the event generation is stated, which evaluates the distance between two trajectories or the distance between a trajectory and a set of positions (Section 4.3).

4.1 Overview

This chapter describes two methods that are utilised by the event-based control units of the agents. First, a prediction method is introduced for the estimation of the uncertain movement of a nearby agent. An execution of the method solves Problem 1.2 on p. 15. The contributions of this chapter are two algorithms, whose execution leads to a feasible inclusion of the future positions of an agent. For a general movement of the agents, Algorithm 11 on p. 88 is given. An execution of Algorithm 12 on p. 92 includes the future positions of an agent for a circular movement. Here, the more strict limitations on the movement are taken into account in order to reduce the conservatism of the estimate. The estimation results are given in Theorem 4.1 on p. 87 for a general movement and in Theorem 4.2 on p. 91 for a circular movement.

Second, a method for the generation of events is introduced that is based on the distance between two agents in order to solve Problem 1.3 on p. 15. The first part of the method considers the general movement of the agents. Events are generated by evaluating the distance between a position and a set of positions in order to invoke communication. The distances between two trajectories or two positions at a certain time instant are determined in order to invoke a change of the trajectory of an agent. The second part of the method takes the circular movement of the agents into account. An evaluation of the phase difference and the height difference between a position and a set of positions leads to an event generation for the invocation of communication. Furthermore, the change of the trajectory of an agent is invoked after an event generation based on the phase difference and the height difference between two trajectories or two positions at a certain time instant.

The chapter provides appropriate event conditions and approaches for a simple evaluation of these conditions.

4.2 Prediction of the movement of mobile agents

4.2.1 *General prediction method*

Motivation for the prediction of the movement of nearby agents. For the event-based control of an agent communicated information from neighbouring agents are only available at discrete points in time. However, in order to fulfil the control aims, the positions of nearby agents must be known continuously, because the trajectories of these agents can change at any time. Predetermined communication time instants can lose their validity due to the changed trajectories. Hence, the movement of neighbouring agents must be predicted. The method developed in this section is utilised by the event-based control unit to solve Problem 1.2.

Exact prediction of the movement of nearby agents. An exact set of possible future positions of a nearby agent that does not overestimate the movement of the agent can be generated by considering the trajectory planning method, which is the same for all agents according to Assumption 1.4 on p. 17. For the prediction, all possible trajectories for the nearby agent have to be determined. Trajectories starting from the last communicated position of the agent to all possible end points $\boldsymbol{w}(t_{k+1})$ for $t_{k+1} > t_k$ within the transition time interval $t_\mathrm{t} = t_{k+1} - t_k$ need to be determined. This can be done by a reformulation of the optimisation problem (3.18) on p. 54. It is searched for all possible end points of the trajectories of the nearby agent in a certain transition time interval t_t, which are furthest away from the start point. The boundary conditions at the start point are given by

$$\boldsymbol{W}(t_k) = \begin{pmatrix} \boldsymbol{w}^\mathrm{T}(t_k) \\ \dot{\boldsymbol{w}}^\mathrm{T}(t_k) \\ \ddot{\boldsymbol{w}}^\mathrm{T}(t_k) \\ \boldsymbol{w}^{(3)\mathrm{T}}(t_k) \\ \boldsymbol{w}^{(4)\mathrm{T}}(t_k) \end{pmatrix} = \begin{pmatrix} w_\mathrm{x}(t_k) & w_\mathrm{y}(t_k) & w_\mathrm{z}(t_k) & w_{,\psi}(t_k) \\ \dot{w}_\mathrm{x}(t_k) & \dot{w}_\mathrm{y}(t_k) & \dot{w}_\mathrm{z}(t_k) & \dot{w}_\psi(t_k) \\ \ddot{w}_\mathrm{x}(t_k) & \ddot{w}_\mathrm{y}(t_k) & \ddot{w}_\mathrm{z}(t_k) & \ddot{w}_\psi(t_k) \\ w_\mathrm{x}^{(3)}(t_k) & w_\mathrm{y}^{(3)}(t_k) & w_\mathrm{z}^{(3)}(t_k) & w_\psi^{(3)}(t_k) \\ w_\mathrm{x}^{(4)}(t_k) & w_\mathrm{y}^{(4)}(t_k) & w_\mathrm{z}^{(4)}(t_k) & w_\psi^{(4)}(t_k) \end{pmatrix}, \tag{4.1}$$

which are contained in the communicated data. The transition time interval t_t increases simultaneously with the ongoing time ($t_{k+1} = t$), since the agent moves on over time and covers longer distances in longer time intervals. Hence, in the reformulated problem the end points $\boldsymbol{w}(t_{k+1})$, which can be reached within t_t subject to the constraints (1.8) – (1.10) are the decision variables:

$$\begin{aligned}
\max_{\boldsymbol{w}(t_{k+1})} \quad & \boldsymbol{J} = \boldsymbol{w}(t_{k+1}) \\
\text{s.t.} \quad & \boldsymbol{W}(t_k) \text{ given by (4.1)}, \ t_{k+1}, \\
& \text{(3.6)}, \ \text{(2.18)}, \ \text{(3.10) in the form (2.15)}, \\
& \text{(3.11), (3.12), (3.15) – (3.17).}
\end{aligned} \tag{4.2}$$

It is a complex problem, because due to the movement of the agents in the 3D airspace it is a multi-dimensional optimisation problem, which needs to be solved for each time step of the ongoing time t to obtain all possible future positions of the agent.

As the cost function $\boldsymbol{J} = \boldsymbol{w}(t_{k+1})$ is a multidimensional linear function, it is convex according to Section 3.2.3. The cost function is exemplarily shown for the x-direction in Fig. 4.1. Since the same constraints are used as in the optimisation problem (3.18) extended by the time instant t_{k+1}, it can be concluded that the constraints are convex as well with the same considerations as in Section 3.2.3. Hence, the optimisation problem (4.2) is convex.

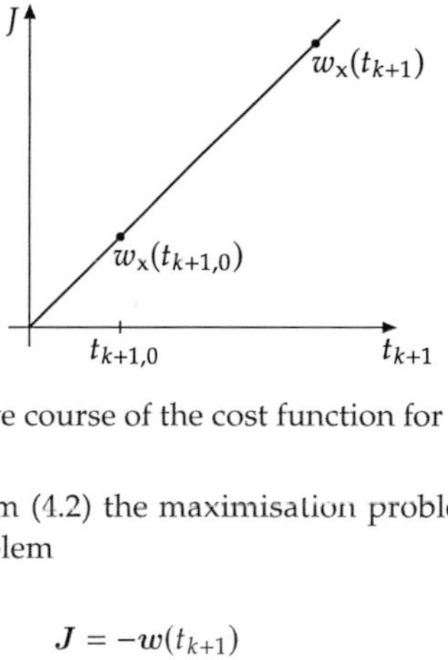

Fig. 4.1: Qualitative course of the cost function for the x-direction.

For the solution of problem (4.2) the maximisation problem is transformed into an equivalent minimisation problem

$$\begin{aligned}
\min_{\boldsymbol{w}(t_{k+1})} \quad & \boldsymbol{J} = -\boldsymbol{w}(t_{k+1}) \\
\text{s.t.} \quad & \boldsymbol{W}(t_k) \text{ given by (4.1)}, \ t_{k+1}, \\
& \text{(3.6)}, \ \text{(2.18)}, \ \text{(3.10) in the form (2.15)}, \\
& \text{(3.11), (3.12), (3.15) – (3.17),}
\end{aligned} \tag{4.3}$$

which can be solved with the gradient descent method [45]. The three-dimensional problem can be handled by fixing two dimensions and optimising for the third dimension. To this aim, Algorithm 1 on p. 39 can be used, in which t_{k+1} is replaced by $\boldsymbol{w}(t_{k+1})$.

Prediction of the movement of nearby agents as an inclusion of positions in a set. A computational simpler approach is to generate a circle with a time-varying radius $r(t_t) = v_{\max} \cdot t_t$ where the communicated position $p_S(t_k) = w(t_k)$ is the centre, as described by [160]. As the maximum speed is bounded by $v_{\max}$, this circle includes all possible future positions, but the position estimation is conservative, because the agent is not able to move with $v_{\max}$ instantaneously in every direction due to its dynamical limitations. Therefore, a position estimation method is proposed, which is computational simpler than solving the optimisation problem (4.3), but less conservative compared to the method from [160], because it considers the dynamics of the agent.

The idea of the position prediction is to generate a set $\mathcal{P}(t, t_k)$ as an ellipsoidal inclusion of the future positions of a nearby agent. An ellipsoidal set is specified in Definition 4.1. The prediction method has to satisfy the limitation (1.10) of Assumption 1.2 on p. 17 and uses the knowledge about the trajectory planning method to generate the set. Due to the limitation, an agent with an initial speed $v(t_k) > 0$ at the time instant t_k cannot move instantaneously from the communicated position $p(t_k)$ in any direction. In the same way, an agent with speed $v(t_k) = 0$ at t_k has to accelerate to the speed $v_{\max}$. The change of the direction and the acceleration of an agent with respect to Assumption 1.2 causes the positions to remain in an ellipsoid with a centre moving in the communicated direction and radii that expand over time.

Definition 4.1. (Ellipsoidal set) [45] An ellipsoidal set is given by

$$\mathcal{A} = \left\{ \boldsymbol{x} \in \mathbb{R}^n : \boldsymbol{x}^{\mathrm{T}} \boldsymbol{P} \boldsymbol{x} \leq 0 \right\}$$

with $\boldsymbol{P} \in \mathbb{R}^{n \times n}$ is symmetric and positive definite.

Definition 4.2. (ε-neighbourhood) [98] Let $\varepsilon > 0$ and $x \in \mathbb{R}$. The ε-neighbourhood of x is defined by

$$U[x] := \{ y \in \mathbb{R} \mid |y - x| < \varepsilon \} =]x - \varepsilon, x + \varepsilon[.$$

The prediction result for the inclusion of the positions of a general moving agent uses Definition 4.2 and Assumptions 1.2 and 1.4 on p. 17. The requirement $w(t_k) = \mathbf{0}$ of Assumption 1.4 can be achieved by a coordinate shift without loss of generality and results in simpler calculations for the prediction.

Theorem 4.1 (Prediction of the movement of an agent). *Consider an agent, that has the position $\mathbf{p}(t_k)$ and the speed $\boldsymbol{v}(t_k)$ at the time instant t_k and moves on trajectories given by Bézier curves (3.19), that satisfy the limitations of Assumptions 1.2 and 1.4. Then, the future positions $\mathbf{p}(t)$, $(t \geq t_k)$ of the agent are included in the set*

$$\mathcal{P}(t,t_k) = \left\{ \mathbf{p}(t) \in \mathbb{R}^3 : \frac{(x(t)-\bar{x}(t,t_k))^2}{r_{\mathrm{x}}^2(t,t_k)} + \frac{(y(t)-\bar{y}(t,t_k))^2}{r_{\mathrm{y}}^2(t,t_k)} + \frac{(z(t)-\bar{z}(t,t_k))^2}{r_{\mathrm{z}}^2(t,t_k)} - 1 \leq 0 \right\} \tag{4.4}$$

with

$$\begin{aligned} \bar{x}(t,t_k) &= x(t_k) + v_{\mathrm{x}}(t_k) \cdot e \cdot (t - t_k) \\ \bar{y}(t,t_k) &= y(t_k) + v_{\mathrm{y}}(t_k) \cdot e \cdot (t - t_k) \\ \bar{z}(t,t_k) &= z(t_k) + v_{\mathrm{z}}(t_k) \cdot e \cdot (t - t_k) \\ r_{\mathrm{x}}(t,t_k) &= r_{\mathrm{y}}(t,t_k) = f \cdot v_{\max} \cdot (t - t_k) \\ r_{\mathrm{z}}(t,t_k) &= v_{\max} \cdot (t - t_k), \end{aligned} \tag{4.5}$$

$e = 1 - f$ *and*

$$f \leq \frac{1}{2} \left[\left[\frac{1}{\max\limits_{U[0.6]} a_1(s)} \cdot \left[\sqrt{1 - \left[\min\limits_{U[0.6]} b(s) \right]^2} - \max\limits_{U[0.6]} a_2(s) \right] \right]^2 + 1 \right]. \tag{4.6}$$

Proof. See Appendix B.2. ∎

The theorem states an interesting result, because it provides an inclusion of the future movement of an agent in a simple way. The computational effort is small compared to other approaches in the literature.

Remark. The set $\mathcal{P}(t,t_k)$ is an ellipsoid with the centre moving as described by the upper three equations of (4.5). The radii expand over time stated by the lower two equations of (4.5). The coefficients e and f are scaling factors, which ensure that the centre of the ellipsoid does not move with the communicated speed $\boldsymbol{v}(t_k)$ and the radii of the ellipsoid do not increase with the maximum speed $v_{\max}$. Hence, these factors lead to a reduction of the conservatism compared to the method proposed by [160]. The operator $\max\limits_{U[x]}$ delivers the maximum of a function in the ε-neighbourhood $U[x]$ of x defined in Definition 4.2.

The functions $a_1(s)$, $a_2(s)$ and $b(s)$ result from the trajectory planning method and are derived in the proof of the theorem. With the substitution $s = \frac{t}{t_t}$ they are given by

$$a_1(t, t_t) = \left(630\left(\frac{t}{t_t}\right)^8 - 2520\left(\frac{t}{t_t}\right)^7 + 3780\left(\frac{t}{t_t}\right)^6 - 2520\left(\frac{t}{t_t}\right)^5 + 630\left(\frac{t}{t_t}\right)^4\right) \tag{4.7}$$

$$a_2(t, t_t) = -315\left(\frac{t}{t_t}\right)^8 + 1240\left(\frac{t}{t_t}\right)^7 - 1820\left(\frac{t}{t_t}\right)^6 + 1176\left(\frac{t}{t_t}\right)^5 - 280\left(\frac{t}{t_t}\right)^4 \tag{4.8}$$

$$b(t, t_t) = -315\left(\frac{t}{t_t}\right)^8 + 1280\left(\frac{t}{t_t}\right)^7 - 1960\left(\frac{t}{t_t}\right)^6 + 1344\left(\frac{t}{t_t}\right)^5 - 350\left(\frac{t}{t_t}\right)^4 + 1. \tag{4.9}$$

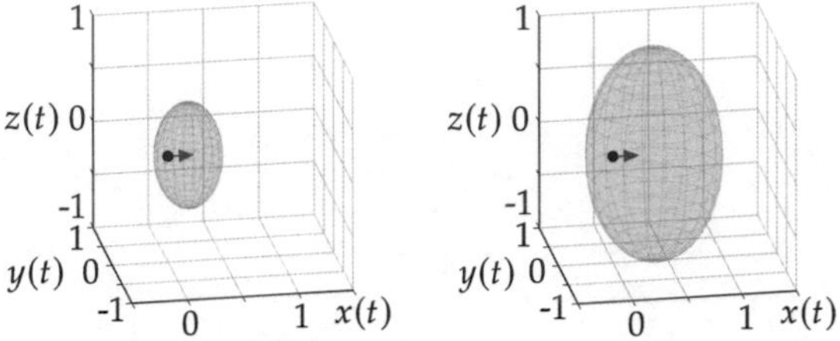

Fig. 4.2: Inclusion for the position $\boldsymbol{p}(t)$ 1 s (left) and 2 s (right) after the time instant t_k. The communicated direction of movement is indicated by the arrow.

An example set $\mathcal{P}(t, t_k)$ for the inclusion of the position of a nearby agent is depicted in Fig. 4.2 as blue ellipsoid. The black dot states the last communicated position $\boldsymbol{p}(t_k)$ with a speed $\boldsymbol{v}(t_k)$ in x-direction. The expansion of the ellipsoid is shown for $t = t_k + 1\,\text{s}$ (Fig. 4.2 left) and $t = t_k + 2\,\text{s}$ (Fig. 4.2 right). The prediction method is summarised in Algorithm 11 that uses Theorem 4.1. The event-based control unit uses the algorithm to solve Problem 1.2.

Algorithm 11 Prediction of the movement of an agent

Given: Position $\boldsymbol{p}(t_k)$ of the agent at a time instant t_k.
Speed $\boldsymbol{v}(t_k)$ of the agent at a time instant t_k.
Parameters e and f.

1: Determine moving centre of the ellipsoid $\bar{x}(t, t_k)$, $\bar{y}(t, t_k)$, $\bar{z}(t, t_k)$ continuously with eqns. (4.5).
2: Determine expanding radii of the ellipsoid $r_x(t, t_k)$, $r_y(t, t_k)$, $r_z(t, t_k)$ continuously with eqns. (4.5).
3: Generate ellipsoid $\mathcal{P}(t, t_k)$ with (4.4).

Result: Ellipsoid $\mathcal{P}(t, t_k)$ that includes the future general movement of an agent.

Example 4.1. The basic idea of Theorem 4.1 is illustrated in Fig. 4.3 in the 2D space at four points in time. The agent starts on the position marked as a black star and is able to move on the blue trajectories. The end points $w(t_{k+1})$ of the trajectories are determined by solving the optimisation problem (4.3) on p. 85. The illustrated trajectories have been planned using the method from Chapter 3. They exemplarily show a selection of all trajectories that the agent could follow. The estimation method includes all possible trajectories in the red circle, which reduces the computational effort considerably. Furthermore, it can be seen that the position estimation is less conservative compared to an inclusion, which has a fixed centre at $p(t_k)$ with radii that expand with $v_{\max}$, shown as the green dotted circle.

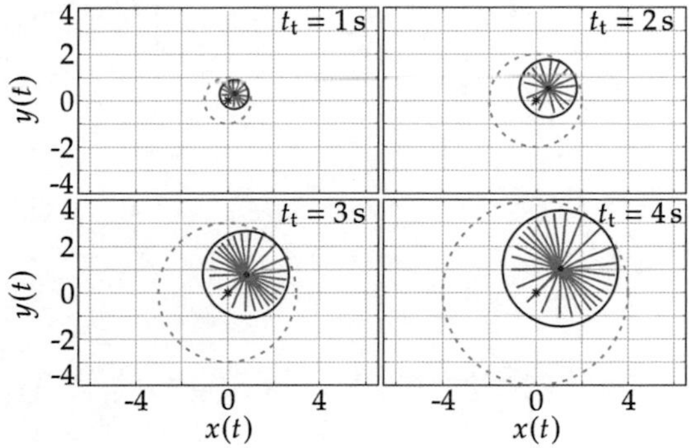

Fig. 4.3: Position estimate with $||v(t_k)|| = 1\,\frac{\mathrm{m}}{\mathrm{s}}$ at four different time instants $t = t_k + t_t$.

The method slightly overestimates the set of possible positions, because only the limitation (1.10) of Assumption 1.2 on p. 17 is taken into account in the estimation. The system limitation (1.8) and the actuator limitation (1.9) are not considered to simplify the calculations. □

Example 4.2. The utilisation of Theorem 4.1 for the prediction of a car movement along a motorway is illustrated in Fig. 4.4. The prediction of the future positions in car applications is simpler compared to other application fields (e.g. UAVs), because the cars are only able to move along their traffic lanes. Hence, the path of their trajectory remains unchanged and the uncertainty about the position arises only from a change of speed along the path.

In the example in Fig. 4.4 the cars are at the positions marked by the black dots at the time instant t_k. They move on the traffic lanes represented by the black lines. The car on the right is on the slip road to the motorway. For a suitable threading, the car on the middle lane predicts the movement of the car on the right. The set $\mathcal{P}_S(t, t_k)$ shown in red increases over time and includes only positions along the lane. As the car has to follow the lane at a positive speed, the set of positions is solely in front of the car. □

Representation of the set $\mathcal{P}(t, t_k)$ in matrix notation. For the event generation described in Section 4.3.1 the set $\mathcal{P}(t, t_k)$ (4.4) needs to be represented in matrix notation. With the vector

$$q(t) = \begin{pmatrix} x(t) & y(t) & z(t) & 1 \end{pmatrix}^{\mathrm{T}}$$

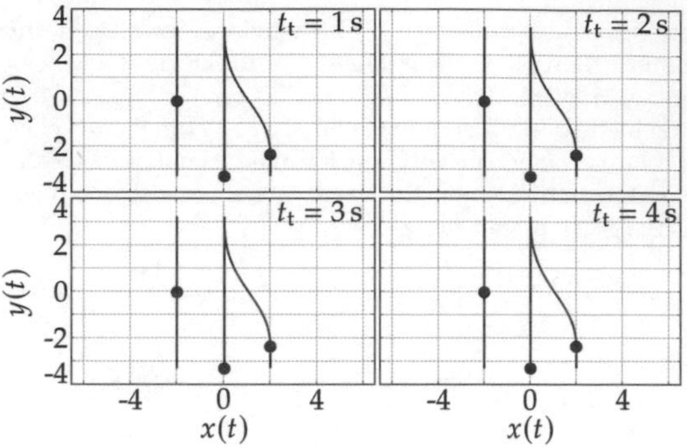

Fig. 4.4: Position estimate for a car moving with $||\boldsymbol{v}(t_k)|| = 1\frac{\mathrm{m}}{\mathrm{s}}$ at four different time instants $t = t_k + t_t$.

and the matrix

$$\boldsymbol{P}(t, t_k) = \begin{pmatrix} \frac{1}{r_x^2(t,t_k)} & 0 & 0 & -\frac{\bar{x}(t,t_k)}{r_x^2(t,t_k)} \\ 0 & \frac{1}{r_y^2(t,t_k)} & 0 & -\frac{\bar{y}(t,t_k)}{r_y^2(t,t_k)} \\ 0 & 0 & \frac{1}{r_z^2(t,t_k)} & -\frac{\bar{z}(t,t_k)}{r_z^2(t,t_k)} \\ -\frac{\bar{x}(t,t_k)}{r_x^2(t,t_k)} & -\frac{\bar{y}(t,t_k)}{r_y^2(t,t_k)} & -\frac{\bar{z}(t,t_k)}{r_z^2(t,t_k)} & p_{4,4}(t,t_k) \end{pmatrix} \tag{4.10}$$

with

$$p_{4,4}(t, t_k) = \frac{\bar{x}^2(t, t_k)}{r_x^2(t, t_k)} + \frac{\bar{y}^2(t, t_k)}{r_y^2(t, t_k)} + \frac{\bar{z}^2(t, t_k)}{r_z^2(t, t_k)} - 1$$

the set $\mathcal{P}(t, t_k)$ is given in matrix notation by

$$\boldsymbol{q}^{\mathrm{T}}(t)\boldsymbol{P}(t, t_k)\boldsymbol{q}(t) = 0.$$

Hence, the set $\mathcal{P}(t, t_k)$ can also be written as

$$\mathcal{P}(t, t_k) = \{\boldsymbol{p}(t) \in \mathbb{R}^3 : \boldsymbol{q}^{\mathrm{T}}(t)\boldsymbol{P}(t, t_k)\boldsymbol{q}(t) = 0\}.$$

4.2.2 *Specific prediction method for a circular movement*

The freedom of movement for agents that are only allowed to move on circular trajectories is strongly restricted by Assumption 1.3 on p. 17 compared to the general movement considered in the last section. Nevertheless, the agent is able to change its speed on the circular path and the height of its circular movement, which makes its trajectory uncertain and thus requires the prediction of the possible future movement.

For the prediction of a circular movement of an agent, the prediction method can be simplified compared to the method for a general movement to reduce its conservatism considerably. Similar to the last section a set $\mathcal{P}_c(t, t_k)$ is generated that is based on the communicated position $\boldsymbol{p}(t_k)$ of the agent at t_k in the cartesian coordinate system. Due to the circular movement it is appropriate to predict the future phases $\Phi(t)$ and the height $z(t)$ of the nearby agent using cylindrical coordinates instead of using cartesian coordinates. Using eqn. (2.3) on p. 30 with the prediction method the phase $\Phi(t_k)$ and the height $z(t_k)$ are determined utilising the communicated position $\boldsymbol{p}(t_k)$. Theorem 4.2 states the specific prediction result for a circular movement of the agents. The resulting set $\mathcal{P}_c(t, t_k)$ consists of two subsets for the prediction of the phase and the prediction of the height of an agent.

Theorem 4.2 (Prediction of the circular movement of an agent). *Consider an agent, that has the phase $\Phi(t_k)$ and the height $z(t_k)$ at time instant t_k and moves on trajectories given by Bézier curves* (3.23) *satisfying the limitations of Assumption* 1.3 *on p.* 17. *Then the future phases $\Phi(t)$ and heights $z(t)$, $(t \geq t_k)$ of the agent are included in the set*

$$\mathcal{P}_c(t, t_k) = \left\{\boldsymbol{p}(t) \in \mathbb{R}^3 : \Phi(t) \in \Delta\Phi(t, t_k),\ z(t) \in \Delta z(t, t_k)\right\} \tag{4.11}$$

where

$$\Delta\Phi(t, t_k) = [\Phi_{\min}(t, t_k), \Phi_{\max}(t, t_k)] \tag{4.12}$$

$$\Delta z(t, t_k) = [z_{\min}(t, t_k), z_{\max}(t, t_k)] \tag{4.13}$$

with

$$\Phi_{\min}(t, t_k) = \mathrm{mod}\ \left(\Phi_S(t_k) + \tfrac{v_{\min}}{2\pi r} \cdot (t - t_k) \cdot 360°, 360°\right) \tag{4.14}$$

$$\Phi_{\max}(t, t_k) = \mathrm{mod}\ \left(\Phi_S(t_k) + \tfrac{v_{\max}}{2\pi r} \cdot (t - t_k) \cdot 360°, 360°\right) \tag{4.15}$$

$$z_{\min}(t, t_k) = \min\left(z_1(t, t_k), z_2(t, t_k)\right) \tag{4.16}$$

$$z_{\max}(t, t_k) = \max\left(z_1(t, t_k), z_2(t, t_k)\right) \tag{4.17}$$

and

$$z_1(t, t_k) = |z_S(t_k) + v_{\max} \cdot (t - t_k)|$$

$$z_2(t, t_k) = |z_S(t_k) - v_{\max} \cdot (t - t_k)|\,.$$

r is the radius of the circle and $\mathrm{mod}\ (a, m)$ *is the modulo operator defined in* (2.4) *on p.* 31.

Proof. The proof of Theorem 4.2 unfolds in two steps. First, the inclusion of the phase is proven. At t_k the phase $\Phi(t_k)$ of an agent is known due to the communicated position

$p(t_k)$. As the agent is only allowed to move with a speed $v_{\min} < v(t) < v_{\max}$ due to Assumption 1.3, its possible covered distance on the circle is between the values

$$\begin{aligned} s_{\min}(t, t_k) &= v_{\min} \cdot (t - t_k) \\ s_{\max}(t, t_k) &= v_{\max} \cdot (t - t_k). \end{aligned} \tag{4.18}$$

As the agent moves on a circular trajectory, the phase covered after the time instant t_k can be calculated from the relation of the distance covered (4.18) and the perimeter of a circle. By adding the phase $\Phi(t_k)$, the limits (4.14) and (4.15) of the set (4.12) are obtained.

Second, the inclusion of the height is proven. Due to Assumption 1.3 the agent changes its height with a maximum speed $v_{\max}$. As the bounds (4.16) and (4.17) of the estimate of the heights change with $v_{\max}$, the heights are included in the set (4.13). The set (4.11) combines the sets (4.12) and (4.13), which proves the theorem. ■

Remark. The set $\mathcal{P}_c(t, t_k)$ is composed of two subsets $\Delta\Phi(t, t_k)$ and $\Delta z(t, t_k)$ that include the possible future phases of the nearby agent and its possible heights, respectively. Both subsets increase over time, indicating the increasing uncertainty of the phase and the height of the moving agent over time.

The prediction method for a circular movement is summarised in Algorithm 12 that uses Theorem 4.2. With the algorithm Problem 1.2 for a circular movement can be solved.

Algorithm 12 Prediction of the movement of an agent for a circular movement

Given: Position $p(t_k)$ of the agent at a time instant t_k.

1: Determine $\Phi(t_k)$ and $z(t_k)$ by using $p(t_k)$.
2: Determine subsets $\Delta\Phi(t, t_k)$ and $\Delta z(t, t_k)$ continuously with (4.12) and (4.13).
3: Generate set $\mathcal{P}_c(t, t_k)$ with (4.11).

Result: Set $\mathcal{P}_c(t, t_k)$ that includes the future circular movement of an agent.

Example 4.3. The prediction method for a circular movement is illustrated in Fig. 4.5 for four time points. For the illustration, two agents $\bar{P}_S$ and $\bar{P}_G$ are considered. Both agents in the figure are located on the circle with radius $r = 1\,\text{m}$ for a movement with the constant speed

$$v_S(t) = v_G(t) = 0.5\, v_{S,\max}$$

where

$$v_{S,\min} = 0\frac{\text{m}}{\text{s}}, \quad v_{S,\max} = \frac{\pi}{4}\frac{\text{m}}{\text{s}} \approx 0.79\frac{\text{m}}{\text{s}}.$$

Hence, in order to cover the complete circle once, the agents need 16 s. At time $t_0 = 0$ the agents are located at positions with the corresponding phases

$$\Phi_G(0) = 0^\circ, \quad \Phi_S(0) = 180^\circ.$$

The position of $\bar{P}_S$ is communicated to $\bar{P}_G$ so that its position is exactly determined. Within the first 2 seconds, both agents cover a distance corresponding to a phase of 45°. The prediction unit of $\bar{P}_G$ determines which positions the agent $\bar{P}_S$ could have reached during this time span. If $\bar{P}_S$

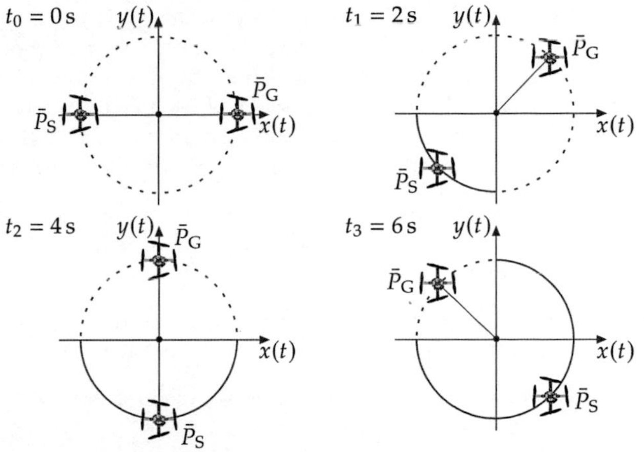

Fig. 4.5: Illustration of the prediction method with $\Delta\Phi_\mathrm{S}(t, t_k)$ shown in red.

had moved at minimum speed $v_{\mathrm{S,min}} = 0\frac{\mathrm{m}}{\mathrm{s}}$ from time $t_0 = 0$ on, it would still be at its initial phase and $\Phi_{\mathrm{S,min}}(2, 0) = 180°$ holds. On the other hand, if it had moved at maximum speed, it would have arrived at the upper bound given by $\Phi_{\mathrm{S,max}}(2, 0) = 270°$. The resulting set $\Delta\Phi_\mathrm{S}(2, 0)$ of $\bar{P}_\mathrm{S}$ is a quarter circle and depicted in red in Fig. 4.5. The set increases over time since the agents cover another distance corresponding to a phase of 45° in the next 2 seconds. The prediction unit estimates the borders of the set $\Delta\Phi_\mathrm{S}(4, 0)$ to $\Phi_{\mathrm{S,min}}(4, 0) = 180°$ and $\Phi_{\mathrm{S,max}}(4, 0) = 270°$, which is a half circle. Hence, the future position of $\bar{P}_\mathrm{S}$ on its circular trajectory becomes more uncertain over time for $\bar{P}_\mathrm{G}$. After $t_3 = 6\,\mathrm{s}$ the uncertainty is a three-quarter circle. Hence, it can be seen that the uncertainty about the phase becomes large in a short time span. This means that communication has to be invoked often enough to ensure the control aims, but a continuous communication link is not necessary. □

4.3 Event generation

4.3.1 *General events*

Overview of the method. In this section the basic method is presented for an appropriate event generation by considering the distance between agents. The method is given for the two moving agents, stand-on agent $\bar{P}_\mathrm{S}$ and give-way agent $\bar{P}_\mathrm{G}$ introduced in Section 1.4. Hence, the corresponding positions and trajectories are denoted with the indices 'S' and 'G'. The event generation with suitable conditions to satisfy the control aims given in Section 1.4 are stated in the Sections 7.3 and 9.4. The distance is evaluated for three different cases:

1. The event generation using the distance between the trajectory $w_{\mathrm{G}}(t)$ of the give-way agent defined in the interval $t_k \leq t \leq t_{\mathrm{end}}$ and the sets $\mathcal{P}_{\mathrm{S}}(t, t_{k+j})$ of positions of the stand-on agent is stated, where t_{end} is the end time instant of the trajectory according to (2.13) on p. 35. With the utilisation of the trajectory of the agent, events at future time instants can be generated. To this aim the sets $\mathcal{P}_{\mathrm{S}}(t, t_{k+j})$ are generated recursively for the future event time instants t_{k+j} based on the positions $w_{\mathrm{G}}(t_{k+j})$.

2. The event generation using the distance between the entire trajectories $w_{\mathrm{G}}(t)$ and $w_{\mathrm{S}}(t)$ of the agents is given for a time interval $t_k \leq t \leq t_{\mathrm{end}}$. Similar to the preceding point, events at future time instants can be triggered.

3. The event generation using the distance between the current positions $p_{\mathrm{G}}(t_k)$ and $p_{\mathrm{S}}(t_k)$ at a time instant t_k is presented.

The method stated in this section is used by the event-based control unit to solve Problem 1.3.

1) Event generation using the distance between $w_{\mathrm{G}}(t)$ and $\mathcal{P}_{\mathrm{S}}(t, t_{k+j})$. In order to generate events in the first case, the distance $\mathrm{dist}(w_{\mathrm{G}}(t), \mathcal{P}_{\mathrm{S}}(t, t_{k+j}))$ between the trajectory $w_{\mathrm{G}}(t)$ and the sets of positions $\mathcal{P}_{\mathrm{S}}(t, t_{k+j})$ is considered for the time interval $t_k \leq t \leq t_{\mathrm{end}}$. Events are generated using the minimum and maximum of the distance $\mathrm{dist}(w_{\mathrm{G}}(t), \mathcal{P}_{\mathrm{S}}(t, t_{k+j}))$ as

$$\begin{aligned} \mathrm{dist}_{\mathrm{min}}(w_{\mathrm{G}}(t), \mathcal{P}_{\mathrm{S}}(t, t_{k+j})) &= \min_{p_{\mathrm{S}}(t) \in \mathcal{P}_{\mathrm{S}}(t, t_{k+j})} (\|w_{\mathrm{G}}(t) - p_{\mathrm{S}}(t)\|) \\ \mathrm{dist}_{\mathrm{max}}(w_{\mathrm{G}}(t), \mathcal{P}_{\mathrm{S}}(t, t_{k+j})) &= \max_{p_{\mathrm{S}}(t) \in \mathcal{P}_{\mathrm{S}}(t, t_{k+j})} (\|w_{\mathrm{G}}(t) - p_{\mathrm{S}}(t)\|) \end{aligned} \tag{4.19}$$

if the condition

$$\text{Event generation}: \begin{cases} \quad \mathrm{dist}_{\mathrm{min}}(w_{\mathrm{G}}(t), \mathcal{P}_{\mathrm{S}}(t, t_{k+j})) \leq \bar{e}_{\mathrm{e,min}} & (I) \\ \vee \;\; \mathrm{dist}_{\mathrm{max}}(w_{\mathrm{G}}(t), \mathcal{P}_{\mathrm{S}}(t, t_{k+j})) \geq \bar{e}_{\mathrm{e,max}} & (II) \end{cases} \tag{4.20}$$

is satisfied at time instants t_{k+j}. The parameters $\bar{e}_{\mathrm{e,min}}$ and $\bar{e}_{\mathrm{e,max}}$ are thresholds, which are chosen appropriately based on the dynamics of the agents in Section 7.3.2 to satisfy the control aims. The problem of generating events infinitely often at a time instant, called Zeno behaviour in the literature of event-based control [63], is excluded for a suitable choice of the thresholds as stated in the analysis in Section 7.5.2. The time instants at which the events are triggered are denoted by t_{k+j}, the ongoing time is given by t. After an event generation the set $\mathcal{P}_{\mathrm{S}}(t, t_{k+j}))$ is reset and newly determined for $t > t_{k+j}$.

The future event time instants t_{k+j}, $(j = 1, \ldots, N)$ are determined by evaluating the distances (4.19) and the condition (4.20) recursively. To this aim, it is assumed that both agents follow their planned trajectories exactly. Based on the position $w_{\mathrm{S}}(t_{k+j})$ and speed $\dot{w}_{\mathrm{S}}(t_{k+j})$ of the stand-on agent on its communicated trajectory at the current event time instant t_{k+j} the set $\mathcal{P}_{\mathrm{S}}(t_{k+j})$ is generated to determine the next event time instant t_{k+j+1} using (4.20). For the examination of the condition, the position $w_{\mathrm{G}}(t_{k+j+1})$ of the give-way agent on its trajectory at the time instants t_{k+j+1} is used.

The method is illustrated in Fig. 4.6, in which the trajectories of the stand-on agent and the give-way agent and the generation of the sets $\mathcal{P}_\mathrm{S}(t, t_{k+j}), (j = 0, \dots, N)$ are shown. For simplicity, events are only generated with condition (I) of (4.20). The event threshold $\bar{e}_{\mathrm{e,min}}$ is marked as a dotted line. The future event time instants are depicted as black beams in the lower part of the figure. The black dots mark the positions of the stand-on agent at the event time instants if it moves on its trajectory.

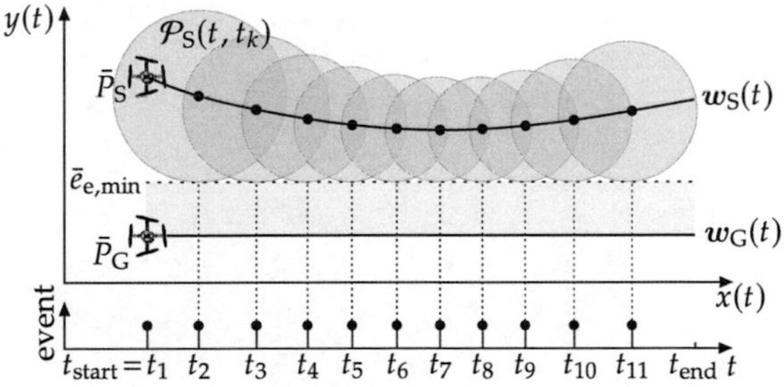

Fig. 4.6: Illustration of the determination of the future event time instants.

Evaluation of condition (I) of (4.20). Condition (I) of (4.20) can be evaluated in a mathematically simple way by considering a ball with the radius $\bar{e}_{\mathrm{e,min}}$ around the position of the give-way agent as the centre. The ball is a special ellipsoid described by

$$\frac{(x(t) - x_\mathrm{G}(t))^2}{r_{\mathrm{G,x}}^2} + \frac{(y(t) - y_\mathrm{G}(t))^2}{r_{\mathrm{G,y}}^2} + \frac{(z(t) - z_\mathrm{G}(t))^2}{r_{\mathrm{G,z}}^2} - 1 = 0 \tag{4.21}$$

in which all radii are equal:

$$r_{\mathrm{G,x}} = r_{\mathrm{G,y}} = r_{\mathrm{G,z}} = \bar{e}_{\mathrm{e,min}}.$$

With the vector

$$\boldsymbol{q}_\mathrm{G}(t) = \begin{pmatrix} x_\mathrm{G}(t) & y_\mathrm{G}(t) & z_\mathrm{G}(t) & 1 \end{pmatrix}^\mathrm{T} \tag{4.22}$$

and the matrix

$$\boldsymbol{P}_\mathrm{G}(t) = \begin{pmatrix} \frac{1}{r_{\mathrm{G,x}}^2} & 0 & 0 & -\frac{x_\mathrm{G}(t)}{r_{\mathrm{G,x}}^2} \\ 0 & \frac{1}{r_{\mathrm{G,y}}^2} & 0 & -\frac{y_\mathrm{G}(t)}{r_{\mathrm{G,y}}^2} \\ 0 & 0 & \frac{1}{r_{\mathrm{G,z}}^2} & -\frac{z_\mathrm{G}(t)}{r_{\mathrm{G,z}}^2} \\ -\frac{x_\mathrm{G}(t)}{r_{\mathrm{G,x}}^2} & -\frac{y_\mathrm{G}(t)}{r_{\mathrm{G,y}}^2} & -\frac{z_\mathrm{G}(t)}{r_{\mathrm{G,z}}^2} & \frac{x_\mathrm{G}^2(t)}{r_{\mathrm{G,x}}^2} + \frac{y_\mathrm{G}^2(t)}{r_{\mathrm{G,y}}^2} + \frac{z_\mathrm{G}^2(t)}{r_{\mathrm{G,z}}^2} - 1 \end{pmatrix} \tag{4.23}$$

the ball (4.21) can be represented in matrix notation as

$$\boldsymbol{q}_\mathrm{G}^\mathrm{T}(t)\boldsymbol{P}_\mathrm{G}(t)\boldsymbol{q}_\mathrm{G}(t) = 0.$$

Considering the set $\mathcal{P}_S(t, t_k)$ in the ellipsoidal representation (4.10), the event condition can be evaluated by analysing the relative position of the two ellipsoids. A multiplication of the matrix $\boldsymbol{P}_G(t)$ with the matrix $\boldsymbol{P}_S(t, t_k)$ of the form (4.10) leads to the matrix

$$\boldsymbol{R}(t, t_k) = \boldsymbol{P}_S^{-1}(t, t_k)\boldsymbol{P}_G(t).$$

Condition (I) of (4.20) is evaluated by an analysis of the eigenvalues λ_i, $(i = 1, \ldots, 4)$ of $\boldsymbol{R}(t, t_k)$. The analysis yields that two eigenvalues (λ_1, λ_2) only depend on the radii of the two ellipsoids. With the other two eigenvalues (λ_3, λ_4) the relative positions of the ellipsoids to each other can be evaluated and the following cases are distinguished, which are shown in Figs. 4.7 and 4.8 [172]:

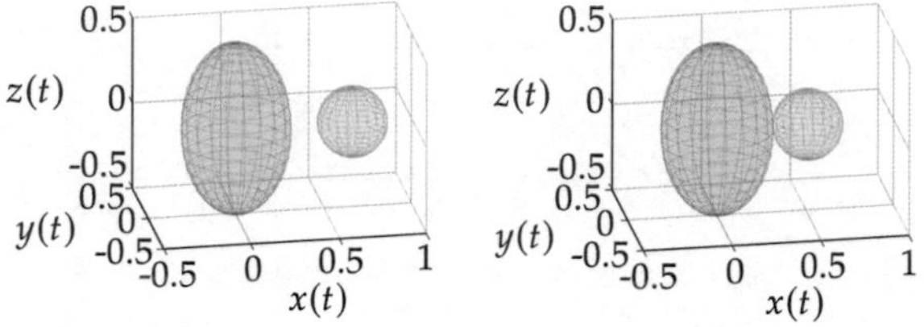

Fig. 4.7: Case (1): Ellipsoids are separated (left); Case (2): ellipsoids touch each other (right).

Lemma 4.1 (Relative positions of two ellipsoids [172]). *The relative positions of two ellipsoids to one another can be evaluated with the eigenvalues of the matrix* $\boldsymbol{R}(t, t_k)$:

- *If for the eigenvalues applies*

$$\lambda_3 \neq \lambda_4, \quad \lambda_3, \lambda_4 \in \mathbb{R}^-,$$

then the ellipsoids are separated (Fig. 4.7 left).

- *If for the eigenvalues applies*

$$\lambda_3 = \lambda_4, \quad \lambda_3, \lambda_4 \in \mathbb{R}^-,$$

then the ellipsoids touch one another (Fig. 4.7 right).

- *If for the eigenvalues applies*

$$\lambda_3 = \lambda_4, \quad \lambda_3, \lambda_4 \in \mathbb{R}^+,$$

then the ellipsoids touch one another at the far side, which means the ellipsoids overlap and touch one another (Fig. 4.8 left).

- *If for the eigenvalues applies*

$$\lambda_3 \neq \lambda_4, \quad \lambda_3, \lambda_4 \in \mathbb{R}^+,$$

then the ellipsoids overlap without touching one another (*Fig.* 4.8 *right*).

As the radius of the ball is equal to the event condition (I) of (4.20), the condition is satisfied and the event is generated whenever the ellipsoids touch (case 2 of Lemma 4.1), because then the minimum distance between the agents is $\bar{e}_{\mathrm{e,min}}$. If the ellipsoids are separated (case 1 of Lemma 4.1), the distance between the agents is larger than the allowed minimum distance.

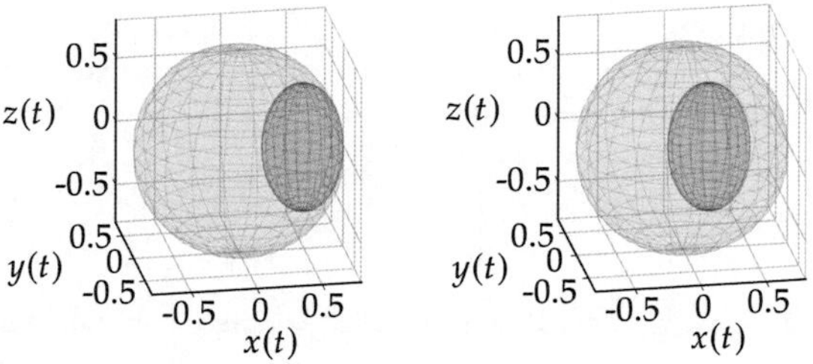

Fig. 4.8: Case (3): Ellipsoids touch one another at the far side (left); Case (4): ellipsoids overlap (right).

Evaluation of condition (II) of (4.20)**.** Condition (II) of (4.20) can be evaluated in the same way as condition (I). The difference is that now the ball (4.21) around the position of the give-way agent has the radius

$$r_{\mathrm{G,x}} = r_{\mathrm{G,y}} = r_{\mathrm{G,z}} = \bar{e}_{\mathrm{e,max}}. \tag{4.24}$$

With (4.22) the matrix $\boldsymbol{P}_{\mathrm{G}}(t)$ in the form (4.23) is formed for the ball (4.21) with radius (4.24) and the matrix $\boldsymbol{R}(t, t_k)$ is calculated. Again an analysis of the eigenvalues gives the relative positions of the two ellipsoids.

As the radius corresponds to the event threshold (II) of (4.20), the condition is fulfilled if the two ellipsoids overlap and touch one another, as shown in Fig. 4.8 (left). This situation corresponds to case 3 of Lemma 4.1. If the ellipsoids overlap without touching one another (case 4 of Lemma 4.1), the stand-on agent does not exceed the maximum distance to the give-way agent.

2) Event generation using the distance between $\boldsymbol{w}_{\mathrm{G}}(t)$ and $\boldsymbol{w}_{\mathrm{S}}(t)$. For the event generation in the second case, the minimum and maximum distances between the entire trajectories $\boldsymbol{w}_{\mathrm{G}}(t)$ and $\boldsymbol{w}_{\mathrm{S}}(t)$ of the agents are considered for the time interval $t_k \leq t \leq t_{\mathrm{end}}$:

$$\begin{aligned} \mathrm{dist}_{\min}(\boldsymbol{w}_{\mathrm{G}}(t), \boldsymbol{w}_{\mathrm{S}}(t)) &= \min_{t \in [\bar{t}_i, \bar{t}_{i+1}]} (||\boldsymbol{w}_{\mathrm{G}}(t) - \boldsymbol{w}_{\mathrm{S}}(t)||) \\ \mathrm{dist}_{\max}(\boldsymbol{w}_{\mathrm{G}}(t), \boldsymbol{w}_{\mathrm{S}}(t)) &= \max_{t \in [\bar{t}_i, \bar{t}_{i+1}]} (||\boldsymbol{w}_{\mathrm{G}}(t) - \boldsymbol{w}_{\mathrm{S}}(t)||). \end{aligned} \tag{4.25}$$

Satisfying the condition at a time instant $t = t_k$

$$\text{Event generation}: \begin{cases} & \text{dist}_{\text{min}}(\boldsymbol{w}_{\text{G}}(t), \boldsymbol{w}_{\text{S}}(t)) \leq \bar{e}_{\text{e,min}} \quad (I) \\ \vee & \text{dist}_{\text{max}}(\boldsymbol{w}_{\text{G}}(t), \boldsymbol{w}_{\text{S}}(t)) \geq \bar{e}_{\text{e,max}} \quad (II) \end{cases} \tag{4.26}$$

leads to an event generation at time instant $t = t_{\bar{e}}$. By generating events with (4.26) no Zeno behaviour can occur, because in the event-based control method, the generation of the event results in a change of the trajectory of the give-way agent. Hence, only the satisfaction of (4.26) at the first time instant is relevant. The fulfilment of condition (4.26) at further time instants is not important, because $\boldsymbol{w}_{\text{G}}(t)$ is already changed to keep the distance requirements between the agents fulfilled. If the condition (4.26) is fulfilled at a time instant t_k, the event is generated at $t_{\bar{e}}$, which is a default transition time interval (3.9) earlier or at least at the current time instant t:

$$\text{Event time instant}: \begin{cases} t_{\bar{e}} = t_k - t_{\text{t,d}} & \text{if } t_k - t \geq t_{\text{t,d}} \\ t_{\bar{e}} = t & \text{if } t_k - t < t_{\text{t,d}}. \end{cases} \tag{4.27}$$

During this time span the give-way agent is able to change its movement smoothly. The method is illustrated in Fig. 4.9, in which the determination of the distance between the trajectories of the stand-on agent and the give-way agent is shown. Condition (4.26) is satisfied at t_k leading to an event generation at $t_{\bar{e}}$.

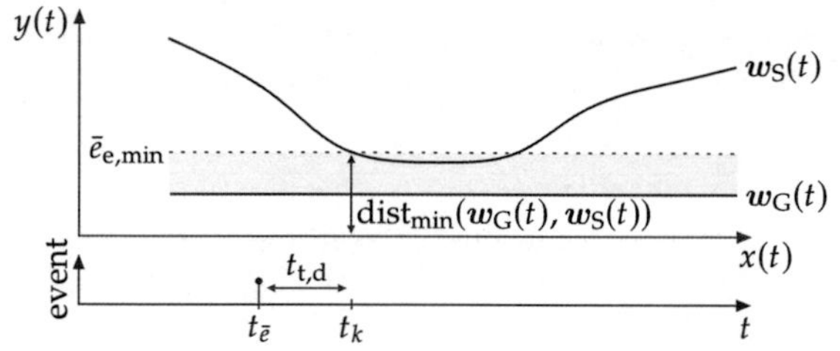

Fig. 4.9: Illustration of the determination of the distance between $\boldsymbol{w}_{\text{G}}(t)$ and $\boldsymbol{w}_{\text{S}}(t)$.

Evaluation of the conditions of (4.26). For a fast determination of the minimum distance between the trajectories, the minimum distance algorithm stated on p. 38 is used. Similarly, the maximum distance between the trajectories is obtained by using the maximum distance algorithm on p. 38.

3) Event generation using the distance between $\boldsymbol{p}_{\text{G}}(t_k)$ and $\boldsymbol{p}_{\text{S}}(t_k)$. The distance between the positions $\boldsymbol{p}_{\text{G}}(t_k)$ and $\boldsymbol{p}_{\text{S}}(t_k)$ of the agents at a time instant t_k needs to be determined for an event generation in the third case:

$$\text{dist}\left(\boldsymbol{p}_{\text{G}}(t_k), \boldsymbol{p}_{\text{S}}(t_k)\right) = ||\boldsymbol{p}_{\text{G}}(t_k) - \boldsymbol{p}_{\text{S}}(t_k)||. \tag{4.28}$$

An event is triggered if the condition

$$\text{Event generation}: \begin{cases} \quad \operatorname{dist}\left(\boldsymbol{p}_{\mathrm{G}}(t_k), \boldsymbol{p}_{\mathrm{S}}(t_k)\right) \leq \bar{e}_{\mathrm{e,min}} & (I) \\ \vee \ \operatorname{dist}\left(\boldsymbol{p}_{\mathrm{G}}(t_k), \boldsymbol{p}_{\mathrm{S}}(t_k)\right) \geq \bar{e}_{\mathrm{e,max}} & (II) \end{cases} \tag{4.29}$$

is fulfilled at a time instant t_k. As the condition is evaluated at time instants t_k, the event is generated at t_k, too.

Evaluation of the condition of (4.29)**.** As both positions are known exactly, the distance between the two points can be determined by using the euclidean distance.

4.3.2 *Specific events for a circular movement*

Overview of the method. The basic method for generating events for a circular movement of the agents $\bar{P}_{\mathrm{S}}$ and $\bar{P}_{\mathrm{G}}$ is similar to the method for a general movement. The difference is that now the phase differences and the height differences between the agents are taken into account. Again three situations are considered that lead to an event generation.

1. Events are generated using the phase difference between the phase $\Phi(\boldsymbol{w}_{\mathrm{G}}(t))$ of the trajectory $\boldsymbol{w}_{\mathrm{G}}(t)$ of the give-way agent and the sets $\Delta\Phi_{\mathrm{S}}(t, t_{k+j})$ of phases of the stand-on agent for a time interval $t_k \leq t \leq t_{\mathrm{end}}$. Additionally, the height difference between the height $w_{\mathrm{G,z}}(t)$ of the trajectory $\boldsymbol{w}_{\mathrm{G}}(t)$ of the give-way agent and the sets $\Delta z_{\mathrm{S}}(t, t_{k+j})$ of heights of the stand-on agent is considered for the event generation for $t_k \leq t \leq t_{\mathrm{end}}$. For an generation of future events the sets $\Delta\Phi_{\mathrm{S}}(t, t_{k+j})$ and $\Delta z_{\mathrm{S}}(t, t_{k+j})$ are generated recursively for the future event time instants t_{k+j}.

2. Events are generated considering the phase difference between the phases $\Phi(\boldsymbol{w}_{\mathrm{G}}(t))$ and $\Phi(\boldsymbol{w}_{\mathrm{S}}(t))$ of the trajectories of the two agents for $t_k \leq t \leq t_{\mathrm{end}}$. In addition, the investigation of the height difference between the heights $w_{\mathrm{G,z}}(t)$ and $w_{\mathrm{S,z}}(t)$ of the trajectories of the agents leads to an event generation.

3. Events are generated using the phase difference between the current phases $\Phi_{\mathrm{G}}(t_k)$ and $\Phi_{\mathrm{G}}(t_k)$ of the agents at time instant t_k as well as using the height difference between the current heights $z_{\mathrm{G}}(t)$ and $z_{\mathrm{S}}(t)$ of the agents at t_k.

The event-based control unit executes the method described in this section to solve Problem 1.3 for a circular movement.

1) Event generation using the differences between $\Phi(\boldsymbol{w}_{\mathrm{G}}(t))$, $\Delta\Phi_{\mathrm{S}}(t, t_{k+j})$ and $w_{\mathrm{G,z}}(t)$, $\Delta z_{\mathrm{S}}(t, t_{k+j})$. For an event generation in the first case, the phase difference $\Phi_{\mathrm{diff}}(t)$ between the phase $\Phi(\boldsymbol{w}_{\mathrm{G}}(t))$ and the sets of phases $\Delta\Phi_{\mathrm{S}}(t, t_{k+j})$ is considered. The minimum and maximum of the phase difference $\Phi_{\mathrm{diff}}(t)$ are determined as

$$\begin{aligned} \Phi_{\mathrm{diff,min}}(t, t_{k+j}) &= \operatorname{mod}\left(\Phi_{\mathrm{S,min}}(t, t_{k+j}) - \Phi_{\mathrm{G}}(t), 360^\circ\right) \\ \Phi_{\mathrm{diff,max}}(t, t_{k+j}) &= \operatorname{mod}\left(\Phi_{\mathrm{S,max}}(t, t_{k+j}) - \Phi_{\mathrm{G}}(t), 360^\circ\right) \end{aligned} \tag{4.30}$$

where $\Phi_{S,min}(t, t_{k+j})$ and $\Phi_{S,max}(t, t_{k+j})$, $(j = 0, \ldots, N)$ are given by (4.14) and (4.15). Furthermore, the minimum possible height difference between the height $w_{G,z}(t)$ and the sets of heights $\Delta z_S(t, t_{k+j})$ is calculated by

$$z_{diff,min}(t, t_{k+j}) = \min\left(z_o(t, t_{k+j}), z_u(t, t_{k+j})\right) \quad (4.31)$$

with

$$\begin{aligned} z_o(t, t_{k+j}) &= \left|z_S(t_{k+j}) + v_{S,max} \cdot (t - t_{k+j}) - z_G(t)\right| \\ z_u(t, t_{k+j}) &= \left|z_S(t_{k+j}) - v_{S,max} \cdot (t - t_{k+j}) - z_G(t)\right|. \end{aligned}$$

The phase differences (4.30) and the height difference (4.31) are determined with respect to the time interval $t_k \leq t \leq t_{end}$.

Example 4.4. The method for the determination of the phase difference is illustrated in Fig. 4.10, in which the two agents $\bar{P}_S$ and $\bar{P}_G$ move in the same way as in Example 4.3. At time $t_0 = 0$ the phase difference $\Phi_{diff}(0) = 180°$ between the phase $\Phi_G(0)$ of $\bar{P}_G$ and the set of phases $\Delta\Phi_S(0, 0)$ of $\bar{P}_S$ is known exactly. Based on the minimum and maximum possible phases $\Phi_{S,min}(2, 0)$, $\Phi_{S,max}(2, 0)$ of $\bar{P}_S$ after 2 s the minimum phase difference is given by $\Phi_{diff,min}(2, 0) = 135°$ (blue line in Fig. 4.10) and the maximum phase difference is $\Phi_{diff,max}(2, 0) = 225°$ (green line in Fig. 4.10). In the same way, after $t_2 = 4$ s the minimum and maximum phase differences are determined to be $\Phi_{diff,min}(4, 0) = 90°$, $\Phi_{diff,max}(4, 0) = 270°$. After 6 s the phase differences increase to $\Phi_{diff,min}(6, 0) = 45°$, $\Phi_{diff,max}(6, 0) = 315°$.

As it can be seen, the difference between the minimum and the maximum phase difference increases over time as the set $\Delta\Phi_S(t, t_k)$ increases over time. □

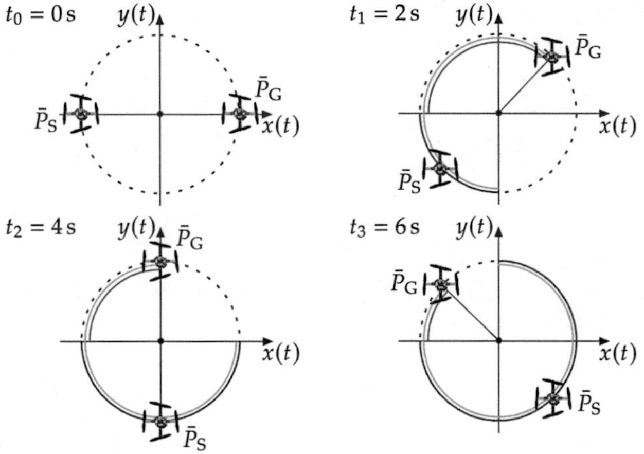

Fig. 4.10: Illustration of the determination of the phase differences $\Phi_{diff,min}(t, t_k)$ (blue) and $\Phi_{diff,max}(t, t_k)$ (green) using the set $\Delta\Phi_S(t, t_k)$ (red).

Events are generated if the condition

$$\text{Event generation}: \begin{cases} & \Phi_{\text{diff,min}}(t, t_{k+j}) \leq \bar{e}_{\text{e,min}} \quad (I) \\ \vee & \Phi_{\text{diff,max}}(t, t_{k+j}) \geq \bar{e}_{\text{e,max}} \quad (II) \\ \vee & z_{\text{diff,min}}(t, t_{k+j}) \leq \bar{e}_{\text{e,z,min}} \quad (III) \end{cases} \tag{4.32}$$

is satisfied at time instants t_{k+j}, where again $\bar{e}_{\text{e,min}}$, $\bar{e}_{\text{e,max}}$ and $\bar{e}_{\text{e,z,min}}$ are thresholds, which are chosen appropriately based on the dynamics of the agents in Section 7.3.6 to satisfy the control aims. The time instants at which the events are triggered are again denoted by t_{k+j}. After an event generation the sets $\Delta\Phi_S(t, t_{k+j})$ and $\Delta z_S(t, t_{k+j})$ are reset and newly determined for $t > t_{k+j}$.

The future event time instants t_{k+j}, $(j = 1, \ldots, N)$ are determined by evaluating the phase difference (4.30) and the height difference (4.31) with the condition (4.32) recursively under the assumption that both agents follow their planned trajectories exactly. Based on the position $w_S(t_{k+j})$ of the stand-on agent at t_{k+j} its phase $\Phi_S(t_{k+j})$ is determined with (2.3) and the sets $\Delta\Phi_S(t, t_{k+j})$ and $\Delta z_S(t, t_{k+j})$ are generated to determine the next event time instant t_{k+j+1} using (4.32). For the examination of the condition, the phase $\Phi_G(t_{k+j})$ of the give-way agent is used.

The method is illustrated in Fig. 4.11. In the upper part of the figure the phase difference $\Phi_{\text{diff}}(t)$ between the trajectories of the agents is shown as the black line. The set $\Delta\Phi_S(t, t_k)$ is illustrated as the grey area, the event thresholds $\bar{e}_{\text{e,min}}$ and $\bar{e}_{\text{e,max}}$ are given by the dotted lines. The future event time instants are determined using only conditions (I) and (II) of (4.32) and are depicted as the black beams. The black dots mark the position of the stand-on agent when it follows its communicated trajectory. After an event generation the set $\Delta\Phi_S(t, t_k)$ is reset. In the lower part of Fig. 4.11 the height difference $z_{\text{diff}}(t)$ is illustrated together with the set $\Delta z_S(t, t_k)$. Events are generated only with condition (III) of (4.32).

Combined, the future event time instants are determined by the condition (I), (II) or (III) of (4.32) that is violated next as illustrated in the lower time line in Fig. 4.11.

Evaluation of the conditions of (4.32)**.** The conditions can be evaluated using eqn. (2.3) to obtain the phase of an agent from its position and by calculating the set $\mathcal{P}_c(t, t_k)$ (4.11) with (4.12) and (4.13).

2) Event generation using the differences between $\Phi(w_G(t))$, $\Phi(w_S(t))$ and $w_{G,z}(t)$, $w_{S,z}(t)$. The event generation in the second case is based on the minimum and maximum phase differences $\Phi_{\text{diff,w,min}}(t, t_k)$ and $\Phi_{\text{diff,w,max}}(t, t_k)$ as well as the minimum height difference

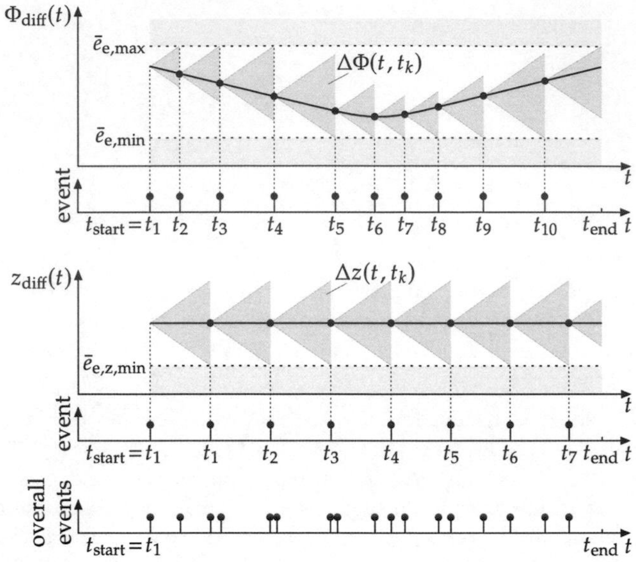

Fig. 4.11: Illustration of the determination of the future event time instants using $\Phi_{\text{diff}}(t)$ (top) and $z_{\text{diff}}(t)$ (bottom).

$z_{\text{diff,w,min}}(t)$ between the entire trajectories $\boldsymbol{w}_{\text{G}}(t)$ and $\boldsymbol{w}_{\text{S}}(t)$ of the agents fot he time interval $t_k \leq t \leq t_{\text{end}}$:

$$\begin{aligned} \Phi_{\text{diff,w,min}}(t) &= \min_{t \in [t_s, t_e]} \left(\text{mod}\left(\Phi(\boldsymbol{w}_{\text{S}}(t)) - \Phi(\boldsymbol{w}_{\text{G}}(t)), 360°\right)\right) \\ \Phi_{\text{diff,w,max}}(t) &= \max_{t \in [t_s, t_e]} \left(\text{mod}\left(\Phi(\boldsymbol{w}_{\text{S}}(t)) - \Phi(\boldsymbol{w}_{\text{G}}(t)), 360°\right)\right) \\ z_{\text{diff,w,min}}(t) &= \min_{t \in [t_s, t_e]} \left(|w_{\text{S,z}}(t) - w_{\text{G,z}}(t)|\right). \end{aligned} \tag{4.33}$$

Events are generated if the condition

$$\text{Event generation}: \begin{cases} & \Phi_{\text{diff,w,min}}(t) \leq \bar{e}_{\text{e,min}} \quad (I) \\ \vee & \Phi_{\text{diff,w,max}}(t) \geq \bar{e}_{\text{e,max}} \quad (II) \\ \vee & z_{\text{diff,w,min}}(t) \leq \bar{e}_{\text{e,z,min}} \quad (III) \end{cases} \tag{4.34}$$

is fulfilled at a time instant $t = t_k$. According to the event generation for the general movement an event generation with (4.26) causes no Zeno behaviour. Again, the generation of this event results in a change of the trajectory of the give-way agent so that further events after the first event are not relevant. If the condition (4.34) is fulfilled at a time instant t_k, the event is generated at $t_{\bar{e}}$, which is a transition time interval (3.25) of a quarter circle earlier or at least at the current time instant t:

$$\text{Event time instant}: \begin{cases} t_{\bar{e}} = t_k - t^i_{\mathrm{t,c}} & \text{if } t_k - t \geq t^i_{\mathrm{t,c}} \\ t_{\bar{e}} = t & \text{if } t_k - t < t^i_{\mathrm{t,c}}. \end{cases}$$

Again, during the time span the give-way agent is able to change its speed or its height smoothly.

Evaluation of the conditions of (4.34). For a fast determination of the minimum and maximum phase differences, first, the phases $\Phi(\boldsymbol{w}_{\mathrm{S}}(t))$ and $\Phi(\boldsymbol{w}_{\mathrm{G}}(t))$ in the time interval $t_k \leq t \leq t_{\mathrm{end}}$ resulting from the trajectories of the agents are derived and second, the phase difference between the phases is calculated. For the determination of the height difference, the height components $w_{\mathrm{S,z}}(t)$ and $w_{\mathrm{G,z}}(t)$ of the trajectories are used. In order to find the minima and maxima of the functions it is sufficient to use a root-finding based approach.

3) Event generation using the differences between $\Phi_{\mathrm{G}}(t_k)$, $\Phi_{\mathrm{S}}(t_k)$ and $z_{\mathrm{G}}(t)$, $z_{\mathrm{S}}(t)$. For the third case the phase difference $\Phi_{\mathrm{diff}}(t_k)$ between the phases $\Phi_{\mathrm{G}}(t_k)$ and $\Phi_{\mathrm{S}}(t_k)$ as well as the height difference $z_{\mathrm{diff}}(t_k)$ between the heights $z_{\mathrm{G}}(t_k)$ and $z_{\mathrm{S}}(t_k)$ of the agents at a time instant t_k are determined:

$$\begin{aligned} \Phi_{\mathrm{diff}}(t_k) &= \mathrm{mod}\,(\Phi_{\mathrm{S}}(t_k) - \Phi_{\mathrm{G}}(t_k), 360^\circ) \\ z_{\mathrm{diff}}(t_k) &= |w_{\mathrm{S,z}}(t_k) - w_{\mathrm{G,z}}(t_k)|\,. \end{aligned} \tag{4.35}$$

An event is triggered if the condition

$$\text{Event generation}: \begin{cases} \quad \Phi_{\mathrm{diff}}(t_k) \leq \bar{e}_{\mathrm{e,min}} & (I) \\ \vee \;\; \Phi_{\mathrm{diff}}(t_k) \geq \bar{e}_{\mathrm{e,max}} & (II) \\ \vee \;\; z_{\mathrm{diff}}(t_k) \leq \bar{e}_{\mathrm{e,z,min}} & (III) \end{cases} \tag{4.36}$$

is satisfied at a time instant t_k. Again, as the condition is evaluated at time instants t_k, the event is generated at t_k.

Evaluation of the conditions of (4.36). As both phases and heights are known exactly, the phase difference and the height difference between the two agents can be determined by using eqn. (2.3) and the euclidean distance.

Properties of the communication network

5

This chapter describes the communication between two mobile agents both over an ideal communication network and an unreliable communication network (Section 5.1). An overview over two common channel models is given for different application fields in Section 5.2. The communication model for UAV applications is stated in detail since UAVs are used as a demonstration example (Section 5.3).

5.1 Communication between mobile agents

5.1.1 *Communication over an ideal network*

This section states the structure of an ideal network between two agents that is used in Part III of this thesis. The agents are connected by a communication network, which provides the channel for a data exchange between the stand-on agent and the give-way agent, as depicted in Fig. 5.1. The connection between the two agents is established automatically once the agents are in the transmission range of one another. The communication over the network is packet based in the sense that only discrete packages containing a finite amount of data are sent across the network at certain instants of time.

In the application scenario in this thesis, the give-way agent sends a request $r_G(t_k)$ at an event time instant t_k to the stand-on agent to demand new information. The stand-on agent sends its information $S_S(t_k)$ after receiving a request $r_G(t_k)$ or autonomously if the next event time instant t_{k+j}, $(j = 1, \ldots, N)$ is reached. The communication flow is described in detail in Section 7.5.

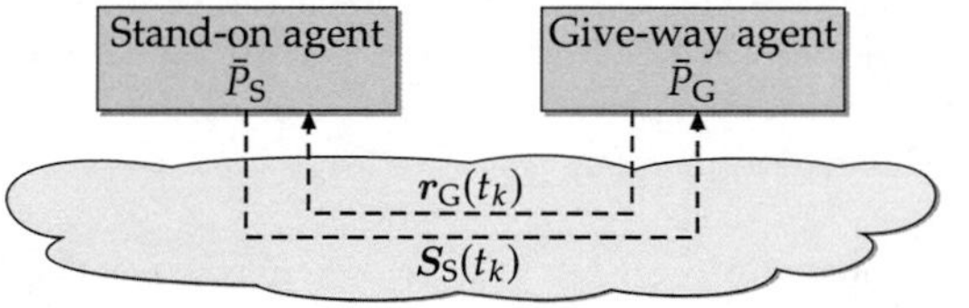

Fig. 5.1: Communication structure of an ideal network between two agents.

Communication takes only place at event time instants t_{k+j}, $(j = 0, \dots, N)$. In between no information transfer occurs. As the communication channel is assumed to be ideal, no transmission delays and packet losses occur. Hence, data that are sent by an agent at t_k are reliably received at t_k by the other agent.

5.1.2 *Communication over an unreliable network*

Communication between the agents. In Part IV of this thesis, the agents are connected by an unreliable communication network with the structure depicted in Fig. 5.2. Again, the communication over the network is packet based and the connection between the two agents is established automatically once the agents are in the transmission range of one another. Networked communication induces inherent imperfections such as time-varying transmission delays or packet losses. This means that a transmission sent by an agent at a communication time instant $t_{c,k}$ is received by the other agent after an unknown delay τ_k at time instant $t_{c,k} + \tau_k$ or never. Hence, now the give-way agent sends a request $\boldsymbol{R}_G(t_{c,k})$ at $t_{c,k}$ and the stand-on agent responds by sending its data $\boldsymbol{S}_S(t_{c,k})$.

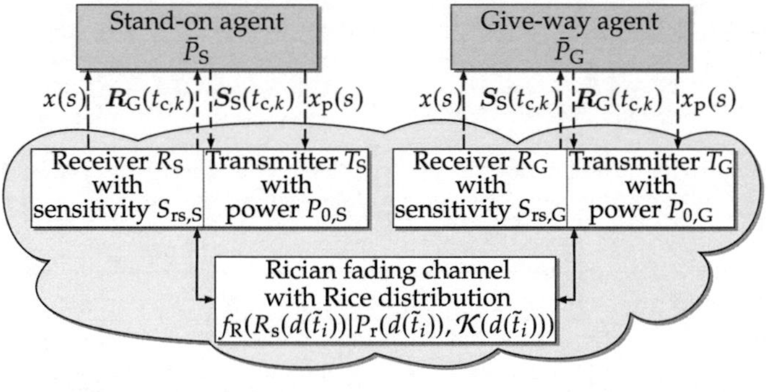

Fig. 5.2: Communication structure of an unreliable network between two agents.

The properties of the channel between the agents are variable due to their motion. As the channel statistics are constant in a time interval $[\tilde{t}_i, \tilde{t}_{i+1}]$ (Assumption 8.1 on p. 175), the properties of the channel depend on the current time instant indicated by $\tilde{t}_i$. Initially at time instant $t = 0$ both agents have to perform a channel estimation by sending a sequence of packets $x_p(s)$ into the network. The analysis of a received observation sequence $x(s)$ is used to initialise the model of the network, which is used by the delay estimators D_S and D_G to estimate the channel properties.

Structure of the communication network. The communication network is split into three parts as shown in Fig. 5.2. First, it consists of the transmitters T_S and T_G of the stand-on agent and the give-way agent, which send their information to the other agent with a transmission power $P_{0,S}$ and $P_{0,G}$, respectively. The value of the transmission power is measured at a reference distance d_0. Second, the information is transmitted over a wireless channel, which is described as a Rician fading channel with the function $f_R(R_s(d(\tilde{t}_i))|P_r(d(\tilde{t}_i)), \mathcal{K}(d(\tilde{t}_i)))$ as described in the next section. Third, the data are received by the receivers R_S and R_G of the stand-on agent and the give-way agent with the corresponding sensitivities $S_{rs,S}$ and $S_{rs,G}$. For receiving an information, the received signal power $P_r(d(\tilde{t}_i))$ has to be larger than the sensitivity of the receiver. The delay estimators D_S and D_G use the transmitters and the receivers of the agents for the channel estimation, as described in Section 5.3.4.

5.2 Two common models of communication networks

5.2.1 *Packet loss models*

Overview. Mathematical models of communication channels are used to describe how the transmitted signal is mapped to the received signal. There exist various types of channel models to represent different effects and to describe the individual layers of a communication system. For the event-based control of the agents an evaluation of the packet loss probability and the transmission delay of a network is required. Hence, in this section a short overview of the common packet loss models to describe packet losses and transmission delays are given. With these models the transmission delay can be determined for a channel with a certain packet loss probability.

Most channel models have been developed at the bit level or the symbol level. Bit level models are computationally complex for network simulations, because the modulation, demodulation and detection has to be executed for every transmitted bit. Hence, in order to reduce the computational complexity, packet level models are suitable for the simulation of communication channels. The computational efficiency results since the simulation has to be executed only for each packet instead of each bit. However, modelling the channels accurately is difficult as packet losses do not only depend on channel properties but also on the characteristics of the source and the distribution of the packet length. Thus, packet level models represent only an approximation of the channel. In general, there are two different approaches to model packet losses for wireless applications:

- A channel model is derived by computing the signal-to-noise-ratio (SNR) based on the channel conditions [134, 165]. These models are optimum because they model the radio propagation accurately between a transmitter and a receiver depending on the chosen model complexity. However, the models are computationally complex as the modulation, transmission, reception and demodulation is taken into account. Furthermore, the performance of these models depends on the accuracy of the model of propagation and the model of the receiver.

- A channel model can be derived with the usage of simulated or collected network traffic traces [65, 72, 84], which model the channel in a simple way but are only an approximate representation. These models contain a set of unknown parameters, which are estimated using network traffic data. The two most popular models for wireless channels are the Bernoulli model, which is an independent channel model and the Gilbert-Elliott model, which is able to represent wireless channels with burst errors more accurately. However, these models are based on the assumption that the packet loss probability is stationary over time. For UAV applications these models are not suitable, because the packet loss probability varies over time, as stated in Section 5.3.

For the UAV application in this thesis, the delay estimators of the agents use a Markov model that extends the Gilbert-Elliott model described in the next paragraph by a radio propagation model. This model is given in the next section and represents the time-varying packet loss probabilities that occur in wireless channels between UAVs. In contrast to the memoryless Bernoulli model, the Gilbert-Elliott model is capable of describing the correlation among packet losses. For the initialisation of the unknown parameters of the model, the delay estimators perform the channel estimation given in Section 5.3.4.

Gilbert-Elliott model. In practice, packet losses are correlated or occur in bursts. If a packet gets lost, it is probable that the next packet gets lost as well. Packet losses are mainly caused by receiver faults and channel conditions such as a degraded link quality and channel congestions. As the channel conditions change slower compared to the transmission rate, the behaviour of packet losses can be represented by Markov models [182]. The Gilbert-Elliott model uses a Hidden Markov Model (HMM) stated in Appendix A.6. An exemplary model is shown in Fig. 5.3. Its background process (upper part of Fig. 5.3) is described by a two-state Markov model, which states are indicated as 'good' and as 'bad' and correspond to specific channel conditions. The states and their transitions are hidden and unobservable. The transition probabilities p_{gg}, p_{bb}, p_{gb} and p_{bg} are constant over time. Each state is assigned with independent packet loss probabilities p_{g} and p_{b} that represent the channel quality and form the sensor model of the HMM (lower part of Fig. 5.3). The outputs of the HMM are the packet loss probabilities.

In the Gilbert-Elliott model the packet loss probabilities are stationary and do not vary over time, which makes the model inappropriate for UAV applications. Hence, in the next section the model is extended by a radio propagation model to make it suitable for UAV applications.

The Markov model contains unknown parameters as the transition probabilities. Different algorithms exist to determine the model parameters as the Baum-Welch algorithm [153, 161]. As the algorithm is computationally complex, in Section 5.3.4 a simpler method based on a channel estimation is presented.

Since the Gilbert-Elliott model contains only two states it may not be suitable if the channel quality varies in a fast way. To overcome this drawback, a finite state Markov

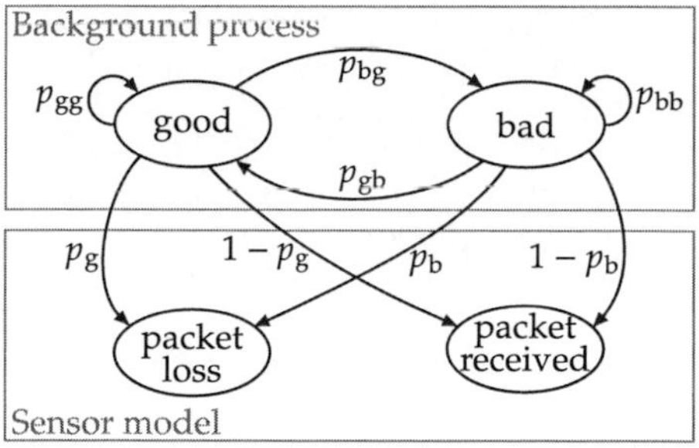

Fig. 5.3: Exemplary Gilbert-Elliott model.

model as described in [165] can be used. However, as the channel between the agents in the application scenarios in this thesis varies depending on the relative movement of the agents, which is limited by the dynamics of the agents, the channel does not change rapidly and a two-state Markov model is suitable.

5.2.2 *Radio propagation models*

Overview. Radio propagation is the behaviour of radio waves as they are propagated from a transmitter to a receiver. Radio waves are affected by different phenomena such as reflection, refraction, diffraction, absorption, polarisation and scattering [132]. Different radio propagation models have been developed to represent the propagation behaviour of radio waves as a function of frequency, distance and other conditions. In wireless communications often fading occurs that is a variation of the attenuation of a signal. Fading is invoked by multipath propagation, weather conditions or shadowing from obstacles. Fading channels can be divided into two types [75]: large-scale fading and small-scale fading, which are shortly summarised in the following paragraphs.

Large-scale fading. It represents the average signal power attenuation or path loss caused by the movements of a transmitter and a receiver over large areas. The large-scale fading is affected by the environment around the transmitter and the receiver. Over long distances (several hundreds or thousands of metres) the signal power decreases steadily. Large-scale fading can be described by several models. The most common models are the following:

- Free-space path loss model: It models the attenuation of radio energy between a transmitter and a receiver when there is only free space between them without any obstacles, called a line-of-sight (LOS) path. Due to its simplicity and the fact that the UAVs in the application scenario move in free airspace, this model is used by the delay estimators to represent the power attenuation of a signal between the agents.

- Log-distance path loss model: It models the path loss of a signal between a transmitter and a receiver inside a building or in densely populated areas.
- Okumara model: The model represents the signal attenuation between a transmitter and a receiver for cities with urban structures but not many tall buildings. The Okumara model is specified for three cases, for urban, suburban and open areas and acts as the base for the Hata model.
- Hata model: It is a path loss model that incorporates the graphical data from the Okumara model and extends it to model the effects of diffraction, reflection and scattering caused by city structures.

Small-scale fading. It represents the rapid changes of the amplitude and phase of a radio signal in a short time interval or a short distance. Every path between moving agents has its own Doppler shift, time delay and path attenuation. When the agents move, multipath propagation results in a time-varying signal. These effects can be described by the following selection of models:

- Rayleigh fading model: It is a statistical model to represent the propagation of a signal, which magnitude is assumed to vary randomly or to fade according to a Rayleigh distribution. It is an appropriate model for urban environments when there is no dominant LOS path. It is a special case of the Rician fading model.
- Rician fading model: It is a stochastic model for the radio propagation when the signals arrive at the receiver from several distinct paths. One path of them is much stronger than all the others and represents a line-of-sight (LOS) connection between transmitter and receiver. The amplitude gain is modelled by a Rice distribution. This model is used by the delay estimators as between the agents good LOS conditions exist.
- Nakagami fading model: The path loss model behaves similar to the Rician fading model near their mean values. The difference is that it is able to model multipath scattering with large time delay spreads. Generally, the phases of individual reflected waves are random, but the time delays are nearly equal for all waves. Hence, the envelope of a cumulated cluster of waves is Rayleigh distributed. If the time delays differ significantly between the clusters, inter-symbol interference can occur and the multipath self-interference approximates the case of co-channel interference through multipath incoherent Rayleigh distributed signals.

Wireless channels in UAV applications. As in the application scenario quadrotors are used, a communication channel for the use in aerospace applications is considered in more detail. The characteristics of wireless channels for UAV applications differ from those of traditional channels. In contrast to commonly used channels in which it is assumed that the transmission delay and the packet loss probability are constant over time, the transmission delay and the packet loss probability are non-stationary and vary with the distance between the UAVs.

This fact can be seen by considering the channel capacity $C(d(\tilde{t}_i))$ of the channel between the agents given by

$$C(d(\tilde{t}_i)) = B \cdot \log(1 + \mathrm{SNR}(d(\tilde{t}_i))), \tag{5.1}$$

which describes the maximum data rate at which information can be transmitted without packet losses over a channel. The achievable bandwidth is denoted by B and states the width of an interval in a frequency spectrum, which is bounded by a lower and an upper cut-off frequency. SNR denotes the signal-to-noise-ratio, which is defined as the ratio of the received signal power P_{signal} to the power P_{noise} of the background noise as

$$\mathrm{SNR} = \frac{P_{\mathrm{signal}}}{P_{\mathrm{noise}}}.$$

The signal-to-noise-ratio is given by

$$\mathrm{SNR}(d(\tilde{t}_i)) = \frac{P_{\mathrm{r}}(d(\tilde{t}_i))}{N},$$

with the received mean signal power $P_{\mathrm{r}}(d(\tilde{t}_i))$ and the assumption of additive white Gaussian noise (AWGN) with zero mean and variance N.

If the agents move parallel to each other the channel does not change, but if the agents are on diverging trajectories the channel properties vary. When the distance between the agents decreases, the received signal power $P_{\mathrm{r}}(d(\tilde{t}_i))$ increases as stated in the next section and hence, the $\mathrm{SNR}(d(\tilde{t}_i))$ increases, which results in a higher channel capacity $C(d(\tilde{t}_i))$. In contrast when the distance between the agents increases, the received signal power $P_{\mathrm{r}}(d(\tilde{t}_i))$ decreases and the $\mathrm{SNR}(d(\tilde{t}_i))$ decreases as well. A lower channel capacity $C(d(\tilde{t}_i))$ results. If the received signal power $P_{\mathrm{r}}(d(\tilde{t}_i))$ is below the receiver sensitivity S_{rs}, no signal can be received and no communication is possible. The change of the channel capacity depending on the movement of the agents and the fact that the amount M of data in Bits stays constant over time results in varying transmission delays and packet losses.

For a reliable information transfer, the data rate $R(d(\tilde{t}_i))$ with which the information is communicated has to be lower or equal to the channel capacity $C(d(\tilde{t}_i))$, $(R(d(\tilde{t}_i)) \leq C(d(\tilde{t}_i)))$, while the received signal power $P_{\mathrm{r}}(d(\tilde{t}_i))$ has to equal at least the receiver sensitivity S_{rs}, $(P_{\mathrm{r}}(d(\tilde{t}_i)) \geq S_{\mathrm{rs}})$.

Furthermore, typically between the UAVs are good line-of-sight conditions due to the movement in free space without obstacles. There is a direct path so that the scattering effects are less relevant. It results that the received signal power caused by the direct path is dominant. Hence, the received signal power $P_{\mathrm{r}}(d(\tilde{t}_i))$ depends on the relative positions of the sending and the receiving UAV. Multipath effects may only exist when the UAVs move in low altitudes or when the traffic density is high with many agents moving in the same airspace. In this case signals are received from different paths due to reflections from the ground or other UAVs.

For the determination of the received signal power the propagation of electromagnetic waves is considered, which depends on many factors such as multipath and shadowing effects and varies with the environment [135], whereby the power attenuates as the electromagnetic waves propagate. Shadowing refers to the power attenuation of signals due to obstacles between the transmitter and the receiver. To arrive at the receiver, signals have to pass through or diffract around the agents. In free airspace, the signal power decays proportionally to the squared distance $d^2(\tilde{t}_i)$ between receiver and transmitter. The mean received signal power is given by

$$P_{\mathrm{r}}(d(\tilde{t}_i)) = P_0 \cdot \left(\frac{d(\tilde{t}_i)}{d_0}\right)^{-\alpha}, \tag{5.2}$$

in which α denotes the path loss exponent to represent a non-ideal environment. For the value of α typically $2 < \alpha < 4$ applies. As good LOS conditions between UAVs are assumed and the traffic density is considered to be low, the value $\alpha = 2$ is used [135], which corresponds to the ideal free-space model. P_0 denotes a received power level at a reference distance d_0.

Wireless channels in vehicular applications. In this paragraph a short overview over channels in vehicle-to-vehicle (V2V) communications is given as the event-based control method is applied to vehicles in some examples in this thesis.

In vehicular communications, the radio propagation is strongly influenced by the type of environment the vehicles are moving in, as buildings, dense traffic and vegetation [163]. In order to model vehicular channels, different aspects have to be considered such as the path loss across space and time, a potentially high Doppler shift and frequency selective fading that is caused by moving or static vehicles. As a complete model that incorporates all aspects is quite complex, the models in literature focus on the modelling of single effects. The most common models are the following:

- The most commonly model used for large-scale propagation for vehicular channels is the log-distance model introduced on p. 110.

- In order to model the small-scale fading, the Nakagami distribution and the Gaussian distribution are commonly used.

Similar to the UAV applications, the propagation characteristics of vehicular channels highly depend on the existence of a LOS path. Hence, the large-scale and the small-scale propagation is usually modelled for LOS and non-LOS links separately [43, 157].

5.3 Communication model for UAV applications

5.3.1 *Model of the wireless channel*

The delay estimators of the agents utilise the model described in this section to generate an estimate of the current transmission delay of the communication channel. The model combines the Rician fading model, which represents the time-varying signal due to the multipath propagation and the Gilbert-Elliott model, which states the correlation of packet losses.

Rician fading model. For modelling the communication channel between the UAVs, it is assumed that they are moving in a high altitude and only two UAVs are flying in the area. Hence, good LOS conditions exist and Rician fading is a suitable model for describing the wireless channel [103]. In a Rician fading channel the amplitude of the received signal envelope $R_s(d(\tilde{t}_i))$ is represented by a Rice distribution with parameters

$$v^2(d(\tilde{t}_i)) = \frac{\mathcal{K}(d(\tilde{t}_i)) \cdot \Omega(d(\tilde{t}_i))}{1 + \mathcal{K}(d(\tilde{t}_i))}, \quad \sigma^2(d(\tilde{t}_i)) = \frac{\Omega(d(\tilde{t}_i))}{2\,(1 + \mathcal{K}(d(\tilde{t}_i)))}, \tag{5.3}$$

where $\Omega(d(\tilde{t}_i)) = E(R_s^2(d(\tilde{t}_i)))$ represents the mean of the total received signal power and $E(\cdot)$ states the expected value introduced in Appendix A.7. $\mathcal{K}(d(\tilde{t}_i))$ denotes the ratio of the received signal power P_{LOS} from the LOS path to the signal power $P_{\mathrm{scattered}}$ from the scattered paths as

$$\mathcal{K}(d(\tilde{t}_i)) = \frac{P_{\mathrm{LOS}}}{P_{\mathrm{scattered}}}$$

and is a measure of fading. When $\mathcal{K}(d(\tilde{t}_i)) = 0$, there is no LOS component and the Rician fading model turns into the Rayleigh fading model. If $\mathcal{K}(d(\tilde{t}_i)) = \infty$, there is no scatter component and the channel exhibits no fading. The parameter $\mathcal{K}(d(\tilde{t}_i))$ can be expressed as [29]

$$\mathcal{K}(d(\tilde{t}_i)) = \frac{\sqrt{1 - \gamma(d(\tilde{t}_i))}}{1 - \sqrt{1 - \gamma(d(\tilde{t}_i))}}$$

with

$$\gamma(d(\tilde{t}_i)) = \frac{\mathrm{V}(R_s^2(d(\tilde{t}_i)))}{(\mathrm{E}(R_s^2(d(\tilde{t}_i))))^2},$$

where $\mathrm{V}(\cdot)$ denotes the variance and $\mathrm{E}(\cdot)$ states the expected value. The probability density function (PDF, introduced in Appendix A.7) of the Rice distribution [137] for the received signal amplitude is given by

$$f_{\mathrm{R}}(R_s(d(\tilde{t}_i))|\Omega(d(\tilde{t}_i)), \mathcal{K}(d(\tilde{t}_i))) = \frac{R_s(d(\tilde{t}_i))}{\sigma^2(d(\tilde{t}_i))} \exp\left(-\frac{R_s^2(d(\tilde{t}_i)) + v^2(d(\tilde{t}_i))}{2\,\sigma^2(d(\tilde{t}_i))}\right) I_0\left(\frac{R_s(d(\tilde{t}_i))\, v(d(\tilde{t}_i))}{\sigma^2(d(\tilde{t}_i))}\right) \tag{5.4}$$

in which $I_0(x)$ denotes the 0-th order modified Bessel function of the first kind given by the analytical expression

$$I_0(x) = \frac{1}{\pi} \int_{t=0}^{\pi} \exp(x \cos(t)) \mathrm{d}t, \quad x \in \mathbb{R}. \tag{5.5}$$

It is a symmetrical function, as shown in Fig. 5.4.

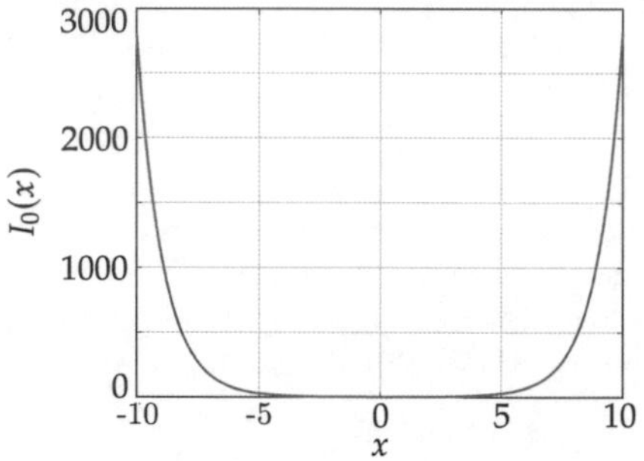

Fig. 5.4: Illustration of the 0-th order modified Bessel function of the first kind $I_0(x)$.

The PDF of the Rice distribution (5.4) depends on the parameters $\Omega(d(\tilde{t}_i))$ and $\mathcal{K}(d(\tilde{t}_i))$, which are contained in the parameters in $\nu(d(\tilde{t}_i))$ and $\sigma(d(\tilde{t}_i))$ as stated in (5.3) and can be rewritten as

$$\begin{aligned} & f_{\mathrm{R}}(R_{\mathrm{s}}(d(\tilde{t}_i))|\Omega(d(\tilde{t}_i)), \mathcal{K}(d(\tilde{t}_i))) \\ & \quad = \underbrace{\frac{R_{\mathrm{s}}(d(\tilde{t}_i))}{\sigma^2(d(\tilde{t}_i))} \exp\left(-\frac{R_{\mathrm{s}}^2(d(\tilde{t}_i))}{2\,\sigma^2(d(\tilde{t}_i))}\right)}_{f_{\mathrm{Ray}}(R_{\mathrm{s}}(d(\tilde{t}_i))|\Omega(d(\tilde{t}_i)))} \underbrace{\exp(-a_0^2(d(\tilde{t}_i))) I_0\left(\frac{R_{\mathrm{s}}(d(\tilde{t}_i))\,\nu(d(\tilde{t}_i))}{\sigma^2(d(\tilde{t}_i))}\right)}_{h(R_{\mathrm{s}}(d(\tilde{t}_i))|\Omega(d(\tilde{t}_i)), \mathcal{K}(d(\tilde{t}_i)))} \end{aligned}$$

with

$$a_0^2(d(\tilde{t}_i)) = \frac{\nu^2(d(\tilde{t}_i))}{2\,\sigma^2(d(\tilde{t}_i))}.$$

It can be seen that the probability density function of the Rice distribution consists of the probability density $f_{\mathrm{Ray}}(R_{\mathrm{s}}(d(\tilde{t}_i))|\Omega(d(\tilde{t}_i)))$ of the Rayleigh distribution multiplied by the function $h(R_{\mathrm{s}}(d(\tilde{t}_i))|\Omega(d(\tilde{t}_i)), \mathcal{K}(d(\tilde{t}_i)))$. Hence, the Rician fading model is a generalisation of the Rayleigh fading model.

In Fig. 5.5 the PDF of the Rice distribution is shown for different values of $a_0^2(d(\tilde{t}_i))$ to illustrate the Rice distribution. It can be seen that the distribution shifts to the right as the coefficient $a_0^2(d(\tilde{t}_i))$ of the random process increases. The value $a_0^2(d(\tilde{t}_i)) = 0$ implies that $\mathcal{K}(d(\tilde{t}_i)) = 0$ holds. Hence, the corresponding curve in Fig. 5.5 shows the PDF of a Rayleigh fading channel. The behaviour of the Rice distribution when only the distance between the agents has changed is illustrated in Fig. 5.8.

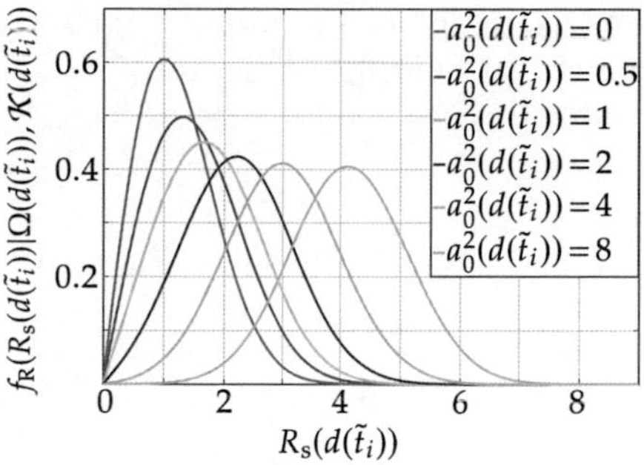

Fig. 5.5: PDF of the Rice distribution for different values of $a_0^2(d(\tilde{t}_i))$ with $\nu(d(\tilde{t}_i)) = 1$.

Determination of the transmission delay of the channel. The transmission delay depends on the properties of the channel, the correlation of packet losses and the distance between the agents, which influences the achievable capacity $C(d(\tilde{t}_i))$ of the channel. The channel capacity (5.1) can be determined in terms of the amount M of data in Bits and the time span t_{data} for the data transmission as

$$C(d(\tilde{t}_i)) = \frac{M}{t_{\text{data}}}. \tag{5.6}$$

As the channel capacity is known due to (5.1) and the amount M of data is known as well and does not change over time, eqn. (5.6) can be solved for t_{data}, which corresponds to the transmission delay $\tau(d(\tilde{t}_i))$ of the channel as

$$\tau(d(\tilde{t}_i)) = \frac{M}{C(d(\tilde{t}_i))}. \tag{5.7}$$

Due to the dependence of the transmission delay on the distance $d(\tilde{t}_i)$ between the agents, the delay changes over time depending on the agents movement. The correlation of packet losses causes them to often occur in bursts, which increases the transmission delay since several packets get lost in a row. Furthermore, additional delays are induced by coding of the information for communication by the transmitters and by decoding the data for the event-based control unit by the receiver.

Determination of the packet loss probability of the channel. Randomly, there might occur packet losses, for example due to possible collisions with packets from other devices communicating over the same network or because the data rate of the channel is too low. The packet loss probability $p(\tilde{t}_i)$ of the channel is determined using the distance $d(\tilde{t}_i)$ between the agents and the Rice distribution defined in (5.4). As the mean received signal power over the direct path is dominant, $P_r(d(\tilde{t}_i))$ is given by

$$\Omega(d(\tilde{t}_i)) = P_r(d(\tilde{t}_i)), \tag{5.8}$$

as stated on p. 113. With the substitution of (5.8) in (5.4) the PDF of the amplitude of the received signal is given by $f_R(R_s(d(\tilde{t}_i))|P_r(d(\tilde{t}_i)), \mathcal{K}(d(\tilde{t}_i)))$ [182]. Then, the packet loss probability can be determined from the cumulative distribution function (CDF, introduced in Appendix A.7) as

$$p(d(\tilde{t}_i)) = \int_{-\infty}^{\sqrt{2 S_{rs}}} f_R(R_s(d(\tilde{t}_i))|P_r(d(\tilde{t}_i)), \mathcal{K}(d(\tilde{t}_i)))\, \mathrm{d}R_s(d(\tilde{t}_i)) \tag{5.9}$$

with S_{rs} as the receiver sensitivity. The receiver sensitivity is defined as the minimum input signal required to produce an output signal. As it is proposed in [129] the packet loss probability can be also stated as

$$p(d(\tilde{t}_i)) = 1 - Q_1\left(\frac{\nu(d(\tilde{t}_i))}{\sigma(d(\tilde{t}_i))}, \frac{\sqrt{2 S_{rs}}}{\sigma(d(\tilde{t}_i))}\right)$$

in which Q_1 denotes the first-order Marcum Q-function given by

$$Q_1(a, b) = \int_b^{\infty} x \cdot \exp\left(-\frac{x^2 + a^2}{2}\right) I_0(ax)\mathrm{d}x \tag{5.10}$$

with $I_0(x)$ as the 0-th order modified Bessel function of the first kind defined in (5.5). The Marcum Q-function is by definition between 0 and 1 and decreases continuously as shown in Fig. 5.6.

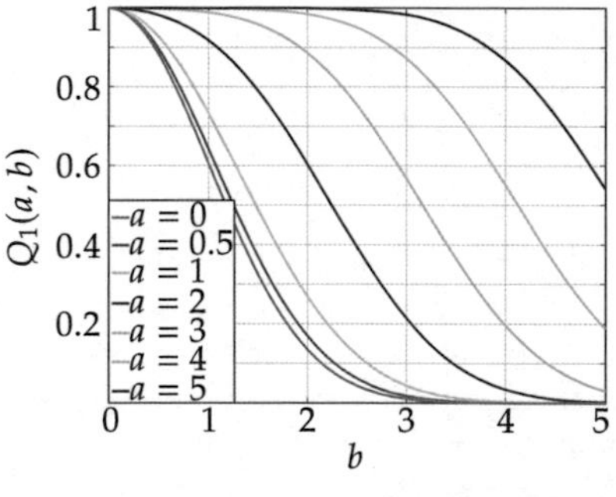

Fig. 5.6: First-order Marcum Q-function $Q_1(a, b)$ for different values of a.

In Fig. 5.7 the Rice distribution is depicted, in which the packet loss probability corresponds to the grey shaded area, which is limited by the receiver sensitivity S_{rs}. The probability of a successful data transmission corresponds to the non-shaded area.

As the Rice distribution depends on the distance between the agents, the curve in Fig. 5.7 shifts to the right if the distance between the agents decreases and the packet loss probability decreases as well, as shown in Fig. 5.8. This results, because for the packet loss probability the integral is determined up to the upper bound of the integration given by

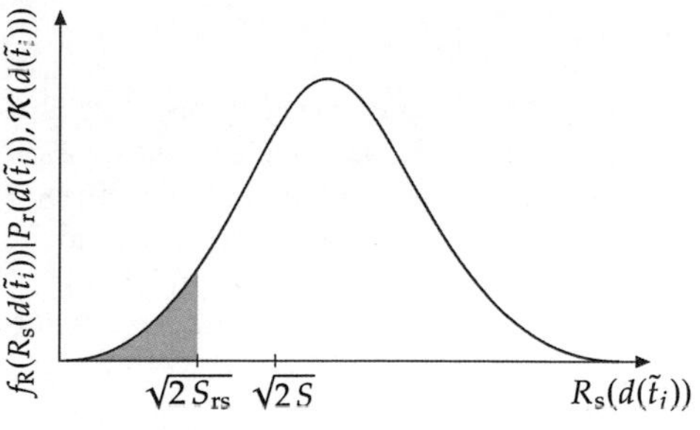

Fig. 5.7: PDF of the Rice distribution stating the packet loss probability.

the fixed receiver sensitivity. As the Rice distribution shifts to the right, the area under the curve decreases and hence $p(d(\tilde{t}_i))$ decreases. If the distance increases, the curve shifts to the left and the packet loss probability $p(d(\tilde{t}_i))$ increases accordingly. In Fig. 5.8 the Rice distribution is shown for four different distances $d(\tilde{t}_i)$ between the agents. The distance influences the mean received signal power $P_r(d(\tilde{t}_i))$ and with eqns. (5.3) and (5.8) the parameters $\nu(d(\tilde{t}_i))$ and $\sigma(d(\tilde{t}_i))$ are changed. It can be seen that with an increasing distance between the agents, the expected value of the curve shifts to the left and the variance of the curve decreases. In the figure an exemplary value of the receiver sensitivity S_{rs} is denoted. It can be seen that with an increasing distance the area under the curve on the left side of the receiver sensitivity increases, which states an increasing packet loss probability.

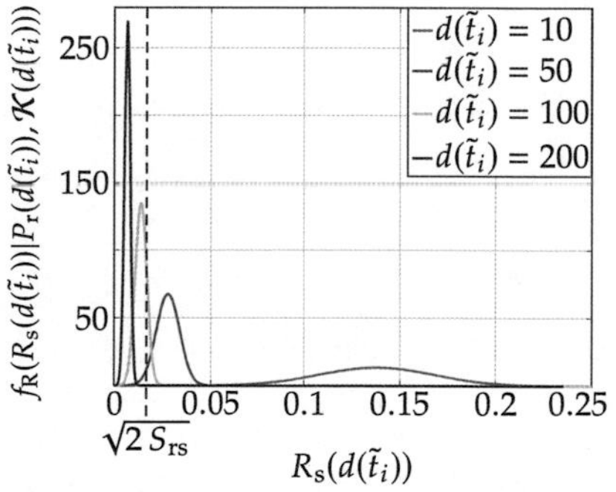

Fig. 5.8: PDF of the Rice distribution for different distances $d(\tilde{t}_i)$.

5.3.2 *Two-state Markov model*

A two-state Hidden Markov Model, as depicted in Fig. 5.9, is used to represent the statistics of the physical wireless channel [182]. Similar to the classical Gilbert-Elliott model reviewed in Section 5.2.1, the upper part of the figure corresponds to the background process of the HMM, which describes the unobservable states and their transitions, while the lower part of the figure conforms to the sensor model with which the output (the packet loss probabilities) of the HMM is modelled. The state vector

$$z = \begin{pmatrix} g & b \end{pmatrix}^{\mathrm{T}}$$

contains two states, which are indicated as 'good' and as 'bad' and correspond to two specific channel conditions. Independent packet loss probabilities $p_{\mathrm{g}}(d(\tilde{t}_i))$ and p_{b} are assigned to each state that represent the channel quality. The probability $p_{\mathrm{g}}(d(\tilde{t}_i))$ of the state 'good' is determined with the Rician fading model described in the preceding section and changes in dependence upon the distance $d(\tilde{t}_i)$ between the agents. It models the dependence of the packet losses on the relative movement of the agents.

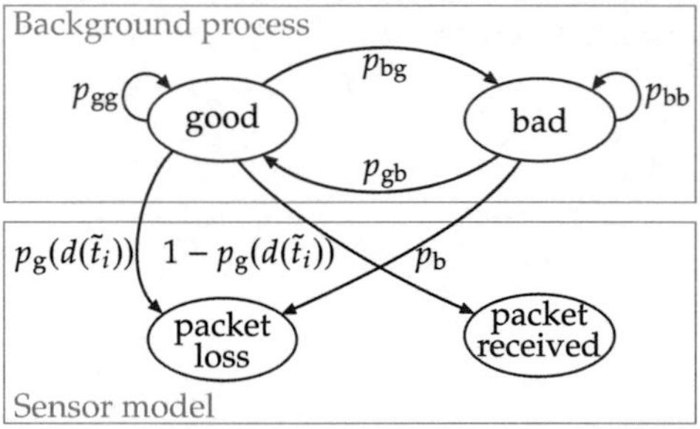

Fig. 5.9: Two-state Markov model of the wireless channel.

The transition probabilities p_{gg}, p_{bb}, p_{gb} and p_{bg} with

$$p_{\mathrm{gg}} = 1 - p_{\mathrm{gb}}, \quad p_{\mathrm{bb}} = 1 - p_{\mathrm{bg}} \tag{5.11}$$

characterise the Markov process and denote the one-step probability to stay in a state or to switch to the other state. The probabilities p_{gb} and p_{bg} are also known as the failure rate and the recovery rate, respectively. These transition probabilities model the correlation of the packet losses. Since the channel conditions, which influence the packet loss correlation are assumed to be constant over time, the probabilities p_{gg}, p_{bb}, p_{gb} and p_{bg} are constant as well and hence are independent from the time interval $[\tilde{t}_i, \tilde{t}_{i+1}]$.

The Hidden Markov Model used in this thesis is similar to the classical Gilbert-Elliott model. The difference is that the packet loss probability $p_{\mathrm{g}}(d(\tilde{t}_i))$ associated with the state 'good' is not constant over time but varies in dependence upon the distance $d(\tilde{t}_i)$ between the agents to cover the channel properties in UAV applications. Since the channel statistics remain constant in the interval $[\tilde{t}_i, \tilde{t}_{i+1}]$ (Assumption 8.1 on p. 175), the probability $p_g(d(\tilde{t}_i))$ is constant in this time interval as well.

The extension of the classical Gilbert-Elliott model by using time-varying packet loss probabilities results in an improved accuracy of the model of the channel states compared to the Gilbert-Elliott model. As the channel statistics stay constant in a time interval $[\tilde{t}_i, \tilde{t}_{i+1}]$ the Markov model is stationary in a time interval. However, the channel realisations can be different within this time interval. Depending on the distances $d(\tilde{t}_i)$ and $d(\tilde{t}_{i+1})$ between the transmitter and the receiver in two consecutive time intervals, the packet loss probability $p_g(d(\tilde{t}_i))$ of the Markov model may be different in these time intervals.

The packet loss probability of the state 'bad' is considered to be

$$p_b = 1, \tag{5.12}$$

which means that packets will always be lost if the channel is in this state. In the state 'good' the packet loss probability is determined by (5.9) using the Rice distribution (5.4) as

$$p_g(d(\tilde{t}_i)) = \int_{-\infty}^{\sqrt{2}\,S_{rs}} f_R(R_s(d(\tilde{t}_i)) | P_r(d(\tilde{t}_i)), \mathcal{K}(d(\tilde{t}_i)))\, \mathrm{d}R_s(d(\tilde{t}_i)). \tag{5.13}$$

As it is stated in Section 5.3.1, the packet loss probability can also be stated in terms of the Marcum Q-function defined in (5.10) as

$$p_g(d(\tilde{t}_i)) = 1 - Q_1\left(\frac{v(d(\tilde{t}_i))}{\sigma(d(\tilde{t}_i))}, \frac{\sqrt{2}\,S_{rs}}{\sigma(d(\tilde{t}_i))}\right). \tag{5.14}$$

With the stationary distributions (5.18) on p. 120 and the packet loss probabilities (5.12) and (5.13) the mean packet loss rate can be evaluated with

$$p_e(d(\tilde{t}_i)) = p_g(d(\tilde{t}_i))\, \pi_g + p_b\, \pi_b = p_g(d(\tilde{t}_i))\, \pi_g + \pi_b. \tag{5.15}$$

5.3.3 *Analysis of the Markov model*

In this section the Markov model is analysed for the properties stated in Appendix A.4. For the analysis of the model (Fig. 5.9) the transition matrix $\boldsymbol{G}_M$ consisting of the constant transition probabilities is built:

$$\boldsymbol{G}_M = \begin{pmatrix} p_{gg} & p_{gb} \\ p_{bg} & p_{bb} \end{pmatrix}. \tag{5.16}$$

As the model represents an unreliable network, the values of the transition probabilities of (5.16) are in the range

$$0 < p_{gg}, p_{bb}, p_{gb}, p_{bg} < 1, \tag{5.17}$$

where the values 0 and 1 are not attained.

Irreducibility. Both states 'good' and 'bad' are reachable from each other, because the transition probabilities p_{gb} and p_{bg} are different from zero due to (5.17). Hence, the two-state Markov model is irreducible.

Aperiodicity. The Markov model is aperiodic if at least for one state i the entry in $\boldsymbol{G}_{\mathrm{M}}$ fulfils $g_{\mathrm{M},ii} > 0$. As for the probabilities $p_{\mathrm{gg}}, p_{\mathrm{bb}}$ of $\boldsymbol{G}_{\mathrm{M}}$, $p_{\mathrm{gg}}, p_{\mathrm{bb}} > 0$ holds, the Markov model is aperiodic.

Ergodicity. The Markov model is ergodic, because it is irreducible and aperiodic and all entries of $\boldsymbol{G}_{\mathrm{M}}$ are strictly positive.

Reversibility. Considering the stationary distributions π_{g} and π_{b} given below in (5.18) and using (A.8) it follows that

$$p_{\mathrm{gb}} \cdot \frac{p_{\mathrm{bg}}}{p_{\mathrm{gb}} + p_{\mathrm{bg}}} = p_{\mathrm{bg}} \cdot \frac{p_{\mathrm{gb}}}{p_{\mathrm{gb}} + p_{\mathrm{bg}}}$$

holds. Hence, the Markov model is reversible.

Sojourn time. The mean time durations that the Markov model remains in the state 'good' and the state 'bad' can be determined by using (A.9) as

$$t_{\mathrm{g}} = \frac{1}{1 - p_{\mathrm{gg}}}, \quad t_{\mathrm{b}} = \frac{1}{1 - p_{\mathrm{bb}}}.$$

Stationary distribution. Since the Markov model is ergodic, a stationary distribution exists. Using (A.15) the stationary distribution can be determined by

$$\pi = \begin{pmatrix} \pi_{\mathrm{g}} \\ \pi_{\mathrm{b}} \end{pmatrix} = \left(\begin{pmatrix} 1 & 0 \\ 0 & 1 \end{pmatrix} - \begin{pmatrix} p_{\mathrm{gg}} & p_{\mathrm{gb}} \\ p_{\mathrm{bg}} & p_{\mathrm{bb}} \end{pmatrix} + \begin{pmatrix} 1 & 1 \\ 1 & 1 \end{pmatrix} \right)^{-1} \begin{pmatrix} 1 \\ 1 \end{pmatrix}$$

and results in

$$\pi_{\mathrm{g}} = \frac{p_{\mathrm{bg}}}{p_{\mathrm{gb}} + p_{\mathrm{bg}}}, \quad \pi_{\mathrm{b}} = \frac{p_{\mathrm{gb}}}{p_{\mathrm{gb}} + p_{\mathrm{bg}}}. \tag{5.18}$$

Equation (5.18) states the mean probability that the Markov model is in the state 'good' or in the state 'bad' when looking at the Markov model at arbitrary points in time. As (5.18) only indicates a mean probability, the Markov model can still be in the other state at a certain point in time. However, the equation only applies after the settling process of the Markov model is over.

Convergence rate towards the stationary distribution. As the Markov model is ergodic and reversible, the convergence rate can be determined by using (A.16) of Theorem A.3 on p. 308. A reformulation of G_{M} with (5.11) leads to

$$G_{\mathrm{M}} = \begin{pmatrix} p_{\mathrm{gg}} & 1 - p_{\mathrm{gg}} \\ 1 - p_{\mathrm{bb}} & p_{\mathrm{bb}} \end{pmatrix}.$$

Then, the eigenvalues of G_{M} are given by

$$\lambda_1 = 1, \quad \lambda_2 = p_{\mathrm{gg}} + p_{\mathrm{bb}} - 1$$

with $\lambda_1 > \lambda_2$, because $p_{\mathrm{gg}}, p_{\mathrm{bb}} < 1$ holds. Hence, the convergence rate towards the stationary distribution (5.18) is given by

$$|g^k_{\mathrm{M},ij} - \pi_j| \leq \frac{1}{\sqrt{\min_{i \in \{\mathrm{g},\mathrm{b}\}} \pi_i}} \cdot |p_{\mathrm{gg}} + p_{\mathrm{bb}} - 1|^k, \quad \forall i,j \in \{\mathrm{g},\mathrm{b}\},\ k \in \mathbb{N}. \tag{5.19}$$

Example 5.1. An exemplary Markov model of a communication channel has the transition probabilities $p_{\mathrm{gg}} = 0.995$ and $p_{\mathrm{bb}} = 0.5$ and is described by the transition matrix

$$G_{\mathrm{M}} = \begin{pmatrix} 0.995 & 0.005 \\ 0.5 & 0.5 \end{pmatrix}.$$

The eigenvalues of G_{M} are given by

$$\lambda_1 = 1, \quad \lambda_2 = 0.495.$$

The stationary distributions are determined with (5.18) as

$$\pi_{\mathrm{g}} = 0.99, \quad \pi_{\mathrm{b}} = 0.01. \tag{5.20}$$

Hence, when looking at the Markov model at arbitrary time instants, the model is with probability 0.99 in the state 'good' and with probability 0.01 in the state 'bad'. The convergence rate towards the stationary distribution is determined with (5.19) and as $\pi_{\mathrm{b}} < \pi_{\mathrm{g}}$ holds given by

$$|g^k_{\mathrm{M},ij} - \pi_j| \leq \frac{1}{\sqrt{\pi_{\mathrm{b}}}} |p_{\mathrm{gg}} + p_{\mathrm{bb}} - 1|^k = \frac{0.495^k}{\sqrt{0.01}}.$$

It takes about $k = 17$ steps to reach the stationary distribution. This means the settling process of the Markov model is finished after 17 steps and the stationary distributions (5.20) are first valid after the settling process is over. □

Due to its properties the Markov model is suitable to model the communication network. The speed of the convergence rate towards the stationary distribution is quite low. An update of the Markov model at runtime would result in an inaccurate estimate of the channel properties. However, since the background process of the Markov model is only initialised once at time instant $t = 0$ and settles to its stationary distribution, this fact is negligible for the scenarios to be considered. These conclusions only concern the Markov model itself. An update of the packet loss probability $p_{\mathrm{g}}(d(\tilde{t}_i))$ due to a changed distance $d(\tilde{t}_i)$ between the agents is possible at runtime without limitations.

5.3.4 Estimation of the parameters of the Markov model

The unknown parameters of the Markov model are the constant transition probabilities p_{gg}, p_{bb}, p_{gb} and p_{bg}, which are estimated using the method presented in [182]. These parameters can be determined initially by a channel estimation at time instant $t = 0$ shown in Fig. 5.2. To this aim, a sequence of N packets $(x_p(1), x_p(2), \ldots, x_p(N))$

$$x_p(s), \quad (s = 1, 2, \ldots, N) \tag{5.21}$$

is sent into the network and the received observation sequence $(x(1), x(2), \ldots, x(N))$

$$x(s), \quad (s = 1, 2, \ldots, N) \tag{5.22}$$

is evaluated in which $x(s) = 0$ indicates a packet loss and $x(s) = 1$ states a successful transmission of the corresponding packet $x_p(s)$. The value N states the total number of observations. The state of the Markov model corresponding to $x(s)$ is denoted by $z(s)$. The state $z(s)$ can take the values 0 or 1 indicating that the Markov model is in the state 'bad' or the state 'good', respectively.

The parameters of the Markov model are estimated based on two statistical parameter estimates, the *average packet loss rate* $\hat{p}_{e,0}$ and the *correlation coefficient* $\hat{c}$ using the given observation sequence consisting of (5.21) and (5.22). With the Markov model the mean packet loss rate can be determined with the stationary distributions (5.18) as

$$p_{e,m} = \frac{1}{N} \sum_{s=1}^{N} \left(\pi_g p_g(s) + \pi_b \right) = p_{g,a}\, \pi_g + \pi_b \tag{5.23}$$

with

$$p_{g,a} = \frac{1}{N} \sum_{s=1}^{N} p_g(s).$$

$p_{g,a}$ states the mean packet loss probability of the state 'good', which is the averaged packet loss probability $p_g(s)$. The mean packet loss rate $\hat{p}_{e,0}$ of the observation sequence is approximated by

$$\hat{p}_{e,0} = 1 - \frac{1}{N} \sum_{s=1}^{N} x(s) = p_{g,a}\, \pi_g + \pi_b, \tag{5.24}$$

which equals the mean packet loss rate (5.23). Equation (5.24) can be written in terms of the probabilities p_{gg} and p_{bb} as

$$p_0\, p_{bb} + (p_0 - 1)\, p_{gg} = 2p_0 - 1 \tag{5.25}$$

with

$$p_0 = \frac{1 - \hat{p}_{e,0}}{1 - p_{g,a}}.$$

As the model parameters p_{gg} and p_{bb} are searched for, a second relation is required. Hence, the expectation $E(x(s)\,x(s+L))$ is considered

$$\begin{aligned} E(x(s)\,x(s+L)) &= E(E(x(s)\,x(s+L)|x(s))) \\ &= E(x(s)E(x(s+L)|x(s))) \\ &= \pi_g(1-p_g(s))(1-p_g(s+L))p_{gg} \end{aligned}$$

in which $x(s)$ and $x(s+L)$ correspond to two adjacent states of the Markov model, where L packets are assumed to be between the states. With the observation sequence the correlation coefficient is approximated as

$$\hat{c} = \frac{1}{N-L}\sum_{s=1}^{N-L} x(s)\,x(s+L) = \pi_g\, p_{gg}\, p_c \tag{5.26}$$

with

$$p_c = \sum_{s=1}^{N-1}(1-p_g(s))(1-p_g(s+1)).$$

With (5.24) and (5.26) a quadratic equation results in terms of p_{gg} as

$$-\frac{p_0\,p_c}{p_0-1}p_{gg}^2 + \frac{p_0\,p_c+\hat{c}}{p_0-1}p_{gg} - \frac{\hat{c}}{p_0-1} = 0.$$

In [182] it is shown that the above equation has two roots, whereby one root is equal to 1. In general, the probability p_{gg} is positive and less than 1. Hence, with (5.25) and the results $\hat{p}_{e,0}$ and $\hat{c}$ of the channel estimation the parameters of the Markov model are estimated. The estimated transition probabilities are denoted by $\hat{p}_{gg}$, $\hat{p}_{bb}$, $\hat{p}_{gb}$ and $\hat{p}_{bg}$.

Result of the estimation:

$$\begin{aligned} \hat{p}_{gg} &= \frac{\hat{c}}{p_0\,p_c} \\ \hat{p}_{bb} &= -\frac{p_0}{p_0-1}\cdot\hat{p}_{gg} + \frac{2\,p_0-1}{p_0-1} \\ \hat{p}_{gb} &= 1-\hat{p}_{gg} \\ \hat{p}_{bg} &= 1-\hat{p}_{bb}. \end{aligned} \tag{5.27}$$

Part III

Event-based control over an ideal communication network

Structure and tasks of the event-based control units

6

This chapter presents the structure of the event-based control units of the agents for an ideal communication network. In Section 6.1 the basic assumptions and the control approach are introduced. The structure and the tasks of the control unit of the stand-on agent are given in Section 6.2. In Section 6.3 the structure and the tasks of the control unit of the give-way agent are stated.

6.1 Introduction

This chapter deals with the event-based control method for the overall system shown in Fig. 1.2 on p. 4. The agents are connected by an ideal communication network as illustrated in Fig. 5.1 on p. 105. The Assumptions 6.1 and 6.2 hold for this part of the thesis.

Assumption 6.1. The ideal communication network does not induce any transmission delays or packet losses. The network is restricted to a structure depicted in Fig. 5.1.

Assumption 6.2. The execution of the control algorithms by the agents does not cause any computational delays.

The control method consists of the event-based control units A_S and A_G of the stand-on agent and the give-way agent, which are able to communicate to one another over the network. The control units have the basic structure shown in Fig. 1.5 on p. 16. As the network is ideal, a delay estimator is not required for the control. The specific structures of the control units of the stand-on agent and the give-way agent are given in the next sections. The control units fulfil the respective local control aims of the agents with respect to their constraints, which leads to the satisfaction of the control aims (A1) – (A3) of the overall system with respect to the constraints (C1) and (C2) introduced in Section 1.4. To this aim the control units execute the methods stated in Chapters 3 and 4.

Remark. Constraint (C3) is neglected, because the computation power onboard the agents is assumed to be appropriate to cause no delays during the execution of the control algorithms (Assumption 6.2).

6.2 Event-based control unit A_S of the stand-on agent

6.2.1 *Structure of the control unit*

The structure of the unit A_S of the stand-on agent is shown in Fig. 6.1. The tasks of the control unit are given in the next sections. Here, also the signals shown in Fig. 6.1 are introduced.

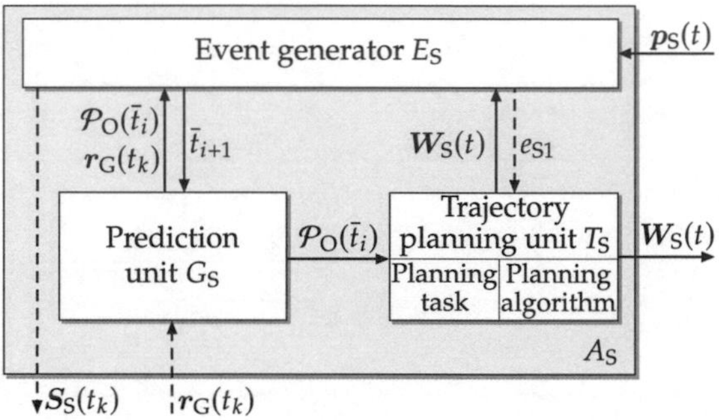

Fig. 6.1: Structure of the event-based control unit A_S of the stand-on agent.

The tasks that the agents have to execute are independent of the form of the movement of the agents. Both, for a general movement in cartesian coordinates and for a circular movement in cylindrical coordinates, the same tasks have to be performed.

6.2.2 *Tasks of the prediction unit G_S*

The prediction unit G_S executes two tasks:

1. It receives the request defined in $r_G(t_k)$ (7.6) on p. 139 of the give-way agent and passes it to the event generator.

2. It possesses a set of positions of obstacles
$$\mathcal{P}_O(\bar{t}_i) = \{p_{O1}(\bar{t}_i), p_{O2}(\bar{t}_i), \ldots, p_{ON}(\bar{t}_i)\}, \tag{6.1}$$
which are located in the area the stand-on agent is moving in. The set is valid in an interval $[\bar{t}_i, \bar{t}_{i+1}]$ of a part $w_S^i(t)$ of its piecewise trajectory $w_S(t)$ and is updated whenever the agent enters a new area, which is indicated by the event generator. For the implementation the event generator sends the time instant $\bar{t}_{i+1}$ at which the stand-on agent reaches the next part $w_S^{i+1}(t)$ of its trajectory to the prediction unit. The set $\mathcal{P}_O(\bar{t}_i)$ is transferred to the event generator and the trajectory planning unit whenever the set is updated in order to check the trajectory for a possible collision with an obstacle and to change the trajectory appropriately.

6.2.3 *Tasks of the event generator E_S*

The event generator E_S acts as a supervisor and monitors the fulfilment of the local control aims (S1) and (S2). At time $t = 0$ the event generator communicates its data $\boldsymbol{S}_S(0)$ to the give-way agent. After that, it uses the request $\boldsymbol{r}_G(t_k)$ to invoke the communication of $\boldsymbol{S}_S(t_k)$. The vector $\boldsymbol{r}_G(t_k)$ given in (7.6) contains a list of future time instants at which communication has to be invoked by the stand-on agent. Whenever a new request is received or the next time instant t_{k+1} in $\boldsymbol{r}_G(t_k)$ is reached the event e_{S0} is generated with the condition

$$\text{Communication event } e_{S0} : \begin{cases} & \text{reception of } \boldsymbol{r}_G(t_k) \\ \vee & t = t_{k+j}, \quad j = 0, 1, \dots, N. \end{cases}$$

As a result the data $\boldsymbol{S}_S(t_k)$ is sent to the give-way agent as a matrix with the structure

$$\boldsymbol{S}_S(t_k) = \begin{pmatrix} u & \\ t_k & \\ \boldsymbol{p}_S(t_k) & \boldsymbol{T}_S \\ \boldsymbol{v}_S(t_k) & \\ \boldsymbol{0}_{n\times 1} & \end{pmatrix}$$

where $\boldsymbol{T}_S$ is given by (3.20) and contains the piecewise trajectory $\boldsymbol{w}_S(t)$ represented by the transition time intervals and the corresponding control points. The vector $\boldsymbol{0}_{n\times 1}$ of length $n = 3(r+1) - 7$ with r as the relative degree of the agent contains only zeros to fill in the missing entries in the matrix. For a trajectory $\boldsymbol{w}_S(t)$ consisting of four pieces as it is considered in the scenarios in this thesis, the communicated information has the following structure:

$$\boldsymbol{S}_S(t_k) = \begin{pmatrix} u & t_{\text{start}} & t_t^1 & t_t^2 & t_t^3 & t_t^4 \\ t_k & b_{S,x}^1 & b_{S,x}^2 & b_{S,x}^3 & b_{S,x}^4 & b_{S,x}^5 \\ \boldsymbol{o}_S(t_k) & b_{S,y}^1 & b_{S,y}^2 & b_{S,y}^3 & b_{S,y}^4 & b_{S,y}^5 \\ \boldsymbol{0}_{8\times 1} & b_{S,z}^1 & b_{S,z}^2 & b_{S,z}^3 & b_{S,z}^4 & b_{S,z}^5 \end{pmatrix} \tag{6.2}$$

with

$$\boldsymbol{o}_S(t_k) = \begin{pmatrix} \boldsymbol{p}_S(t_k) & \boldsymbol{v}_S(t_k) \end{pmatrix}^{\mathrm{T}}.$$

The trajectory is communicated for the time interval $t_k \le t \le t_{\text{end}}$, where t_{end} is the end time instant of the trajectory according to (2.13). The first column of the data is composed of the current time instant t_k, the position $\boldsymbol{p}_S(t_k)$ and speed $\boldsymbol{v}_S(t_k)$ of the stand-on agent at t_k and a variable u. It indicates whether the stand-on agent changed its trajectory:

$$u = \begin{cases} 0, & \text{no change of the trajectory of the stand-on agent} \\ 1, & \text{change of the trajectory of the stand-on agent.} \end{cases} \tag{6.3}$$

The four pieces $w_S^i(t)$, $(i = 1, \ldots, 4)$ of the trajectory $w_S(t)$ are derived using the transition time intervals t_t^i, $(i = 1, \ldots, 4)$ and the control points $b_{S,x}^i$, $b_{S,y}^i$ and $b_{S,z}^i$, $(i = 1, \ldots, 4)$ according to Section 3.2.4. The time instant t_{start} represents the start time instant of the first piece $w_S^1(t)$ with the control points $b_{S,x}^1$, $b_{S,y}^1$ and $b_{S,z}^1$ at the start point. As communication occurs when the stand-on agent is moving on the trajectory, the time instant t_{start} may be in the past ($t_{start} < t_k$). Nevertheless, the communication of t_{start} is necessary to enable the give-way agent to determine the trajectory $w_S(t)$ for an event generation. The control point vectors consist of five elements since five control points are required to define the start point and the end point of the trajectory as the agents in the quadrotor example have the relative degree given by (2.21). The vectors are given by

$$b_{S,j}^1 = \begin{pmatrix} b_{S,0,j}^1 & b_{S,1,j}^1 & b_{S,2,j}^1 & b_{S,3,j}^1 & b_{S,4,j}^1 \end{pmatrix}^T, \quad j \in \{x, y, z\}$$

and

$$b_{S,j}^i = \begin{pmatrix} b_{S,5,j}^i & b_{S,6,j}^i & b_{S,7,j}^i & b_{S,8,j}^i & b_{S,9,j}^i \end{pmatrix}^T, \quad j \in \{x, y, z\},\ i = 1, \ldots, 4.$$

Furthermore, the event generator decides when it is necessary to change the trajectory of the stand-on agent to avoid a collision with an obstacle or to keep its individual task fulfilled. It monitors the distance between the trajectory $w_S(t)$ of the stand-on agent provided by the trajectory planning unit and the positions of the obstacles $p_{Oj}(\bar{t}_i)$, $(j = 1, \ldots, N)$ supplied by the prediction unit as

$$\text{dist}_{\min}\left(w_S(t), p_{Oj}(\bar{t}_i)\right) = \min_{t \in \bar{t}_i}\left(\left\|w_S(t) - p_{Oj}(\bar{t}_i)\right\|\right), \quad \forall j.$$

In addition, it checks the maximum distance $\text{dist}_{\max}(w_S(t), w_d(t))$ between $w_S(t)$ and the desired trajectory $w_d(t)$ using (4.25) on p. 97 to verify if a change of the trajectory due to an obstacle leads to a violation of its individual task. The fulfilment of the task can be neglected when it is necessary to avoid a collision with an obstacle. An event e_{S1} is generated if the condition

$$\text{Local event } e_{S1}: \begin{cases} & \text{dist}_{\min}\left(w_S(t), p_{Oj}(\bar{t}_i)\right) \leq \bar{e}_S, \quad \forall j \\ \vee & \text{dist}_{\max}(w_S(t), w_d(t)) \geq \bar{e}_S \end{cases}$$

is satisfied. After the generation of the event the trajectory $w_S(t)$ is changed to keep the control aim (S1) fulfilled.

6.2.4 *Task of the trajectory planning unit T_S*

The unit T_S is split into the planning task section and the planning algorithm section. In the planning task section the boundary requirements $W_S^i(\bar{t}_i)$, $W_S^i(\bar{t}_{i+1})$ in the form (3.2) on p. 46 of the collision-free trajectory are defined by considering the list of obstacles $\mathcal{P}_O(\bar{t}_i)$

(6.1), so that the local control aim (S1) is satisfied. In the planning algorithm section the trajectory is planned with Algorithm 3 on p. 57 for a trajectory for a general movement, Algorithm 4 on p. 67 for a circular trajectory and Algorithm 10 on p. 80 for a transition segment, respectively. The unit generates the matrix

$$\boldsymbol{W}_{\mathrm{S}}(t) = \begin{pmatrix} \boldsymbol{w}_{\mathrm{S}}^{\mathrm{T}}(t) \\ \dot{\boldsymbol{w}}_{\mathrm{S}}^{\mathrm{T}}(t) \\ \ddot{\boldsymbol{w}}_{\mathrm{S}}^{\mathrm{T}}(t) \\ \boldsymbol{w}_{\mathrm{S}}^{(3)\mathrm{T}}(t) \\ \boldsymbol{w}_{\mathrm{S}}^{(4)\mathrm{T}}(t) \end{pmatrix} = \begin{pmatrix} w_{\mathrm{S,x}}(t) & w_{\mathrm{S,y}}(t) & w_{\mathrm{S,z}}(t) & w_{\mathrm{S},\psi}(t) \\ \dot{w}_{\mathrm{S,x}}(t) & \dot{w}_{\mathrm{S,y}}(t) & \dot{w}_{\mathrm{S,z}}(t) & \dot{w}_{\mathrm{S},\psi}(t) \\ \ddot{w}_{\mathrm{S,x}}(t) & \ddot{w}_{\mathrm{S,y}}(t) & \ddot{w}_{\mathrm{S,z}}(t) & \ddot{w}_{\mathrm{S},\psi}(t) \\ w_{\mathrm{S,x}}^{(3)}(t) & w_{\mathrm{S,y}}^{(3)}(t) & w_{\mathrm{S,z}}^{(3)}(t) & w_{\mathrm{S},\psi}^{(3)}(t) \\ w_{\mathrm{S,x}}^{(4)}(t) & w_{\mathrm{S,y}}^{(4)}(t) & w_{\mathrm{S,z}}^{(4)}(t) & w_{\mathrm{S},\psi}^{(4)}(t) \end{pmatrix}. \tag{6.4}$$

It consists of the trajectory $\boldsymbol{w}_{\mathrm{S}}(t)$ and its first four derivatives, because the feedforward controller of the quadrotor requires reference variables until the fourth derivative as stated in Section 10. In addition to plan the trajectory, the heading angle $\psi(t)$ of the quadrotor must be specified. In this thesis, the heading angle is set to $\psi(t) = 0$ constantly. Hence $w_{\mathrm{S},\psi}(t) = \dot{w}_{\mathrm{S},\psi}(t) = \ddot{w}_{\mathrm{S},\psi}(t) = w_{\mathrm{S},\psi}^{(3)}(t) = w_{\mathrm{S},\psi}^{(4)}(t) = 0$ results.

Remark. The method can be used in the same way for other types of agents (e.g. cars). To this aim, only the different dynamics of the agents have to be considered as well as the relative degree of the agents for the generation of the matrix in (6.4).

6.3 Event-based control unit A_G of the give-way agent

6.3.1 *Structure of the control unit*

The structure of the unit A_G of the give-way agent is depicted in Fig. 6.2. In the next sections the tasks of the control unit are stated together with an introduction of the signals.

It can be seen that the control unit has the same structure with the same parts as the unit A_S of the stand-on agent. This is necessary as both agents should be able to fulfil both assignments 'stand-on' and 'give-way'. Depending on their respective assignment, the parts of the units execute different tasks to fulfil their specific local control aims.

6.3.2 *Tasks of the prediction unit G_G*

The prediction unit G_G executes the following two tasks to solve Problem 1.2:

1. Data processing: The communicated data $\boldsymbol{S}_{\mathrm{S}}(t_k)$ of the stand-on agent is received by the prediction unit. It evaluates the variable u (6.3) to check if the trajectory of

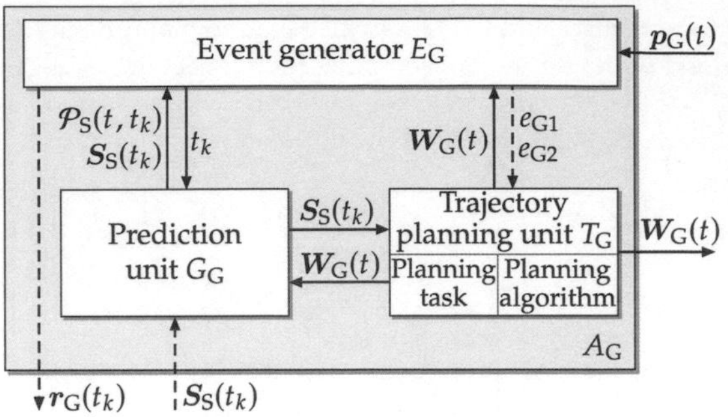

Fig. 6.2: Structure of the event-based control unit A_G of the give-way agent.

the stand-on agent has been changed. Furthermore, it transfers the received data to the event generator and to the trajectory planning unit for a verification whether a violation of a control aim threatens and whether the trajectory of the give-way agent needs to be changed.

2. Inclusion of the movement of the stand-on agent: Based on the trajectory planning technique and the position $p_S(t_k)$ and speed $v_S(t_k)$ of the stand-on agent communicated at the last event time instant t_k the prediction unit of the give-way agent generates a set $\mathcal{P}_S(t, t_k)$ that includes all possible future positions of the stand-on agent:

$$p_S(t) \in \mathcal{P}_S(t, t_k), \quad \forall t \geq t_k.$$

In order to determine the inclusion of the future movement of the stand-on agent in the time intervals $[t_{k+j}, t_{k+j+1}], (j = 1, \ldots, N)$ the prediction method is initialised with the position $w_S(t_{k+j})$ and the speed $\dot{w}_S(t_{k+j})$ of the stand-on agent on its trajectory at the time instant t_{k+j} under the assumption that the stand-on agent followed its trajectory exactly in the interval before.

6.3.3 *Tasks of the event generator E_G*

The event generator E_G acts as supervisor and solves Problem 1.3. Its tasks are split into an initialisation phase and an execution phase:

1) Initialisation phase. At time $t = 0$ the give-way agent sends a request to the stand-on agent to obtain its initial information. The stand-on agent responds by sending the requested data $S_S(0)$.

2) Execution phase. The event generator monitors the distance between the current position $p_G(t)$ of the give-way agent and the set $\mathcal{P}_S(t, t_k)$ of predicted positions of the stand-on agent as well as between the trajectories $w_G(t)$, $w_S(t)$ of the two agents. To this aim it determines the trajectory $w_S(t)$ and the sequence of time instants $\bar{t}_i$ (2.13) of the piecewise trajectory recursively as

$$\bar{t}_{i+1} = \bar{t}_i + t_t^i, \quad i = 1, \ldots, 4 \tag{6.5}$$

using the communicated data $S_S(t_k)$ (6.2). In (6.5) $\bar{t}_1 = t_{start}$ applies.

The event generator determines the event time instants over the entire time course at which communication must be invoked and checks if the trajectory of the give-way agent needs to be changed to fulfil the control aims. If a change of the trajectory is necessary, it triggers the planning of a *proactive trajectory* or a *reactive trajectory* to keep the control aims fulfilled. The planning of these trajectories is stated in detail in Section 7.4. The following three events are generated to fulfil the control aims:

- With the *communication event* e_{G0} communication to the stand-on agent is invoked.
- The *proactive event* e_{G1} triggers the planning of a proactive trajectory $w_{G,p}(t)$ of the give-way agent.
- With the *reactive event* e_{G2} the generation of a reactive trajectory $w_{G,r}(t)$ of the give-way agent is invoked.

6.3.4 *Task of the trajectory planning unit T_G*

The task of the trajectory planning unit T_G is to solve Problem 1.4. The unit is subdivided into the planning task section and the planning algorithm section. In the planning task section the boundary conditions on the start point and the end point of each part of the trajectory are defined in a way to fulfil control aim (G1). In the planning algorithm section the initial trajectory $w_G(t)$, the proactive trajectory $w_{G,p}(t)$ and the reactive trajectory $w_{G,r}(t)$ of the give-way agent are planned using the result of the planning task section and Algorithm 3 on p. 57 for a general movement, Algorithm 4 on p. 67 for a circular movement and Algorithm 10 on p. 80 for a transition segment. The unit generates the matrix

$$W_G(t) = \begin{pmatrix} w_G^T(t) \\ \dot{w}_G^T(t) \\ \ddot{w}_G^T(t) \\ w_G^{(3)T}(t) \\ w_G^{(4)T}(t) \end{pmatrix} = \begin{pmatrix} w_{G,x}(t) & w_{G,y}(t) & w_{G,z}(t) & w_{G,\psi}(t) \\ \dot{w}_{G,x}(t) & \dot{w}_{G,y}(t) & \dot{w}_{G,z}(t) & \dot{w}_{G,\psi}(t) \\ \ddot{w}_{G,x}(t) & \ddot{w}_{G,y}(t) & \ddot{w}_{G,z}(t) & \ddot{w}_{G,\psi}(t) \\ w_{G,x}^{(3)}(t) & w_{G,y}^{(3)}(t) & w_{G,z}^{(3)}(t) & w_{G,\psi}^{(3)}(t) \\ w_{G,x}^{(4)}(t) & w_{G,y}^{(4)}(t) & w_{G,z}^{(4)}(t) & w_{G,\psi}^{(4)}(t) \end{pmatrix},$$

which has the same structure as the matrix in (6.4).

According to Section 1.4, the stand-on agent is only responsible to follow its trajectory exactly (control aims (A2) and (A3) of the overall system). In contrast, the give-way agent is responsible to plan its trajectory so as to fulfil the requirements (1.1) and (1.2) of control aim (A1). Hence, in the next chapter the parts of the event-based control unit A_G of the give-way agent are presented in more detail.

Components of the event-based control unit

7

This chapter presents the components of the event-based control unit A_G of the give-way agent for an ideal communication network as it is responsible to satisfy collision avoidance and a maximum spatial separation between the agents. The chapter gives a detailed description of the prediction unit G_G (Section 7.2), the event generator E_G (Section 7.3) and the trajectory planning unit T_G (Section 7.4). Furthermore, the event-based communication scheme is described in Section 7.5. The event-based control method is summarised by an algorithm in Section 7.6.

7.1 Overview

This chapter describes the components of the event-based control unit of the give-way agent with the structure shown in Fig. 6.2 on p. 132. It is assumed that the agents are connected over an ideal communication network that satisfies Assumption 6.1 on p. 127. As the give-way agent is responsible for the fulfilment of the requirements (1.1) and (1.2) of control aim (G1) the components solve the Problems 1.2 – 1.4. To this aim they use the methods stated in Chapters 3 and 4 for trajectory planning, prediction and event generation.

The main contribution of this chapter is the combination of the algorithms from Chapters 3 and 4 to obtain Algorithm 14 on p. 168, which execution leads to a guaranteed satisfaction of control aim (G1). This main result is stated in Theorem 7.5 on p. 170. Furthermore, it is shown that between two consecutive events a minimum time span is ensured to exclude Zeno behaviour. In Theorem 7.3 on p. 165 this result is given for a general movement of the agents and in Theorem 7.4 on p. 168 it is stated for a circular movement. Important results for the event generation are summarised in Theorem 7.1 on p. 140 and Theorem 7.2 on p. 147. In the following sections the components are stated in detail.

7.2 Prediction unit G_G

7.2.1 General prediction

The prediction unit G_G uses Algorithm 11 on p. 88 to generate an inclusion of the possible future positions of the stand-on agent for a general movement. It results in the set

$$\mathcal{P}_S(t,t_k)=\left\{p_S(t)\in\mathbb{R}^3: \frac{(x_S(t)-\bar{x}_S(t,t_k))^2}{r^2_{S,x}(t,t_k)}+\frac{(y_S(t)-\bar{y}_S(t,t_k))^2}{r^2_{S,y}(t,t_k)}+\frac{(z_S(t)-\bar{z}_S(t,t_k))^2}{r^2_{S,z}(t,t_k)}-1\leq 0\right\}, \tag{7.1}$$

which was introduced in eqn. (4.4) in Theorem 4.1 on p. 87 for a general movement of an arbitrary agent. In order to generate the inclusion (7.1), the prediction unit requires the position $p_S(t_k)$ and the speed $v_S(t_k)$ of the stand-on agent at the time instant t_k. In addition, the parameters e and f are required, which are derived in Theorem 4.1 depending on the dynamics of the stand-on agent. The maximum speed $v_{S,\max}$ is used to parametrise the values in (4.5).

The set $\mathcal{P}_S(t,t_k)$ can also include the positions of the stand-on agent for $t > t_{k+1}$. To this aim the set is parametrised with the positions $w_S(t_{k+j})$ and the speeds $\dot{w}_S(t_{k+j})$ of the stand-on agent on its communicated trajectory at the future time instants t_{k+j}, $(j = 2,\dots,N)$ that are determined by the event generator.

Theorem 4.1 incorporating eqn. (7.1) states a solution to Problem 1.2 for a general movement of the agents.

7.2.2 Specific prediction for a circular movement

In order to generate an inclusion of all possible future phases and heights of the stand-on agent for a circular movement the prediction unit executes Algorithm 12 on p. 92. As a result, the set

$$\mathcal{P}_{S,c}(t,t_k)=\left\{p_S(t)\in\mathbb{R}^3:\ \Phi_S(t)\in\Delta\Phi_S(t,t_k),\ z_S(t)\in\Delta z_S(t,t_k)\right\} \tag{7.2}$$

with

$$\begin{aligned}\Delta\Phi_S(t,t_k)&=[\Phi_{S,\min}(t,t_k),\Phi_{S,\max}(t,t_k)]\\ \Delta z_S(t,t_k)&=[z_{S,\min}(t,t_k),z_{S,\max}(t,t_k)]\end{aligned} \tag{7.3}$$

is generated, which corresponds to the set introduced in (4.11) of Theorem 4.2 on p. 91 for a circular movement of an arbitrary agent. The values in (7.3) are determined with (4.14) – (4.17). The generation of the set (7.2) is based on the phase $\Phi_S(t_k)$, which can be derived from the position $p_S(t_k)$ with (2.3) and the height $z_S(t_k)$ of the stand-on agent.

The future phases and heights of the stand-on agent for $t > t_{k+1}$ can also be included in $\mathcal{P}_{S,c}(t,t_k)$ when the set is determined based on the phases $\Phi_S(t_{k+j})$ and the heights

$z_S(t_{k+j})$ that are derived from the positions $w_S(t_{k+j})$ of the stand-on agent given by its communicated trajectory at the future time instants t_{k+j}, $(j = 2, \dots, N)$.

Problem 1.2 is solved by Theorem 4.2 incorporating eqn. (7.2) for a circular movement of the agents.

7.3 Event generator E_G

7.3.1 *Overview of the general event generation*

This section summarises the situations that lead to an event generation for a general movement of the agents and the reactions to the events. The summary is given in Tab. 7.1. The generation method for the events is stated in more detail in the Sections 7.3.2 – 7.3.4.

Table 7.1: Summary of the event generation for a general movement.

Situation	Event	Reaction
Position of agent $\bar{P}_S$ becomes too uncertain.	e_{G0}	$\bar{P}_G$ sends $r_G(t_k)$, $\bar{P}_S$ responds by sending $S_S(t_k)$ if $w_S(t)$ has changed. $\bar{P}_S$ sends $S_S(t_k)$ autonomously if $w_S(t)$ is unchanged.
$\bar{P}_S$ or $\bar{P}_G$ changed their trajectories and a violation of control aim (G1) is detected in the future.	e_{G1}	$\bar{P}_G$ plans a proactive trajectory $w_{G,p}(t)$.
$\bar{P}_S$ changed its trajectory without notification and a violation of control aim (G1) is imminent.	e_{G2}	$\bar{P}_G$ plans a reactive trajectory $w_{G,r}(t)$.

Remark. The event generation for other types of agents (e.g. cars) is realised in the same way. The only difference is that the event threshold has to be adapted to the dynamics of the agents.

7.3.2 *Communication event e_{G0}*

Situation. Whenever the uncertainty about the movement of the stand-on agent becomes too large so that new information of the stand-on agent is required to ensure control aim (G1), the event e_{G0} is generated.

Event generation. The generation of the communication event is split into two phases. In the initialisation phase, at time $t = 0$, it is assumed that the agents are located such that the control aim (A1) is satisfied. The give-way agent generates an event e_{G0} to obtain the initial information of the stand-on agent.

In the execution phase data are exchanged depending on the movement of the agents. In the following the event conditions and the reaction after triggering an event are stated. For the generation of event e_{G0} the maximum and minimum distances between the sets $\mathcal{P}_S(t, t_{k+j})$, $(j = 0, \ldots, N)$ of positions of the stand-on agent and the position $w_G(t)$ of the give-way agent defined in (4.19) on p. 94 are considered. The event condition (7.4) stated below is only evaluated if one of the following situations occurs:

- the give-way agent has received the initial information $S_S(0)$ of the stand-on agent.
- the give-way agent has received the data $S_S(t_k)$ and the stand-on agent has changed its trajectory indicated by the variable $u = 1$ according to (6.3).
- the give-way agent has changed its trajectory.

Otherwise there is no need to determine new event time instants, because the agents move in a way for which the already determined event time instants are valid.

As stated in Section 4.3.1, the event condition is evaluated recursively for the time interval $t_k \leq t \leq t_{\text{end}}$ that is included in the communicated information $S_S(t_k)$. After the determination of an event time instant at t_{k+j}, the set $\mathcal{P}_S(t, t_{k+j})$ is reset and newly generated to determine the next event time instants at t_{k+j+1}. The events are generated if the condition

$$\text{Communication event } e_{G0}: \begin{cases} \quad \text{dist}_{\min}(w_G(t), \mathcal{P}_S(t, t_{k+j})) \leq \underline{s} + \bar{e}_G & (I) \\ \vee \;\; \text{dist}_{\max}(w_G(t), \mathcal{P}_S(t, t_{k+j})) \geq \bar{s} - \bar{e}_G & (II) \\ \wedge \;\; t - t_{k+j} \geq t_{\text{com}} & (III) \end{cases} \tag{7.4}$$

for $t_k \leq t \leq t_{\text{end}}$ is satisfied at time instants t_{k+j+1}. The decision if the parts (I) and (II) or part (III) of condition (7.4) are used depends on the distance $\text{dist}(w_G(t), \mathcal{P}_S(t, t_{k+j}))$:

$$\text{Event generated with}: \begin{cases} \begin{matrix} \text{part } (I) \\ \text{part } (II) \end{matrix} & \text{for} \quad \text{dist}(w_G(t), \mathcal{P}_S(t, t_{k+j})) \in \left[\underline{s} + \bar{e}_G, \bar{s} - \bar{e}_G\right] \\[2ex] \text{part } (III) & \text{for} \quad \begin{matrix} \text{dist}(w_G(t), \mathcal{P}_S(t, t_{k+j})) \in \left]\underline{s}, \underline{s} + \bar{e}_G\right[\\ \vee \, \text{dist}(w_G(t), \mathcal{P}_S(t, t_{k+j})) \in \left]\bar{s} - \bar{e}_G, \bar{s}\right[. \end{matrix} \end{cases}$$

The parameters $\underline{s}$ and $\bar{s}$ correspond to the requirements of collision avoidance (1.1) and a maximum separation (1.2) of control aim (G1). The parameter $\bar{e}_G$ is a threshold, which depends on the dynamics of the agents and is derived in Theorem 7.1 on p. 140 to satisfy the control aim. The parameter t_{com} is utilised to avoid a possible Zeno behaviour if the distance between the agents violates a time span condition (I) or (II) of (7.4) due to the dynamics of the agents, as determined in Section 7.5.2.

Evaluation of condition (7.4)**.** Conditions (I) and (II) are evaluated with the method presented in Section 4.3.1 by adapting the radii of the ball. The third condition of (7.4) is evaluated by using a clock, which measures the time span that is elapsed since the last event time instant t_k and triggers the event after the time span t_{com}.

Remark. The agents use clocks to track the ongoing time. When the clocks run exactly, a synchronisation of the clocks is not necessary.

Event time instant. The event e_{G0} is generated at the next event time instants t_{k+j+1}, $(j = 0, \ldots, N)$ at which condition (7.4) is fulfilled:

$$\text{Event time instant} : \begin{cases} \quad t_{k+j+1} = \arg\min_{t > t_{k+j}} \left\{ \text{dist}(\boldsymbol{w}_{\text{G}}(t), \mathcal{P}_{\text{S}}(t, t_{k+j})) \leq \underline{s} + \bar{e}_{\text{G}} \right\} \\ \vee \; t_{k+j+1} = \arg\min_{t > t_{k+j}} \left\{ \text{dist}(\boldsymbol{w}_{\text{G}}(t), \mathcal{P}_{\text{S}}(t, t_{k+j})) \geq \bar{s} - \bar{e}_{\text{G}} \right\} \\ \wedge \; t_{k+j+1} = \arg\min_{t > t_{k+j}} \left\{ t - t_{k+j} \geq t_{\text{com}} \right\} \end{cases} \tag{7.5}$$

Reaction. As a result, communication between the agents is invoked. If a situation occurs in which condition (7.4) is satisfied, a request is generated that combines all determined time instants (7.5) at which an event e_{G0} has to be generated in a vector

$$\boldsymbol{r}_{\text{G}}(t_k) = \begin{pmatrix} t_k & t_{k+1} & \ldots & t_{k+N} \end{pmatrix}^{\text{T}}. \tag{7.6}$$

The request is sent to the stand-on agent so that it is able to send its information at t_{k+j}, $(j = 1, \ldots, N)$ autonomously. The give-way agent checks the newly received information for a change of the trajectory of the stand-on agent.

Determination of the event threshold $\bar{e}_{\text{G}}$. The event threshold $\bar{e}_{\text{G}}$ must be determined appropriately so that after an event generation there is sufficient space for the give-way agent to change its trajectory and to adjust the distance to the stand-on agent under consideration of its movement. The threshold $\bar{e}_{\text{G}}$ is derived in Theorem 7.1, where the worst-case scenario is considered that both agents move directly towards each other with their maximum speeds.

Theorem 7.1 (Event threshold $\bar{e}_G$). *Consider a stand-on agent and a give-way agent that satisfy Assumptions* 1.1, 1.2 *and* 1.4 *on p.* 16 *and that are connected over a communication network fulfilling Assumption* 6.1 *on p.* 127. *The trajectory of the give-way agent can be changed appropriately to fulfil requirements* (1.1) *and* (1.2) *of control aim* (A1) *if the following holds:*

- *The give-way agent uses the threshold*

$$\bar{e}_G \geq \frac{13}{9} v_{max}\, t_{t,min} \tag{7.7}$$

for the event generation, where $t_{t,min}$ is the minimum transition time interval of the give-way agent.

- *The end point of the first phase of the reactive trajectory $w_{G,r}(t)$ is given by*

$$w_G(t_{e1}) = \begin{pmatrix} x_G(t_{s1}) + f_x(t_{t,min}, v_G, v_{G,max}) \\ y_G(t_{s1}) + f_y(t_{t,min}, v_G, v_{G,max}) \\ z_G(t_{s1}) + f_z(t_{t,min}, v_G, v_{G,max}) \end{pmatrix} \tag{7.8}$$

using the functions defined in (7.9), *which results in a distance of*

$$d_{SG} \geq \bar{s} + \varepsilon_d\, \bar{e}_G$$

between the agents after the time interval $t \geq t_{t,min}$ with ε_d given by (B.8).

Proof. See Appendix B.3. ■

The theorem gives an event threshold, which is used by the give-way agent to generate appropriate events in order to fulfil requirements (1.1) and (1.2) of control aim (A1). Hence, the theorem states the solution to Problem 1.3.

Remark. For the determination of the event threshold the worst-case scenario is considered. The time interval $t_{t,min}$ is the minimum transition time interval of the trajectory of the give-way agent, which changes the movement of the give-way agent for the worst-case scenario. The time interval is determined by solving optimisation problem (3.18) with $p_G(t_{s1})$ as the start point, where t_{s1} is given by (7.16) on p. 144 and $p_G(t_{e1})$ defined in (7.8) as the end point of the movement.

The value v_G denotes the speed of the give-way agent at the event time instant t_k and $v_{G,max} = 1.5\, v_{S,max}$ is the maximum speed of the agent according to Assumption 1.2 on p. 17. The position of the end point affects the transition time interval. With the point (7.8) it is ensured that the transition time interval is as short as possible while the increase of the distance is as large as possible. Furthermore, the agent satisfies the system and actuator limitations (1.8) – (1.10) on p. 17. In the second phase of the reactive trajectory the desired distances $\underline{s} + 2\,\bar{e}_G$ or $\bar{s} - 2\,\bar{e}_G$ are obtained. The functions used in (7.8) are given by

$$\begin{aligned} f_x(t_{t,\min}, v_G, v_{G,\max}) &= \tfrac{4}{9}\, t_{t,\min}(\mu_x\, v_G + \nu_x\, v_{G,\max}) \\ f_y(t_{t,\min}, v_G, v_{G,\max}) &= \tfrac{4}{9}\, t_{t,\min}(\mu_y\, v_G + \nu_y\, v_{G,\max}) \\ f_z(t_{t,\min}, v_G, v_{G,\max}) &= \tfrac{4}{9}\, t_{t,\min}(\mu_z\, v_G + \nu_z\, v_{G,\max}). \end{aligned} \tag{7.9}$$

The scaling factors $\boldsymbol{\mu} = (\mu_x\ \mu_y\ \mu_z)^T$, $||\boldsymbol{\mu}|| = 1$ and $\boldsymbol{\nu} = (\nu_x\ \nu_y\ \nu_z)^T$, $||\boldsymbol{\nu}|| = 1$ are used to force the give-way agent to adjust the distance to the stand-on agent in the right direction depending from which direction the stand-on agent approaches.

Example 7.1. The dynamics of the stand-on agent and the give-way agent are described by the linear model (10.8), (10.10) on p. 232. If the speeds of the agents are limited by

$$v_{S,\max} = 2\frac{\text{m}}{\text{s}}, \quad v_{G,\max} = 3\frac{\text{m}}{\text{s}}$$

and they have to keep the distances to one another described by the bounds

$$\underline{s} = 0.4\,\text{m}, \quad \bar{s} = 9\,\text{m},$$

then the threshold $\bar{e}_G$ and the parameter t_{com} are determined to be

$$\bar{e}_G = 1.1\,\text{m}, \quad t_{com} = 0.43\,\text{s}.$$

The event threshold is about three times larger than the safety distance. This indicates that the agents need a considerable space to change their direction of movement. At the same time it can be seen at the parameter t_{com} that a certain time span is between two consecutive events. Hence, even in worst-case scenario the amount of communication is reduced compared to a continuos communication scheme. □

7.3.3 *Proactive event* e_{G1}

Situation. The two agents move on trajectories that cause a violation of the requirements (1.1) or (1.2) of control aim (G1) in the future. The upcoming violation of (G1) is detected at an event time instant t_k.

Event generation. For triggering event e_{G1} the maximum and minimum distances between the trajectories $\boldsymbol{w}_G(t)$ and $\boldsymbol{w}_S(t)$ of the agents stated in (4.25) on p. 97 are evaluated for the time interval $t_k \le t \le t_{end}$. The event condition (7.11) given below is checked

- at $t = t_k$ if the stand-on agent changed its trajectory or
- at a time instant at which the give-way agent changes its trajectory.

Otherwise no violation of control aim (G1) can occur as the distance between the trajectories was already checked before. The event e_{G1} is triggered if the distance between the agents is in the interval

$$\mathrm{dist}(\boldsymbol{w}_{\mathrm{G}}(t), \boldsymbol{w}_{\mathrm{S}}(t)) \in \left[\underline{s} + \bar{e}_{\mathrm{G}}, \bar{s} - \bar{e}_{\mathrm{G}}\right]. \tag{7.10}$$

and the condition

$$\text{Proactive event } e_{\mathrm{G1}}: \begin{cases} \quad \mathrm{dist}_{\mathrm{min}}(\boldsymbol{w}_{\mathrm{G}}(t), \boldsymbol{w}_{\mathrm{S}}(t)) \leq \underline{s} + 2\,\bar{e}_{\mathrm{G}} & (I) \\ \vee \ \ \mathrm{dist}_{\mathrm{max}}(\boldsymbol{w}_{\mathrm{G}}(t), \boldsymbol{w}_{\mathrm{S}}(t)) \geq \bar{s} - 2\,\bar{e}_{\mathrm{G}} & (II) \end{cases} \tag{7.11}$$

is fulfilled at a time instant $t = t_{\mathrm{EG1}}$ determined in (7.12).

Remark. Event e_{G1} is only triggered within the range of distance specified in (7.10). Outside this interval, the condition is fulfilled, but then the condition for generating event e_{G2} is also fulfilled, as stated in the next section and event e_{G2} is triggered.

The event threshold in (7.11) is different from the threshold in (7.4). This difference is necessary to guarantee a minimum time span between two consecutive events, as analysed in Section 7.5.2.

Event time instant. Whenever condition (7.11) is satisfied at a time instant t_k according to (4.27), the event e_{G1} is generated at

$$\text{Event time instant}: \begin{cases} t_{\mathrm{EG1}} = t_k - t_{\mathrm{t,d}} & \text{if } t_k - t \geq t_{\mathrm{t,d}} \\ t_{\mathrm{EG1}} = t & \text{if } t_k - t < t_{\mathrm{t,d}} \end{cases} \tag{7.12}$$

with $t_{\mathrm{t,d}}$ as the default transition time interval (3.9).

Remark. In case the value for the minimum or the maximum distance (4.25) occurs several times, only the value at the earliest point in time is relevant, because starting from this time instant, the give-way agent changes its trajectory in order to keep control aim (G1) satisfied. Hence, event e_{G1} is generated if condition (7.11) is satisfied for the first time.

The reason for triggering the event earlier than the time instant t_k at which condition (7.11) is violated is, that the resulting time span is long enough so that the trajectory of the give-way agent can be changed smoothly after the event generation.

Return time instant. According to control aim (G1), the give-way agent is required to return to its initial trajectory in order to reach its destination. To this aim, whenever event e_{G1} was generated at time instant t_{EG1}, the event generator checks the distance

$$\mathrm{dist}_{\mathrm{return}}(\boldsymbol{w}_{\mathrm{G}}(t), \boldsymbol{w}_{\mathrm{S}}(t)) = ||\boldsymbol{w}_{\mathrm{G}}(t) - \boldsymbol{w}_{\mathrm{S}}(t)||, \ \text{for } t > t_{\mathrm{EG1}} \tag{7.13}$$

between the trajectories of the agents for the time interval $t_{\mathrm{EG1}} < t \leq t_{\mathrm{end}}$. It determines the time instant t_{ret} at which the violation of control aim (G1) is over with

$$\text{Return time instant}: \begin{cases} \quad t_{\mathrm{ret}} = \arg\min\limits_{t > t_{\mathrm{EG1}}} \left\{\mathrm{dist}_{\mathrm{return}}(\boldsymbol{w}_{\mathrm{G}}(t), \boldsymbol{w}_{\mathrm{S}}(t)) > \underline{s} + 2\,\bar{e}_{\mathrm{G}}\right\} \\ \vee \ \ t_{\mathrm{ret}} = \arg\min\limits_{t > t_{\mathrm{EG1}}} \left\{\mathrm{dist}_{\mathrm{return}}(\boldsymbol{w}_{\mathrm{G}}(t), \boldsymbol{w}_{\mathrm{S}}(t)) < \bar{s} - 2\,\bar{e}_{\mathrm{G}}\right\}. \end{cases} \tag{7.14}$$

From this time instant on the give-way agent can follow its initial trajectory again to reach its individual destination according to control aim (G1).

Remark. Due to the movement of the stand-on agent, it may not be possible for the give-way agent to find a time instant t_{ret} with (7.14). In this case a further movement of the give-way agent along its initial trajectory is not possible, while fulfilling requirements (1.1) and (1.2).

Reaction. After the generation of the event, a proactive trajectory $w_{G,p}(t)$ for the give-way agent is planned to keep the control aim (G1) satisfied, as described in Section 7.4.1. At time instant t_{EG1} the give-way agent changes its movement to the proactive trajectory. Due to Assumption 1.2 on p. 17 the give-way agent has a faster dynamic compared to the stand-on agent, which ensures that the give-way agent is able to adjust the distance between the agents. At time instant t_{ret} the give-way agent starts the return to its initial trajectory, as stated in Section 7.4.1.

7.3.4 *Reactive event* e_{G2}

Situation. If the stand-on agent changed its trajectory directly after the last event time instant t_k without notification and deviates from its communicated trajectory, it can change the distance to the give-way agent unnoticed during the time span $t_{k+1} - t_k$ of no communication so that a violation of the requirements (1.1) or (1.2) of control aim (G1) is imminent.

Event generation. The event e_{G2} is generated based on the evaluation of the distance between the current positions $p_G(t_k)$ and $p_S(t_k)$ of the agents at an event time instant t_k with (4.28) on p. 98. The event condition (7.15) is checked at every event time instant t_k. If the condition

$$\text{Reactive event } e_{G2}: \begin{cases} & \text{dist}(p_G(t_k), p_S(t_k)) \leq \underline{s} + \bar{e}_G \quad (I) \\ \vee & \text{dist}(p_G(t_k), p_S(t_k)) \geq \bar{s} - \bar{e}_G \quad (II) \end{cases} \tag{7.15}$$

is satisfied at time t_k, the event is generated.

Remark. If the situation occurs that event e_{G2} is generated, the distance between the agents reduces for a time span to the interval $]\underline{s}, \underline{s} + \bar{e}_G[$ or increases to the interval $]\bar{s} - e_G, \bar{s}[$. This is caused by the dynamics of the agents. When they approach one another or increase the distance to one another, it takes a short time span to change the movement of the give-way agent so as to keep the control aim (G1) fulfilled. In the two intervals, conditions (I) and (II) of (7.4) are always satisfied, which would cause Zeno behaviour. Hence, in these intervals communication is invoked using the parameter t_{com} in condition (III) of (7.4).

Event time instant. Event e_{G2} is generated at a time instant

$$t_{\mathrm{EG2}} = t_k \tag{7.16}$$

when condition (7.15) is fulfilled.

Return time instant. After the generation of event e_{G2} at t_{EG2} the event generator evaluates the distance between the agents with (7.13) for $t_{\mathrm{EG2}} < t \leq t_{\mathrm{end}}$. It determines the time instant t_{ret} at which the give-way agent can return to its initial trajectory with condition (7.14).

Remark. As the changed trajectory of the stand-on agent is communicated in the matrix $\boldsymbol{S}_{\mathrm{S}}(t_k)$ at the event time instant t_k, condition (7.14) can be used to determine the time instant t_{ret} even if an event e_{G2} was triggered.

Reaction. After the generation of the event, a reactive trajectory $\boldsymbol{w}_{\mathrm{G,r}}(t)$ for the give-way agent is planned to keep the control aim (G1) satisfied, as described in Section 7.4.1. At time instant t_{EG2} the give-way agent changes its movement to the reactive trajectory. As the give-way agent has a faster dynamic compared to the stand-on agent, it is able to adjust the distance between the agents. At time instant t_{ret} the give-way agent starts the return to its initial trajectory, as stated in Section 7.4.1.

7.3.5 *Overview of the specific event generation for a circular movement*

This section summarises the situations in which an event has to be generated for a circular movement of the agents and the reactions after triggering an event. The same types of events are triggered as for a general movement of the agents. However, the event conditions are adapted to the circular movement by considering the phase difference and the height difference between the agents. The summary of the event generation is given in Tab. 7.2. The method for triggering the events is stated in greater detail in the Sections 7.3.6 – 7.3.8.

Table 7.2: Summary of the event generation for a circular movement.

Situation	Event	Reaction
Phase and height of agent $\bar{P}_{\mathrm{S}}$ become too uncertain.	e_{G0}	$\bar{P}_{\mathrm{G}}$ sends $\boldsymbol{r}_{\mathrm{G}}(t_k)$, $\bar{P}_{\mathrm{S}}$ responds by sending $\boldsymbol{S}_{\mathrm{S}}(t_k)$ if $\boldsymbol{w}_{\mathrm{S}}(t)$ has changed. $\bar{P}_{\mathrm{S}}$ sends $\boldsymbol{S}_{\mathrm{S}}(t_k)$ autonomously if $\boldsymbol{w}_{\mathrm{S}}(t)$ is unchanged.

Table 7.2: Summary of the event generation for a circular movement. (continued)

Situation	Event	Reaction
$\bar{P}_S$ or $\bar{P}_G$ changed their trajectories and a violation of control aim (G1) for a circular movement is detected in the future.	e_{G1}	$\bar{P}_G$ plans a proactive trajectory $w_{G,p}(t)$.
$\bar{P}_S$ changed its trajectory without notification and a violation of control aim (G1) for a circular movement is imminent.	e_{G2}	$\bar{P}_G$ plans a reactive trajectory $w_{G,r}(t)$.

7.3.6 *Communication event* e_{G0} *for a circular movement*

Situation. Whenever the uncertainty about the phase and the height of the stand-on agent become too large so that new information of the stand-on agent is required to ensure control aim (G1) for a circular movement, the event e_{G0} is triggered.

Event generation. According to the general movement of the agents, the event e_{G0} is generated in two phases. In the initialisation phase, at time $t = 0$, the agents are located so as to fulfil requirements (1.3) and (1.7) of control aim (A1) for a circular movement. The give-way agent generates an event e_{G0} to receive the initial data of the stand-on agent.

In the execution phase communication is invoked depending on the movement of the agents. The situations that lead to an event generation and the reaction to it are described in the following. For triggering event e_{G0} the minimum and maximum phase difference between the phase $\Phi(w_G(t))$ of the give-way agent and the sets of phases $\Delta\Phi_S(t, t_{k+j})$, $(j = 0, \dots, N)$ of the stand-on agent given by (4.30) on p. 99 as well as the minimum height difference between the height $w_{G,z}(t)$ and the sets of heights $\Delta z_S(t, t_{k+j})$ defined in (4.31) are considered. Again the event condition (7.17) is evaluated only if

- the give-way agent receives the initial data $\boldsymbol{S}_S(0)$.
- the give-way agent receives the information $\boldsymbol{S}_S(t_k)$ and the stand-on agent changed its trajectory indicated by the variable $u = 1$ according to (6.3).
- the give-way agent changed its trajectory.

Events are triggered if the condition

$$\text{Communication event } e_{\mathrm{G0}}: \begin{cases} \quad \Phi_{\mathrm{diff,min}}(t, t_{k+j}) \leq \underline{s} + \bar{e}_{\Phi} & (I) \\ \vee \quad \Phi_{\mathrm{diff,max}}(t, t_{k+j}) \geq \bar{s} - \bar{e}_{\Phi} & (II) \\ \vee \quad z_{\mathrm{diff,min}}(t, t_{k+j}) \leq \bar{z} - \bar{e}_{\mathrm{z}} & (III) \\ \wedge \quad t - t_{k+j} \geq t_{\mathrm{com}} & (IV) \end{cases} \tag{7.17}$$

for $t_k \leq t \leq t_{\mathrm{end}}$ is fulfilled at time instants t_{k+j+1}. By checking the condition recursively for the entire time course that is included in $\boldsymbol{S}_{\mathrm{S}}(t_k)$, the future event time instants are predetermined. The decision if the parts (I) and (II) or part (III) or part (IV) of condition (7.4) are used depends on the phase difference $\Phi_{\mathrm{diff}}(t)$ and the height difference $z_{\mathrm{diff}}(t)$:

$$\text{Event generated with}: \begin{cases} \begin{matrix}\text{part } (I) \\ \text{part } (II)\end{matrix} & \text{for} \quad \Phi_{\mathrm{diff}}(t) \in \left[\underline{s} + \bar{e}_{\Phi}, \bar{s} - \bar{e}_{\Phi}\right] \\ \text{part } (III) & \text{for} \quad z_{\mathrm{diff}}(t) \in [\bar{z} - \bar{e}_{\mathrm{z}}, \bar{z}] \\ & \qquad \Phi_{\mathrm{diff}}(t) \in \left]\underline{s}, \underline{s} + \bar{e}_{\Phi}\right[\\ \text{part } (IV) & \text{for} \ \vee \Phi_{\mathrm{diff}}(t) \in \left]\bar{s} - \bar{e}_{\Phi}, \bar{s}\right[\\ & \qquad \vee \ z_{\mathrm{diff}}(t) \in [0, \bar{z} - \bar{e}_{\mathrm{z}}[\,. \end{cases}$$

The parameters $\underline{s}$ and $\bar{s}$ correspond to the lower bound (1.4) and the upper bound (1.5) of the phase difference (1.3). The parameter $\bar{z}$ given by (1.7) is the desired height difference of control aim (G1) for a circular movement. The thresholds $\bar{e}_{\Phi}$ and $\bar{e}_{\mathrm{z}}$ depend on the dynamics of the agents and are derived in Theorem 7.2 on p. 147 to satisfy the control aim. The parameter t_{com} is utilised to avoid a possible Zeno behaviour, if the phase difference or the height difference between the agents violates a time span conditions (I), (II) or (III) of (7.17) due to the dynamics of the agents, as determined in Section 7.5.3.

Event time instant. The event e_{G0} is generated at the next event time instants t_{k+j+1}, $(j = 0, \dots, N)$ at which condition (7.17) is fulfilled:

$$\text{Event time instant}: \begin{cases} \quad t_{k+j+1} = \arg\min\limits_{t>t_{k+j}} \left\{\Phi_{\mathrm{diff,min}}(t, t_{k+j}) \leq \underline{s} + \bar{e}_{\Phi}\right\} \\ \vee \quad t_{k+j+1} = \arg\min\limits_{t>t_{k+j}} \left\{\Phi_{\mathrm{diff,max}}(t, t_{k+j}) \geq \bar{s} - \bar{e}_{\Phi}\right\} \\ \vee \quad t_{k+j+1} = \arg\min\limits_{t>t_{k+j}} \left\{z_{\mathrm{diff,min}}(t, t_{k+j}) \leq \bar{z} - \bar{e}_{\mathrm{z}}\right\} \\ \wedge \quad t_{k+j+1} = \arg\min\limits_{t>t_{k+j}} \left\{t - t_{k+j} \geq t_{\mathrm{com}}\right\} \end{cases} \tag{7.18}$$

Reaction. As a result, communication between the agents is invoked. If a situation occurs in which condition (7.17) is evaluated, a request is generated by combining the event time instants (7.18) in the vector (7.6) and sent to the stand-on agent. After that, the stand-on agent sends its data autonomously at the defined time instants. The give-way agent checks the newly received information for a change of the trajectory of the stand-on agent.

Determination of the event thresholds $\bar{e}_\Phi$ and $\bar{e}_z$. The event thresholds $\bar{e}_\Phi$ and $\bar{e}_z$ need to be determined so that after an event generation there is sufficient space for the give-way agent to change its speed or its height of the circular movement with respect to the movement of the stand-on agent. For the derivation of the thresholds in Theorem 7.2 the worst-case scenarios are considered:

- Both agents move with the speed $v_{S,max}$ and $\bar{P}_S$ reduces its speed to $v_{S,min}$ in the shortest possible time span. Then, $\bar{P}_G$ has to reduce its speed to $v_{G,min}$ as fast as possible.
- Both agents move with the speed $v_{S,min}$ and $\bar{P}_S$ accelerates as fast as possible to $v_{S,max}$. Then, $\bar{P}_G$ has to accelerate to $v_{G,max}$ as quickly as possible.

Theorem 7.2 (Event thresholds $\bar{e}_\Phi$, $\bar{e}_z$). *Consider a stand-on agent and a give-way agent that satisfy Assumptions* 1.1, 1.2 *and* 1.4 *on p.* 16 *and that are connected over a communication network fulfilling Assumption* 6.1 *on p.* 127. *The trajectory of the give-way agent can be changed appropriately to fulfil requirements* (1.3) *and* (1.7) *of control aim* (A1) *for a circular movement if events are generated using the thresholds*

$$\begin{aligned} \bar{e}_\Phi &= \max\left(\Delta\Phi_{add,a}, \Delta\Phi_{add,b}\right) \\ 0 &< \bar{e}_z < \bar{z}. \end{aligned} \tag{7.19}$$

Proof. See Appendix B.4. ∎

Remark. The values $\Phi_{add,a}$ and $\Phi_{add,b}$ are phase differences that arise from the movement of the agents with different speeds and are derived in (B.11) and (B.12).

The theorem states event thresholds, which are utilised by the give-way agent to generate events in order to fulfil requirements (1.3) and (1.7) of control aim (A1) for a circular movement. Hence, the theorem provides the solution to Problem 1.3 for a circular movement of the agents.

Example 7.2. The dynamics of the stand-on agent and the give-way agent are described by the linear model (10.8), (10.10) on p. 232. The radius of the circular trajectories is chosen to be $r = 1\,\mathrm{m}$ and the tolerance range is given by $\gamma = 90°$. If the speeds of the agents are limited by

$$v_{S,min} = 0.2\frac{\mathrm{m}}{\mathrm{s}}, \quad v_{G,min} = 0\frac{\mathrm{m}}{\mathrm{s}}$$
$$v_{S,max} = 1\frac{\mathrm{m}}{\mathrm{s}}, \quad v_{G,max} = 1.2\frac{\mathrm{m}}{\mathrm{s}}$$

and they have to keep the phase difference and the height difference to one another

$$\Phi_{\text{diff}}(t) \in \left[\underline{s}, \overline{s}\right], \quad z_{\text{diff}}(t) = \overline{z}$$

with

$$\underline{s} = 90°, \quad \overline{s} = 270° \text{ and } \overline{z} = 1\,\text{m},$$

then the thresholds $\bar{e}_\Phi$ and $\bar{e}_z$ and the parameter t_{com} are determined to be

$$\bar{e}_\Phi = 4.58° \mathrel{\widehat{=}} 0.08\,\text{m}, \quad \bar{e}_z = 0.25\,\text{m}, \quad t_{\text{com}} = 1.41\,\text{s}.$$

The event thresholds are small compared to the bounds of the phase difference and the height difference. In particular they are smaller compared to the threshold derived in Example 7.1. The reason for this is that the movement of the agents on the circular path is much more restricted compared to the general movement and hence the agents do not need as much space to change their movement. Consequently, the time span between two events in the worst-case scenario is significantly longer than in Example 7.1. □

7.3.7 *Proactive event e_{G1} for a circular movement*

Situation. The two agents move on trajectories that cause a violation of the requirements (1.4), (1.5) or (1.7) of control aim (G1) for a circular movement in the future. The upcoming violation of (G1) for a circular movement is detected at an event time instant t_k.

Event generation. The minimum and maximum phase differences as well as the minimum height difference between the trajectories $w_{\text{G}}(t)$ and $w_{\text{S}}(t)$ of the agents given by (4.33) on p. 102 are evaluated for the time interval $t_k \leq t \leq t_{\text{end}}$ for generating event e_{G1}. The event condition is checked

- at $t = t_k$ if the stand-on agent changed its trajectory or
- at a time instant at which the give-way agent changes its trajectory.

The event is generated if the phase difference and the height difference are in the intervals

$$\begin{aligned} \Phi_{\text{diff}}(t) &\in \left[\underline{s} + 2\,\bar{e}_\Phi, \overline{s} - 2\,\bar{e}_\Phi\right], \\ z_{\text{diff}}(t) &\in \left[\overline{z} - 2\,\bar{e}_z, \overline{z}\right] \end{aligned} \tag{7.20}$$

and the condition

$$\text{Proactive event } e_{\text{G1}} : \begin{cases} \quad \Phi_{\text{diff,w,min}}(t) \leq \underline{s} + 2\,\bar{e}_\Phi & (I) \\ \vee \;\; \Phi_{\text{diff,w,max}}(t) \geq \overline{s} - 2\,\bar{e}_\Phi & (II) \\ \vee \;\; z_{\text{diff,w,min}}(t) \leq \overline{z} - 2\,\bar{e}_z & (III) \end{cases} \tag{7.21}$$

is satisfied at a time instant $t = t_{\text{EG1}}$.

Remark. Event e_{G1} is only triggered within the range of distance specified in (7.20). Outside this interval, the condition is fulfilled, but then the condition for generating event e_{G2} is also fulfilled, as stated in the next section and event e_{G2} is triggered.

In accordance with the event generation for the general movement, the event threshold in (7.21) differs from the threshold in (7.17). The difference is again required to guarantee a minimum time span between two consecutive events, as analysed in Section 7.5.3.

Event time instant. Event e_{G1} is generated when condition (7.21) is satisfied for the first time. Again, in the case the same value for the minimum or the maximum phase difference or the minimum height difference (4.33) occurs several times, only the value at the earliest point in time is relevant. If condition (7.21) is satisfied at a time instant t_k, the event time instant is given by (7.12).

Remark. For the event generation for the circular movement no return time instant needs to be determined. As the movement of the give-way agent directly depends on the movement of the stand-on agent, the give-way agent moves on a proactive trajectory or a reactive trajectory once it had to leave its initial trajectory.

Reaction. After the generation of the event, at $t = t_{EG1}$ a proactive trajectory $w_{G,p}(t)$ for the give-way agent is planned to keep the control aim (G1) for a circular movement satisfied, as described in Section 7.4.5. Due to Assumption 1.3 on p. 17 the give-way agent has a faster dynamic compared to the stand-on agent, which ensures that the give-way agent is able to adjust the phase difference and the height difference between the agents.

7.3.8 *Reactive event* e_{G2} *for a circular movement*

Situation. If the stand-on agent changed its trajectory directly after the last event time instant t_k without notification, the phase difference and the height difference between the agents can change unnoticed so that a violation of the requirements (1.4), (1.5) or (1.7) of control aim (G1) for a circular movement is imminent.

Event generation. For the generation of the event e_{G2} the phase difference and the height difference between the agents at an event time instant t_k is checked with (4.35) on p. 103 at every event time instant. If the condition

$$\text{Reactive event } e_{G2}: \begin{cases} \quad \Phi_{\text{diff}}(t_k) \leq \underline{s} + \bar{e}_\Phi & (I) \\ \vee \quad \Phi_{\text{diff}}(t_k) \geq \bar{s} - \bar{e}_\Phi & (II) \\ \vee \quad z_{\text{diff}}(t_k) \leq \bar{z} - \bar{e}_z & (III) \end{cases} \tag{7.22}$$

is fulfilled at time t_k, the event is triggered.

Remark. If the situation occurs that event e_{G2} is generated, the phase difference and the height difference between the agents reduces for a time span to the intervals $]\underline{s}, \underline{s} + \bar{e}_\Phi[$, $]0, \bar{z} + \bar{e}_z[$ or increases to the interval $]\bar{s} - e_\Phi, \bar{s}[$ due to the dynamics of the agents. In these intervals the conditions (I), (II) and (III) of (7.17) are always satisfied, which would cause Zeno behaviour. Hence, in the intervals communication is invoked using the parameter t_{com} in condition (IV) of (7.17).

Event time instant. The event is directly triggered when condition (7.22) is satisfied at the time instant given by (7.16).

Reaction. After the generation of the event at $t = t_{EG2}$ a reactive trajectory $w_{G,r}(t)$ for the give-way agent is planned to keep the control aim (G1) for a circular movement satisfied, as described in Section 7.4.5.

7.4 Trajectory planning unit T_G

7.4.1 *General planning tasks*

As the trajectory planning method is already described in Chapter 3, this section focusses on the planning tasks of the give-way agent for the proactive trajectory $w_{G,p}(t)$ and the reactive trajectory $w_{G,r}(t)$ for a general movement of the agents. These trajectories are planned in three phases:

- **Phase 1: Trajectory to adjust the distance to the stand-on agent.** After the occurrence of the event e_{G1} with condition (7.11) a trajectory $w_{G,p1}(t)$ is planned to adjust the distance to the communicated future trajectory of the stand-on agent to $\underline{s} + 2\,\bar{e}_G$ or $\bar{s} - 2\,\bar{e}_G$ in order to keep control aim (G1) satisfied. The generation of the event e_{G2} with condition (7.15) triggers a trajectory $w_{G,r1}(t)$, which exploits the dynamic limitations of the give-way agent to maintain control aim (G1) by adjusting the distance to the stand-on agent. Triggering events e_{G1} or e_{G2} for the first time forces the give-way agent to leave its initially planned trajectory.

- **Phase 2: Trajectory to keep the adjusted distance to the stand-on agent.** As long as the movement of the stand-on agent does not allow a return of the give-way agent to its initially planned trajectory without violating control aim (G1), a trajectory $w_{G,p2}(t)$ or $w_{G,r2}(t)$ is planned to keep the distances $\underline{s} + 2\,\bar{e}_G$ or $\bar{s} - 2\,\bar{e}_G$ maintained.

- **Phase 3: Trajectory to return to the initial trajectory.** As soon as the return to the initial trajectory is possible for the give-way agent at the time instant t_{ret}, a trajectory $w_{G,p3}(t)$ or $w_{G,r3}(t)$ is planned to lead the agent back to this trajectory.

Normally the phase 1 is directly followed by the phase 2. If the stand-on agent changes its trajectory several times in a row, the give-way agent can also adjust the distance to the stand-on agent multiple times by executing phase 1 repeatedly. In the same way, if the give-way agent cannot return to its initial trajectory, phase 2 can be performed several times in a row.

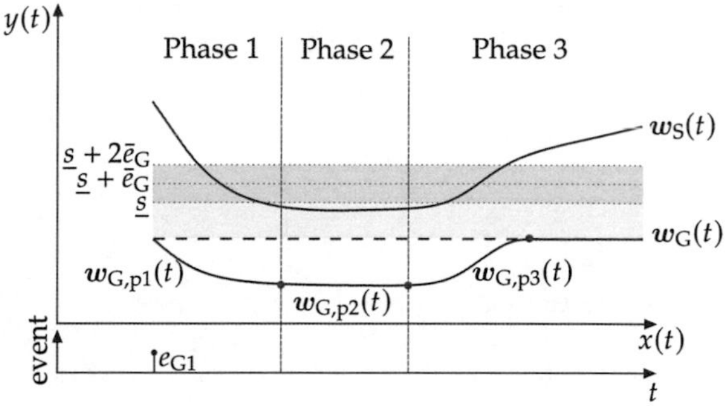

Fig. 7.1: Three phases of planning the proactive trajectory.

The combination of the three phases of planning are exemplarily shown in the 2D space after a generation of the event e_{G1} with condition (I) of (7.11) in Fig. 7.1 and after a generation of event e_{G2} in Fig. 7.2 with condition (I) of (7.15). The initially planned trajectory $w_G(t)$ is indicated by the dashed line. After the generation of event e_{G1}, the trajectory $w_{G,p1}(t)$ is determined prematurely to adjust the distance to the stand-on agent, as shown by the solid line in the left part of the figure. After the required distance to the stand-on agent has been obtained, the trajectory $w_{G,p2}(t)$ of the give-way agent is aligned to the trajectory of the stand-on agent, as shown by the solid line in the middle part of the figure. In the last phase, the give-way agent returns to its initial trajectory over the trajectory $w_{G,p3}(t)$. The red dots in the figure illustrate the end points of the trajectory of each planning phase.

In Fig. 7.2 an event e_{G2} triggers the determination of the trajectory $w_{G,r1}(t)$ that adapts the distance to the stand-on agent in a minimum time, as illustrated by the solid line. While the planning of the trajectories $w_{G,p1}(t)$ and $w_{G,r1}(t)$ differs, the trajectories $w_{G,p2}(t)$ and $w_{G,r2}(t)$ as well as $w_{G,p3}(t)$ and $w_{G,r3}(t)$ are planned identically.

The planning tasks for the three phases of the proactive trajectory and the reactive trajectory are described in the following sections.

Remark. The trajectory planning for other types of agents (e.g. cars) is carried out in the same way. However, the boundary conditions for the planning have to be adjusted to the dynamics of the agents.

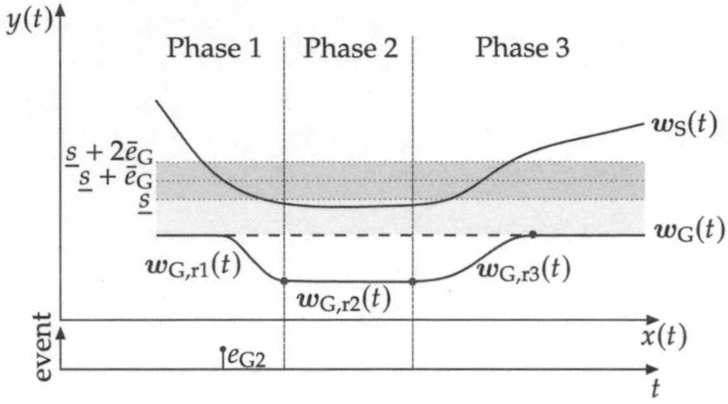

Fig. 7.2: Three phases of planning the reactive trajectory.

7.4.2 *Phase 1: Trajectory to adjust the distance to the stand-on agent*

Planning task. The planning task for the first phase of the trajectories $w_{G,p}(t)$ and $w_{G,r}(t)$ is given by

Given:	• Conditions on the start point $\boldsymbol{W}_{G,p}(t_{s1})$ or $\boldsymbol{W}_{G,r}(t_{s1})$ given by (7.23) and the end point $\boldsymbol{W}_{G,p}(t_{e1})$ or $\boldsymbol{W}_{G,r}(t_{e1})$ given by (7.24) or (7.25). • Transition time interval t_t. • Data of the stand-on agent $\boldsymbol{S}_S(t_k)$. • Parameters $\underline{s}$ and $\bar{s}$ used in the requirements (1.1) and (1.2) of control aim (G1). • Event threshold $\bar{e}_G$ (7.7).
Find:	Trajectory $w_{G,p1}(t)$ or $w_{G,r1}(t)$ that guarantees control aim (G1).
Boundary conditions:	• System and actuator limitations given by (1.8), (1.9). • Maximum speed (1.10).

1) Start conditions of the proactive trajectory and the reactive trajectory. For a smooth transition between the current trajectory and the first part of the proactive trajectory or the reactive trajectory the conditions on the start point of the trajectory according to (3.2) on p. 46 are given by

$$
\boldsymbol{W}_{G,p}(t_{s1}) = \boldsymbol{W}_{G,r}(t_{s1}) = \begin{pmatrix} x_G(t_{s1}) & y_G(t_{s1}) & z_G(t_{s1}) & 0 \\ \dot{x}_G(t_{s1}) & \dot{y}_G(t_{s1}) & \dot{z}_G(t_{s1}) & 0 \\ \ddot{x}_G(t_{s1}) & \ddot{y}_G(t_{s1}) & \ddot{z}_G(t_{s1}) & 0 \\ x_G^{(3)}(t_{s1}) & y_G^{(3)}(t_{s1}) & z_G^{(3)}(t_{s1}) & 0 \\ x_G^{(4)}(t_{s1}) & y_G^{(4)}(t_{s1}) & z_G^{(4)}(t_{s1}) & 0 \end{pmatrix}. \tag{7.23}
$$

The start conditions are composed of the position of the give-way agent at the time instant t_{s1} and the heading angle, which is set to $\psi(t) = 0$. The start time instant is set to $t_{s1} = t_{EG1}$ after the generation of event e_{G1}, where t_{EG1} is given by (7.12). After triggering event e_{G2} $t_{s1} = t_{EG2}$ holds, where t_{EG2} is determined by (7.16).

2) End conditions of the proactive trajectory. For the derivation of the end point of the proactive trajectory the data $S_S(t_k)$ are used in order to adjust the distance to the stand-on agent to keep control aim (G1) fulfilled. With the communicated control points $b^i_{S,j}$, $(j \in \{x, y, z\}, i = 0, \ldots, 4)$ of $S_S(t_k)$ and eqn. (3.5) the boundary conditions on the trajectory of the stand-on agent are calculated. These conditions are used to determine the conditions on the end point of the proactive trajectory of the give-way agent. The end conditions are given by

$$
\boldsymbol{W}_{G,p}(t_{e1}) = \begin{pmatrix} w_{1,1}(t_{e1}) & w_{1,2}(t_{e1}) & w_{1,3}(t_{e1}) & 0 \\ \dot{w}_{S,x}(t_{e1}) & \dot{w}_{S,y}(t_{e1}) & \dot{w}_{S,z}(t_{e1}) & 0 \\ \ddot{w}_{S,x}(t_{e1}) & \ddot{w}_{S,y}(t_{e1}) & \ddot{w}_{S,z}(t_{e1}) & 0 \\ w_{S,x}^{(3)}(t_{e1}) & w_{S,y}^{(3)}(t_{e1}) & w_{S,z}^{(3)}(t_{e1}) & 0 \\ w_{S,x}^{(4)}(t_{e1}) & w_{S,y}^{(4)}(t_{e1}) & w_{S,z}^{(4)}(t_{e1}) & 0 \end{pmatrix} \tag{7.24}
$$

with

$$
\begin{aligned}
w_{1,1}(t_{e1}) &= w_{S,x}(t_{e1}) + \varepsilon_x(\underline{s} + 2\,\bar{e}_G) \\
w_{1,2}(t_{e1}) &= w_{S,y}(t_{e1}) + \varepsilon_y(\underline{s} + 2\,\bar{e}_G) \\
w_{1,3}(t_{e1}) &= w_{S,z}(t_{e1}) + \varepsilon_z(\underline{s} + 2\,\bar{e}_G)
\end{aligned}
$$

when the planning of the proactive trajectory is triggered with condition (I) of (7.11) and

$$
\begin{aligned}
w_{1,1}(t_{e1}) &= w_{S,x}(t_{e1}) + \varepsilon_x(\bar{s} - 2\,\bar{e}_G) \\
w_{1,2}(t_{e1}) &= w_{S,y}(t_{e1}) + \varepsilon_y(\bar{s} - 2\,\bar{e}_G) \\
w_{1,3}(t_{e1}) &= w_{S,z}(t_{e1}) + \varepsilon_z(\bar{s} - 2\,\bar{e}_G)
\end{aligned}
$$

when the planning of the trajectory is initiated with condition (II) of (7.11). A scaling factor $\boldsymbol{\varepsilon} = (\varepsilon_x \;\; \varepsilon_y \;\; \varepsilon_z)^T$, $||\boldsymbol{\varepsilon}|| = 1$ is used so that the give-way agent adjusts the distance to the stand-on agent by the distance $\underline{s} + 2\,\bar{e}_G$ or $\bar{s} - 2\,\bar{e}_G$ in a suitable direction. According to

(7.11) these distances are appropriate for an event generation. The conditions (7.24) force the give-way agent to move with the same speed as the stand-on agent for $t > t_{\mathrm{e1}}$. The end time instant is given by $t_{\mathrm{e1}} = t_{\mathrm{s1}} + t_{\mathrm{t}}^{i}$, where t_{t}^{i} is the transition time interval of the stand-on agent.

3) End conditions of the reactive trajectory. Again the data $S_{\mathrm{S}}(t_k)$ are used to derive the boundary conditions of the trajectory of the stand-on agent, which are utilised to determine the conditions on the end point of the reactive trajectory given by

$$\boldsymbol{W}_{\mathrm{G,r}}(t_{\mathrm{e1}}) = \begin{pmatrix} w_{1,1}(t_{\mathrm{s1}}) & w_{1,2}(t_{\mathrm{s1}}) & w_{1,3}(t_{\mathrm{s1}}) & 0 \\ \dot{w}_{\mathrm{S,x}}(t_{\mathrm{e1}}) & \dot{w}_{\mathrm{S,y}}(t_{\mathrm{e1}}) & \dot{w}_{\mathrm{S,z}}(t_{\mathrm{e1}}) & 0 \\ \ddot{w}_{\mathrm{S,x}}(t_{\mathrm{e1}}) & \ddot{w}_{\mathrm{S,y}}(t_{\mathrm{e1}}) & \ddot{w}_{\mathrm{S,z}}(t_{\mathrm{e1}}) & 0 \\ w_{\mathrm{S,x}}^{(3)}(t_{\mathrm{e1}}) & w_{\mathrm{S,y}}^{(3)}(t_{\mathrm{e1}}) & w_{\mathrm{S,z}}^{(3)}(t_{\mathrm{e1}}) & 0 \\ w_{\mathrm{S,x}}^{(4)}(t_{\mathrm{e1}}) & w_{\mathrm{S,y}}^{(4)}(t_{\mathrm{e1}}) & w_{\mathrm{S,z}}^{(4)}(t_{\mathrm{e1}}) & 0 \end{pmatrix} \tag{7.25}$$

with

$$\begin{aligned} w_{1,1}(t_{\mathrm{s1}}) &= x_{\mathrm{G}}(t_{\mathrm{s1}}) + f_{\mathrm{x}}(t_{\mathrm{t,min}}, v_{\mathrm{G}}, v_{\mathrm{G,max}}) \\ w_{1,2}(t_{\mathrm{s1}}) &= y_{\mathrm{G}}(t_{\mathrm{s1}}) + f_{\mathrm{y}}(t_{\mathrm{t,min}}, v_{\mathrm{G}}, v_{\mathrm{G,max}}) \\ w_{1,3}(t_{\mathrm{s1}}) &= z_{\mathrm{G}}(t_{\mathrm{s1}}) + f_{\mathrm{z}}(t_{\mathrm{t,min}}, v_{\mathrm{G}}, v_{\mathrm{G,max}}) \end{aligned}$$

and $t_{\mathrm{e1}} = t_{\mathrm{s1}} + t_{\mathrm{t,min}}$. Since the violation of the control aim is imminent, it is required to adjust the distance to the stand-on agent in the minimum transition time interval $t_{\mathrm{t,min}}$ determined by Algorithm 2 on p. 56 in a way to keep control aim (G1) satisfied. The functions $f_{\mathrm{x}}(t_{\mathrm{t,min}}, v_{\mathrm{G}}, v_{\mathrm{G,max}})$, $f_{\mathrm{y}}(t_{\mathrm{t,min}}, v_{\mathrm{G}}, v_{\mathrm{G,max}})$ and $f_{\mathrm{z}}(t_{\mathrm{t,min}}, v_{\mathrm{G}}, v_{\mathrm{G,max}})$ depend on the minimum transition time interval $t_{\mathrm{t,min}}$, the speed v_{G} at t_{s1} and the maximum speed $v_{\mathrm{G,max}}$ of the give-way agent and are specified in (7.9) on p. 141 for the dynamics of the agents. After the time span $t_{\mathrm{t,min}}$ the agents have the distances $\underline{s} + \varepsilon_{\mathrm{d}}\,\bar{e}_{\mathrm{G}}$ or $\bar{s} - \varepsilon_{\mathrm{d}}\,\bar{e}_{\mathrm{G}}$ to one another depending on which condition ((I) or (II)) of (7.15) is generated according to Theorem 7.1 on p. 140. The factor $0 < \varepsilon_{\mathrm{d}}(f_{\mathrm{x}}, f_{\mathrm{y}}, f_{\mathrm{z}}) \le 1$ is derived by (B.8).

Example 7.3. The method is exemplarily shown in Fig. 7.3 for a movement of the agents in the 2D space. The dynamics of the stand-on agent and the give-way agent are described by the linear model (10.8), (10.10) on p. 232. The reactive trajectory is triggered by condition (I) of (7.15). At the start time instant $t_{\mathrm{s1}} = t_{\mathrm{EG2}}$ the agents have the distance $\underline{s} + \bar{e}_{\mathrm{G}}$ to one another. After the time span $t_{\mathrm{t,min}}$ the distance is given by $\underline{s} + \varepsilon_{\mathrm{d}}\,\bar{e}_{\mathrm{G}}$ with $\varepsilon_{\mathrm{d}} = 0.8285$, which is derived by (B.8). Hence, the distance remains nearly constant despite the movement of the stand-on agent. After the avoidance manoeuvre for $t > t_{\mathrm{EG2}} + t_{\mathrm{t,min}}$ the desired distance $\underline{s} + 2\,\bar{e}_{\mathrm{G}}$ is recovered and the trajectory of the give-way agent is aligned to the trajectory of the stand-on agent with this distance. □

4) Planning of the trajectories $w_{\mathrm{G,p1}}(t)$ or $w_{\mathrm{G,r1}}(t)$. The trajectory for the first phase is planned with Algorithm 3 on p. 57. The transition time interval t_{t} of the trajectory after the

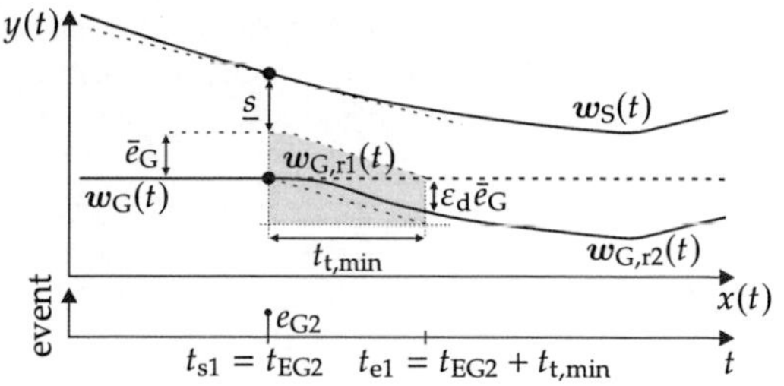

Fig. 7.3: Planning of the reactive trajectory $w_{G,r1}(t)$ after triggering condition (I) of (7.15).

generation of event e_{G1} is given by the transition time interval t_t^i of the stand-on agent in this interval. After the generation of event e_{G2} the transition time interval of the trajectory is determined by solving optimisation problem (3.18) on p. 54.

7.4.3 *Phase 2: Trajectory to keep the adjusted distance to the stand-on agent*

As long as it is not possible for the give-way agent to return to its initial trajectory without violating control aim (G1), a trajectory is planned to maintain the adjusted distance to the stand-on agent, as shown by the dashed line in the middle part of Figs. 7.1 and 7.2. The planning task is as follows:

Given:
- Conditions on the start point $\boldsymbol{W}_{G,p}(t_{s2})$ or $\boldsymbol{W}_{G,r}(t_{s2})$ given by (7.23) and the end point $\boldsymbol{W}_{G,p}(t_{e2})$ or $\boldsymbol{W}_{G,r}(t_{e2})$ given by (7.24).
- Transition time interval t_t.
- Data of the stand-on agent $\boldsymbol{S}_S(t_k)$.
- Parameters $\underline{s}$ and $\bar{s}$ used in the requirements (1.1) and (1.2) of control aim (G1).
- Event threshold $\bar{e}_G$ (7.7).

Find: Trajectory $w_{G,p2}(t)$ or $w_{G,r2}(t)$, which keeps the distance to the stand-on agent to maintain control aim (G1).

Boundary conditions:
- System and actuator limitations given by (1.8), (1.9).
- Maximum speed (1.10).

1) Start conditions of the trajectory. The second phase of the proactive trajectory and the reactive trajectory are planned in the same way. The conditions on the start point of the trajectory are obtained by (7.23) with the start time instant $t_{s2} = t_{e1}$, where t_{e1} is the end time instant of the trajectory of the first phase.

2) End conditions of the trajectory. The conditions on the end point of the trajectory are given by (7.24) with $t_{e2} = t_{s2} + t_t$. This leads to a trajectory for the give-way agent, which is aligned to the trajectory of the stand-on agent and maintains control aim (G1).

3) Planning of the trajectory. The trajectory is generated with Algorithm 3 on p. 57. The transition time interval t_t is given by the transition time interval t_t^i of the stand-on agent in this interval. If the phase 2 is executed for the first time and the give-way agent is at position (7.25), with this trajectory the distance to the stand-on agent is adjusted so that the distances $\underline{s} + \bar{e}_G$ or $\bar{s} - \bar{e}_G$ are recovered.

7.4.4 *Phase 3: Trajectory to return to the initial trajectory*

As soon as the give-way agent can return to its initial trajectory, a trajectory is planned to bring the give-way agent back on this trajectory, as illustrated by the dashed line in the right part of Figs. 7.1 and 7.2. The planning task is given by

Given:
- Conditions on the start point $\boldsymbol{W}_{G,p}(t_{s3})$ or $\boldsymbol{W}_{G,r}(t_{s3})$ given by (7.23) and the end point $\boldsymbol{W}_{G,p}(t_{e3})$ or $\boldsymbol{W}_{G,r}(t_{e3})$ given by (7.26).
- Transition time interval t_t.
- Initial trajectory $w_G(t)$.

Find: Trajectory $w_{G,p3}(t)$ or $w_{G,r3}(t)$, which leads the give-way agent back to its initial trajectory.

Boundary conditions:
- System and actuator limitations given by (1.8), (1.9).
- Maximum speed (1.10).

1) Start conditions of the trajectory. Similarly to the second phase of the proactive and reactive trajectory, the third phase of both trajectories is planned in the same way. The conditions on the start point of the trajectory are given by (7.23) with the start time instant $t_{s3} = t_{ret}$, where t_{ret} is the time instant at which the give-way agent can return to its initial trajectory determined by (7.14).

2) End conditions of the trajectory. The conditions on the end point of the trajectory are given by the conditions of the initial trajectory $w_G(t_{e3})$ of the give-way agent at the corresponding time instant $t_{e3} = t_{s3} + t_t$ as

$$W_{G,p}(t_{e3}) = W_{G,p}(t_{e3}) = \begin{pmatrix} w_{G,x}(t_{e3}) & w_{G,y}(t_{e3}) & w_{G,z}(t_{e3}) & 0 \\ \dot{w}_{G,x}(t_{e3}) & \dot{w}_{G,y}(t_{e3}) & \dot{w}_{G,z}(t_{e3}) & 0 \\ \ddot{w}_{G,x}(t_{e3}) & \ddot{w}_{G,y}(t_{e3}) & \ddot{w}_{G,z}(t_{e3}) & 0 \\ w^{(3)}_{G,x}(t_{e3}) & w^{(3)}_{G,y}(t_{e3}) & w^{(3)}_{G,z}(t_{e3}) & 0 \\ w^{(4)}_{G,x}(t_{e3}) & w^{(4)}_{G,y}(t_{e3}) & w^{(4)}_{G,z}(t_{e3}) & 0 \end{pmatrix}. \tag{7.26}$$

3) Planning of the trajectory. The trajectory is determined with Algorithm 3. The transition time interval t_t is again given by the transition time interval t_t^i of the stand-on agent in this segment. Afterwards, the give-way agent continues moving on its initial trajectory.

The derivation of the planning tasks for the three phases of the proactive trajectory or the reactive trajectory for a general movement of the agents is summarised in Algorithm 13. The execution of the algorithm together with the determination of the trajectories solves Problem 1.4.

Algorithm 13 Derivation of the planning tasks for the three phases of the proactive trajectory or the reactive trajectory

---

Given: Events e_{G1}, e_{G2}.
Start time instant t_{s1} of the trajectory $w_{G,p}(t)$ or $w_{G,r}(t)$.
Return time instant t_{ret} (7.14).
Position $p_G(t_{s1})$ of the give-way agent.
Trajectory $w_G(t)$, $(t_{s1} \leq t \leq t_{end})$ of the give-way agent.
Data $S_S(t_k)$ of the stand-on agent.
Parameters $\underline{s}$ (1.1), $\bar{s}$ (1.2) of control aim (G1).
Event threshold $\bar{e}_G$ (7.7).

1: **If** event e_{G1} or e_{G2} occurred and $t = t_{s1}$:
Determine boundary conditions of the trajectory on the start point at $t = t_{s1}$ with (7.23) and on the end point at $t = t_{e1}$ with (7.24) or (7.25) to adjust the distance to the stand-on agent.
Use the communicated transition time intervals t_t^i of the stand-on agent or determine the transition time interval with Algorithm 2 on p. 56.

2: **If** $t_{e1} < t < t_{ret}$:
Determine boundary conditions of the trajectory on the start point at $t = t_{s2}$ with (7.23) and on the end point at $t = t_{e2}$ with (7.24) to keep the adjusted distance to the stand-on agent.
Use the communicated transition time intervals t_t^i of the stand-on agent.

3: **If** $t \geq t_{ret}$:
Determine boundary conditions of the trajectory on the start point at $t = t_{s3}$ with (7.23) and on the end point at $t = t_{e3}$ with (7.26) to return to the initial trajectory.

Use the communicated transition time intervals t_t^i of the stand-on agent.

Result: Conditions on the start points and the end points for the three phases of the proactive trajectory or the reactive trajectory.

7.4.5 *Specific planning tasks for a circular movement*

This section states the planning task for the proactive trajectory $w_{G,pc}(t)$ and the reactive trajectory $w_{G,rc}(t)$ for a circular movement of the agents. The trajectories are planned in two phases:

- **Phase 1: Trajectory to adjust the phase difference or the height difference.** After the generation of the event e_{G1} with condition (7.21) a trajectory $w_{G,pc1}(t)$ is planned to adjust the phase difference to the communicated future trajectory of the stand-on agent to the middle of the interval $[\underline{s}, \overline{s}]$ or to maintain the height difference $\overline{z}$ in order to keep control aim (G1) for a circular movement satisfied. The generation of the event e_{G2} with condition (7.22) invokes the planning of a trajectory $w_{G,cr1}(t)$, which exploits the dynamic limitations of the give-way agent to maintain control aim (G1) for a circular movement.

- **Phase 2: Trajectory to keep the adjusted phase difference or the height difference.** As the movement of the give-way agent depends on the movement of the stand-on agent, a trajectory $w_{G,pc2}(t)$ or $w_{G,rc2}(t)$ is planned to keep the phase difference and the height difference of control aim (G1) for a circular movement maintained.

Remark. In contrast to the general movement of the agents, the proactive trajectory and the reactive trajectory are planned in two phases. The third phase, the return of the give-way agent to its initial trajectory is not necessary, because now its movement directly depends on the movement of the stand-on agent. Hence, it has permanently to adapt its movement to the movement of the stand-on agent.

The planning tasks for the two phases of the proactive trajectory and the reactive trajectory are described in the following sections.

7.4.6 *Phase* 1: *Trajectory to adjust the phase or the height to the stand-on agent*

Planning task. The planning task for the first phase of the trajectories $w_{G,pc}(t)$ and $w_{G,rc}(t)$ is given by

Given:
- Conditions on the start point $\boldsymbol{W}_{G,pc}(t_{s1})$ or $\boldsymbol{W}_{G,rc}(t_{s1})$ given by (7.27) and the end point $\boldsymbol{W}_{G,pc}(t_{e1})$ or $\boldsymbol{W}_{G,rc}(t_{e1})$ given by (7.28).
- Transition time interval t_t.
- Data of the stand-on agent $\boldsymbol{S}_S(t_k)$.
- Parameters $\underline{s}$, $\bar{s}$ and $\bar{z}$ used in the requirements (1.3) and (1.7) of control aim (G1) for a circular movement.

Find: Trajectory $\boldsymbol{w}_{G,pc1}(t)$ or $\boldsymbol{w}_{G,rc1}(t)$ that guarantees control aim (G1) for a circular movement.

Boundary conditions:
- System and actuator limitations given by (1.8), (1.9).
- Maximum speed (1.11).

1) Start conditions of the trajectory. The proactive trajectory and the reactive trajectory have the same start conditions and end conditions, but they use different transition time intervals for the planning. According to (3.27) the start conditions of the proactive trajectory and the reactive trajectory are given by

$$\boldsymbol{W}_{G,pc}(t_{s1}) = \boldsymbol{W}_{G,rc}(t_{s1}) = \begin{pmatrix} r\cos(\Phi_G(t_{s1})) & r\sin(\Phi_G(t_{s1})) & z_G(t_{s1}) \\ -\omega\, r\sin(\Phi_G(t_{s1})) & \omega\, r\cos(\Phi_G(t_{s1})) & \dot{z}_G(t_{s1}) \\ -\omega^2 r\cos(\Phi_G(t_{s1})) & -\omega^2 r\sin(\Phi_G(t_{s1})) & \ddot{z}_G(t_{s1}) \\ \omega^3 r\sin(\Phi_G(t_{s1})) & -\omega^3 r\cos(\Phi_G(t_{s1})) & z_G^{(3)}(t_{s1}) \\ \omega^4 r\cos(\Phi_G(t_{s1})) & \omega^4 r\sin(\Phi_G(t_{s1})) & z_G^{(4)}(t_{s1}) \end{pmatrix} \quad (7.27)$$

for a smooth transition between the trajectories. The start conditions are composed of the phase $\Phi_G(t_{s1})$ and the height $z_G(t_{s1})$ of the give-way agent and the corresponding first four derivatives at the start time instant t_{s1}. The start time instant is set to $t_{s1} = t_{EG1}$ after the generation of event e_{G1}, where t_{EG1} is given by (7.12). After triggering event e_{G2} $t_{s1} = t_{EG2}$ holds, where t_{EG2} is determined by (7.16).

2) End conditions of the trajectory. For the derivation of the conditions on the end point of the proactive trajectory or the reactive trajectory the data $\boldsymbol{S}_S(t_k)$ of the stand-on agent is used. As it is desired that the give-way agent is opposite to the stand-on agent ($\Phi_{diff}(t) = 180°$) and on a different height, the end conditions are given by

$$\boldsymbol{W}_{G,pc}(t_{e1}) = \boldsymbol{W}_{G,rc}(t_{e1}) = \begin{pmatrix} -r\cos(\Phi_S(t_{e1})) & -r\sin(\Phi_S(t_{e1})) & w_{1,3}(t_{e1}) \\ \omega\, r\sin(\Phi_S(t_{e1})) & -\omega\, r\cos(\Phi_S(t_{e1})) & -\dot{z}_S(t_{e1}) \\ \omega^2 r\cos(\Phi_S(t_{s1})) & \omega^2 r\sin(\Phi_S(t_{e1})) & -\ddot{z}_S(t_{e1}) \\ -\omega^3 r\sin(\Phi_S(t_{e1})) & \omega^3 r\cos(\Phi_S(t_{e1})) & -z_S^{(3)}(t_{e1}) \\ -\omega^4 r\cos(\Phi_S(t_{e1})) & -\omega^4 r\sin(\Phi_S(t_{e1})) & -z_S^{(4)}(t_{e1}) \end{pmatrix} \quad (7.28)$$

with

$$w_{1,3}(t_{e1}) = \begin{cases} z_h & \text{for } z_S(t_{e1}) = z_l \\ z_l & \text{for } z_S(t_{e1}) = z_h. \end{cases} \tag{7.29}$$

The conditions are composed of the phase $\Phi_S(t_{e1})$ and the height $z_S(t_{e1})$ of the stand-on agent and the corresponding first four derivatives at the time instant t_{e1}. The negation of the entries of the matrix (7.28) compared to the matrix (7.27) gives the position opposite to the stand-on agent at t_{e1}. The parameter (7.29) forces the give-way agent to be on the other height than the stand-on agent. The end time instant is determined to be $t_{e1} = t_{s1} + t_t$, where t_t is chosen depending on which event (e_{G1} or e_{G2}) was generated, as stated in the next paragraph.

3) Planning of the trajectory. The trajectory is planned with Algorithm 10 on p. 80 for a change of the speed or the height to recover the desired phase difference and the height difference of control aim (G1) for a circular movement. The transition time interval t_t of the trajectory after the generation of event e_{G1} is chosen to be equal to the transition time interval t_t of the stand-on agent. After triggering event e_{G2} the transition time interval is given by the solution of optimisation problem (3.18) on p. 54.

7.4.7 *Phase 2: Trajectory to keep the adjusted phase or the height to the stand-on agent*

In order to keep the phase difference and the height difference of control aim (G1) for a circular movement maintained, a trajectory is generated with the following planning task:

Given:
- Conditions on the start point $\boldsymbol{W}_{G,pc}(t_{s2})$ or $\boldsymbol{W}_{G,rc}(t_{s2})$ given by (7.27) and the end point $\boldsymbol{W}_{G,pc}(t_{e2})$ or $\boldsymbol{W}_{G,rc}(t_{e2})$ given by (7.28).
- Transition time interval t_t.
- Data of the stand-on agent $\boldsymbol{S}_S(t_k)$.
- Parameters $\underline{s}$, $\bar{s}$ and $\bar{z}$ used in the requirements (1.3) and (1.7) of control aim (G1) for a circular movement.

Find: Trajectory $\boldsymbol{w}_{G,pc2}(t)$ or $\boldsymbol{w}_{G,rc2}(t)$ that keeps control aim (G1) for a circular movement satisfied.

Boundary conditions:
- System and actuator limitations given by (1.8), (1.9).
- Maximum speed (1.11).

1) Start conditions of the trajectory. The second phase of the proactive trajectory or the reactive trajectory are planned in the same way. The conditions on the start point of the trajectory are given by (7.27) with the start time instant $t_{s2} = t_{e1}$, where t_{e1} is the end time instant of the trajectory of the first phase.

2) End conditions of the trajectory. The conditions on the end point of the trajectory are given by (7.28) with $t_{e2} = t_{s2} + t_t$. The transition time interval t_t is given by the transition time interval t_t^i of the stand-on agent in this segment. This leads to a trajectory for the give-way agent, which is aligned to the trajectory of the stand-on agent and maintains control aim (G1) for a circular movement.

3) Planning of the trajectory. In order to maintain the desired phase difference and the height difference of control aim (G1) for a circular movement, a periodic continuation of the trajectory is planned with Algorithm 4 on p. 67 and Algorithm 5 on p. 68.

For the derivation of the planning tasks for the three phases of the proactive trajectory or the reactive trajectory for a circular movement of the agents Algorithm 13 on p. 157 can be used with small modifications:

- Replace requirements (1.1) and (1.2) by (1.3) and (1.7).
- Replace conditions (7.23) by (7.27) and (7.24), (7.25) by (7.28).

The third step of Algorithm 13 is not required. An execution of the modified algorithm together with the generation of the trajectories solves Problem 1.4 for a circular movement.

7.5 Communication method

7.5.1 *Communication flow*

The communication flow of the event-based control method is shown in the flow chart in Fig. 7.4. The time instants at which communication is invoked are written at the top of an execution step. The communication is split into an initialisation phase and an execution phase. Initially at time $t = 0$ the give-way agent sends a request $\boldsymbol{r}_G(0)$ to the stand-on agent to obtain its initial data $\boldsymbol{S}_S(0)$ including its initial trajectory.

In the execution phase, communication is as follows:

- The give-way agent determines based on $\boldsymbol{S}_S(0)$ the future event time instants t_{k+j}, $j = 1, \dots, N$ at which communication must be invoked and sends them with a request $\boldsymbol{r}_G(t_k)$ (7.6) to the stand-on agent.
- The stand-on agent responds by sending its current data $\boldsymbol{S}_S(t_k)$.
- After reception of the data, the give-way agent checks whether the stand-on agent is on its communicated trajectory as $\boldsymbol{p}_S(t_k) = \boldsymbol{w}_S(t_k)$ and if it did not change its future trajectory. If this is the case, the give-way agent just waits for the information of the stand-on agent at the next predicted event time instant t_{k+1}. On the other hand, if the give-way agent detects at an event time instant t_k that the stand-on agent deviated

from its trajectory as $p_S(t_k) \neq w_S(t_k)$ or it has changed its future trajectory $w_S(t)$, $(t > t_k)$ or the give-way agent has changed its future trajectory $w_G(t)$, $(t > t_k)$, the give-way agent determines new future event time instants t_{k+j}, $(j = 1, \ldots, N)$ at the time instant t_k and sends them as a request $r_G(t_{k+1})$ to the stand-on agent.

- The stand-on agent responds by sending its new data $S_S(t_{k+1})$.

The advantage of sending future event time instants compared to a communication scheme where the give-way agent sends a request $r_G(t_k)$ at each event time instant t_k and the stand-on agent just responds to these requests, is a reduction of the communication effort. As the stand-on agent sends the data at the predicted times autonomously, the give-way agent does not have to send a request at all event time instants.

Between two event time instants there is a minimum time span so that Zeno behaviour cannot occur in the communication flow of the event-based control method, as it is analysed in the next sections. The communication flow is applied for every movement of the agents.

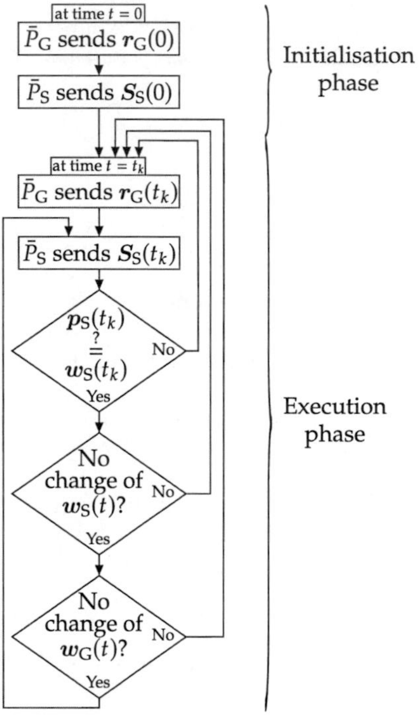

Fig. 7.4: Communication flow of the event-based control method.

7.5.2 General communication frequency

This section gives a bound on the frequency of the data communication within the event-based scheme for a general movement of the agents. For the determination of the minimum time span $t_{\mathrm{min,com}}$ between two communication events the following two time spans have to be considered:

- the time span t_{sep} between two consecutive events when the distance $s(t)$ between the agents is in the interval $\left[\underline{s} + \bar{e}_{\mathrm{G}}, \bar{s} - \bar{e}_{\mathrm{G}}\right]$ stated in Lemma 7.1,
- the parameter t_{com} stating the time span between two events when the distance $s(t)$ between the agents is in the intervals $\left]\underline{s}, \underline{s} + \bar{e}_{\mathrm{G}}\right[$ or $\left]\bar{s} - \bar{e}_{\mathrm{G}}, \bar{s}\right[$ given in Lemma 7.2.

Minimum time span between two events e_{G0} in the interval $\left[\underline{s} + \bar{e}_{\mathrm{G}}, \bar{s} - \bar{e}_{\mathrm{G}}\right]$. The minimum time span t_{sep} between two communication events in the interval $\left[\underline{s} + \bar{e}_{\mathrm{G}}, \bar{s} - \bar{e}_{\mathrm{G}}\right]$ is stated in Lemma 7.1.

Lemma 7.1 (Time span t_{sep} between two events e_{G0}). *Consider a stand-on agent and a give-way agent fulfilling Assumptions* 1.1, 1.2 *and* 1.4 *on p.* 16, *which are connected over a communication network satisfying Assumption* 6.1 *on p.* 127. *The minimum time span between two communication events e_{G0} when the distance between the agents is in the interval $s(t) \in \left[\underline{s} + \bar{e}_{\mathrm{G}}, \bar{s} - \bar{e}_{\mathrm{G}}\right]$, is bounded from below by*

$$t_{\mathrm{sep}} \geq \frac{\bar{e}_{\mathrm{G}}}{v_{\mathrm{SG,max}}}. \tag{7.30}$$

Proof. See Appendix B.5. ■

The speed $v_{\mathrm{SG,max}}$ is the maximum speed difference between the agents given by

$$v_{\mathrm{SG,max}} = v_{\mathrm{S,max}} + v_{\mathrm{G,max}} \tag{7.31}$$

if both agents move directly towards each other at their maximum speeds $v_{\mathrm{S,max}}$ and $v_{\mathrm{G,max}}$.

The lemma states that the event-based control method requires significantly less communication compared to a continuous communication scheme to maintain control aim (G1) when the distance between the agents is in the interval $s(t) \in \left[\underline{s} + \bar{e}_{\mathrm{G}}, \bar{s} - \bar{e}_{\mathrm{G}}\right]$.

Example 7.4. Considering agents with the dynamics introduced in Example 7.1 on p. 141, the time span t_{sep} between two consecutive events is given by

$$t_{\mathrm{sep}} = 0.22\,\mathrm{s}.$$

A comparison of the value of t_{sep} with t_{com} derived in Example 7.1 gives that t_{sep} is about half as large as t_{com}. This seems remarkable at first sight. If in the worst-case scenario the parameter t_{com} is used for communication, the agents are close to the limits $\underline{s}$ or $\bar{s}$. However, due to Assumption 1.2 on p. 17, this limit cannot be violated. The parameter t_{com} is determined in a way that the other bound cannot be violated unnoticed if the stand-on agent changes its movement significantly. Hence, the parameter t_{com} can be larger than t_{sep} depending on the chosen bounds $\underline{s}$ and $\bar{s}$. □

Parameter t_{com}. The necessity of using the parameter t_{com} after triggering event e_{G2} is illustrated in Fig. 7.5 where a situation is shown in which e_{G2} is generated with condition (I) of (7.15). The following considerations apply in the same way for an event generation with condition (II) of (7.15).

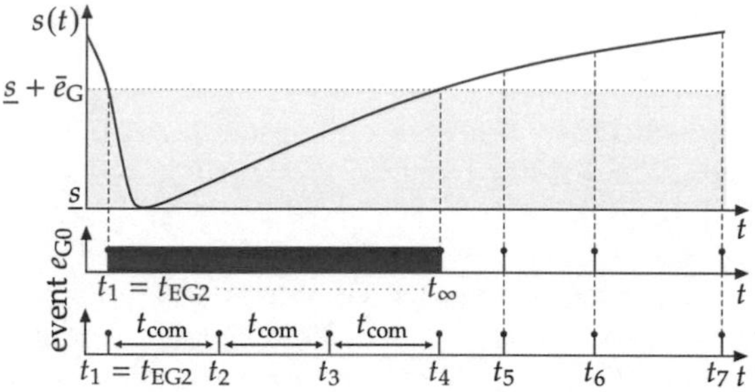

Fig. 7.5: Distance $s(t)$ between the agents after an event generation with condition (I) of (7.15).

At the event time instant t_{EG2} the distance between the agents equals the threshold for the communication event e_{G0}. Due to the dynamics of the agents the distance between them reduces in the worst-case scenario to $\underline{s}$. The threshold $\bar{e}_G$ (7.7) has been specified for this purpose. Hence, condition (I) of (7.4) is permanently fulfilled, which causes an event generation infinitely often (Zeno behaviour) shown in the upper timeline in Fig. 7.5. In order to avoid this behaviour the parameter t_{com} is used (lower timeline in Fig. 7.5). The parameter is determined with Lemma 7.2 so that control aim (G1) is always fulfilled. To this aim the worst-case scenario is considered that the stand-on agent reverses its movement, which causes a violation of (G1) at the other boundary $\bar{s}$ if communication is invoked not frequently enough. Hence, communication has to be invoked at the latest when the distance $\underline{s} + 2\,\bar{e}_G$ is reached between the agents.

Lemma 7.2 (Parameter t_{com}). *Consider a stand-on agent and a give-way agent fulfilling Assumptions* 1.1, 1.2 *and* 1.4 *on p.* 16, *which communicate over a communication network satisfying Assumption* 6.1 *on p.* 127. *When the distance between the agents is in the intervals*

$$\begin{aligned} & s(t) \in \left]\underline{s}, \underline{s} + 2\,\bar{e}_G\right[\\ \vee\, & s(t) \in \left]\bar{s} - 2\,\bar{e}_G, \bar{s}\right[\end{aligned} \tag{7.32}$$

and event e_{G2} *is generated, control aim* (G1) *is satisfied if communication is invoked with the parameter*

$$t_{com} = \varepsilon_t \cdot t_{t,min}. \tag{7.33}$$

The factor $\varepsilon_t = 1.07$ *depends on the dynamics of the agents and* $t_{t,min}$ *is the minimum transition time interval determined with the optimisation problem* (3.18) *on p.* 54.

Proof. See Appendix B.6. ∎

The lemma gives the interesting result that even in the worst-case scenario no continuous communication is necessary to fulfil control aim (G1). The communication frequency is still significantly reduced compared to a continuous communication scheme. An exemplary value for t_{com} can be found in Example 7.1 on p. 141.

Minimum time span between two events e_{G0}. The minimum time span $t_{min,com}$ between two communication events is stated in Theorem 7.3 using Lemma 7.1 and 7.2.

Theorem 7.3 (Minimum time span between two events e_{G0}). *Consider a stand-on agent and a give-way agent that satisfy Assumptions* 1.1, 1.2 *and* 1.4 *on p.* 16 *and that are connected over a communication network fulfilling Assumption* 6.1 *on p.* 127. *The minimum time span between two consecutive communication events* e_{G0} *is bounded from below by*

$$t_{min,com} \geq \min\left(t_{sep}, t_{com}\right)$$

where t_{sep} *is given by* (7.30) *and* t_{com} *is specified in* (7.33).

Proof. The proof is a combination of the proofs of Lemma 7.1 and 7.2. ∎

The theorem combines the results of Lemma 7.1 and 7.2 and states that there is a certain time span between two consecutive events e_{G0} at any time.

7.5.3 *Specific communication frequency for a circular movement*

This section states an upper bound on the frequency of the event-based communication for a circular movement of the agents. For the determination of the minimum time span between two consecutive communication events e_{G0} the following two time spans are taken into account:

- the time span $t_{c,sep}$ between two events when the phase difference $\Phi_{diff}(t)$ is in the interval $\left[\underline{s} + \bar{e}_{\Phi}, \bar{s} - \bar{e}_{\Phi}\right]$ and the height difference $z_{diff}(t)$ is in the interval $[\bar{z} - \bar{e}_z, \bar{z}]$ given in Lemma 7.3.
- the parameter t_{com} stating the time span between two events when the phase difference $\Phi_{diff}(t)$ is in the intervals $\left]\underline{s}, \underline{s} + \bar{e}_{\Phi}\right[$ or $]\bar{s} - \bar{e}_{\Phi}, \bar{s}[$ and the height difference $z_{diff}(t)$ is in the interval $]0, \bar{z} - \bar{e}_z[$ stated in Lemma 7.4.

Minimum time span between two events e_{G0} in the intervals $\left[\underline{s} + \bar{e}_{\Phi}, \bar{s} - \bar{e}_{\Phi}\right], [\bar{z} - \bar{e}_z, \bar{z}]$. The minimum time span $t_{c,sep}$ between two communication events in the intervals $\left[\underline{s} + \bar{e}_{\Phi}, \bar{s} - \bar{e}_{\Phi}\right]$ and $[\bar{z} - \bar{e}_z, \bar{z}]$ is stated in Lemma 7.3.

Lemma 7.3 (Time span $t_{c,sep}$ between two events e_{G0} for a circular movement). *Consider a stand-on agent and a give-way agent satisfying Assumptions* 1.1, 1.2 *and* 1.4 *on p.* 16, *which communicate over a communication network fulfilling Assumption* 6.1 *on p.* 127 *and move on circular trajectories in different heights. The minimum time span between two communication events e_{G0} for a circular movement when the phase difference and the height difference between the agents are in the intervals*

$$\Phi_{diff}(t) \in \left[\underline{s} + \bar{e}_{\Phi}, \bar{s} - \bar{e}_{\Phi}\right]$$
$$z_{diff}(t) \in [\bar{z} - \bar{e}_z, \bar{z}]$$

is bounded from below by

$$t_{c,sep} = \min\left(t_{\Phi,sep}, t_{z,sep}\right) \tag{7.34}$$

with

$$t_{\Phi,sep} \geq \frac{\bar{e}_{\Phi}\, 2\,\pi\, r}{v_{SG,max}}, \quad t_{z,sep} \geq \frac{\bar{e}_z}{v_{SG,z,max}}.$$

Proof. See Appendix B.7. ■

In accordance to Lemma 7.1, the lemma states that also for a circular movement of the agents the communication frequency is reduced significantly compared to a continuous communication while fulfilling control aim (G1) for a circular movement.

Example 7.5. Considering agents with the dynamics introduced in Example 7.2 on p. 148, the time spans $t_{\Phi,sep}$ and $t_{z,sep}$ between two consecutive events are given by

$$t_{\Phi,sep} = 0.51\,\text{s}, \quad t_{z,sep} = 0.25\,\text{s}.$$

Again these time spans are small compared to t_{com} for the same reasons as in the case of a general movement of the agents. It can be seen that the height changes cause the most communication, but the communication effort is still reduced to a continuous communication scheme. □

Parameter t_{com}. The usage of the parameter t_{com} is already motivated in the preceding section. It is stated in Lemma 7.4 and applied after the generation of event e_{G2}, which is triggered if one of the following worst-case scenarios occur:

- Both agents move with the speed $v_{\text{S,min}}$ and $\bar{P}_{\text{S}}$ accelerates as fast as possible to $v_{\text{S,max}}$. Then, $\bar{P}_{\text{G}}$ has to accelerate to $v_{\text{G,max}}$ as quickly as possible.
- Both agents move with the speed $v_{\text{S,max}}$ and $\bar{P}_{\text{S}}$ reduces its speed to $v_{\text{S,min}}$ in the shortest possible time span. Then, $\bar{P}_{\text{G}}$ has to reduce its speed to $v_{\text{G,min}}$ as fast as possible.
- Both agents move on their circular trajectories with arbitrary speeds and $\bar{P}_{\text{S}}$ changes its height with its maximum speed $v_{\text{S,max}}$. Then, $\bar{P}_{\text{G}}$ has to change its height with speed $v_{\text{G,max}}$.

Lemma 7.4 (Parameter t_{com} for a circular movement)**.** *Consider a stand-on agent and a give-way agent fulfilling Assumptions* 1.1, 1.2 *and* 1.4 *on p.* 16, *which communicate over a communication network satisfying Assumption* 6.1 *on p.* 127 *and move on circular trajectories in different heights. When the phase difference and the height difference are in the intervals*

$$\begin{aligned} &\Phi_{\text{diff}}(t) \in \left]\underline{s}, \underline{s} + \bar{e}_{\Phi}\right[\\ \vee\, &\Phi_{\text{diff}}(t) \in \left]\bar{s} - \bar{e}_{\Phi}, \bar{s}\right[\\ \vee\, &z_{\text{diff}}(t) \in \left]0, \bar{z} - \bar{e}_{\text{z}}\right[\end{aligned}$$

and event e_{G2} is generated, control aim (G1) *for a circular movement is fulfilled if communication is invoked with the parameter*

$$t_{\text{com}} = \min\left(\tilde{t}_{I}, \tilde{t}_{II}, \tilde{t}_{III}\right) \tag{7.35}$$

with

$$\begin{aligned} \tilde{t}_{I} &= \frac{\frac{\gamma - \bar{e}_{\Phi} + \Delta\Phi_{\text{G}}}{360^\circ} \cdot 2\,\pi\, r - t_{\text{t,min}} \cdot v_{\text{S,max}}}{v_{\text{S,max}} - v_{\text{G,min}}} \\ \tilde{t}_{II} &= \frac{\frac{-\gamma + \bar{e}_{\Phi} + \Delta\Phi_{\text{G}}}{360^\circ} \cdot 2\,\pi\, r - t_{\text{t,min}} \cdot v_{\text{S,min}}}{v_{\text{S,min}} - v_{\text{G,max}}} \\ \tilde{t}_{III} &= \frac{\bar{e}_{\text{z}} - v_{\text{G,max}} \cdot t_{\text{t,min}}}{v_{\text{S,max}}}. \end{aligned}$$

Proof. See Appendix B.8. ∎

Similar to Lemma 7.2 the lemma states the result that in the worst-case scenario no continuous communication is necessary to fulfil control aim (G1) for a circular movement. An exemplary value for t_{com} can be found in Example 7.2 on p. 148.

Minimum time span between two events e_{G0}. The minimum time span $t_{c,min,com}$ is given in Theorem 7.4 using Lemma 7.3 and 7.4.

Theorem 7.4 (Minimum time span between two events e_{G0} for a circular movement). *Consider a stand-on agent and a give-way agent that satisfy Assumptions* 1.1, 1.2 *and* 1.4 *on p.* 16, *communicate over a network that fulfils Assumption* 6.1 *on p.* 127 *and move on circular trajectories in different heights. The minimum time span between two consecutive communication events e_{G0} is bounded from below by*

$$t_{c,min,com} \geq \min\left(t_{c,sep}, t_{com}\right)$$

where $t_{c,sep}$ is specified in (7.34) *and t_{com} is given by* (7.35).

Proof. The proof is composed of the proofs of Lemma 7.3 and 7.4. ∎

The theorem combines the results of Lemma 7.3 and 7.4 and gives a certain time span between two consecutive events at any time for a circular movement.

7.6 Algorithms for the event-based control of mobile agents for an ideal network

General movement of the agents. This section provides an algorithm that is executed by the two agents and ensures the satisfaction of control aim (A1) using only local information and communicated data at event time instants. The algorithm summarises the event-based control method proposed in this chapter. For the implementation of the algorithm, the variable e_{count} is used acting as an event counter, which ensures that the events are only generated if the corresponding conditions are fulfilled for the first time. The counter is reset whenever the trajectories of the give-way agent or the stand-on agent have changed.

Algorithm 14 Event-based control method

Given: Position $\boldsymbol{p}_G(0)$ of the give-way agent $\bar{P}_G$.
Trajectory $\boldsymbol{w}_G(t)$, $(t_{start} \leq t \leq t_{end})$ of the give-way agent $\bar{P}_G$.
Event threshold $\bar{e}_G$ (7.7).
Parameters $\underline{s}$ (1.1) and $\bar{s}$ (1.2) of control aim (G1).
Event counter $e_{count} = 0$.

1: $\bar{P}_G$ requests initial information of $\bar{P}_S$ by sending the request $\boldsymbol{r}_G(0)$.
$\bar{P}_S$ sends data $\boldsymbol{S}_S(0)$.
2: $\bar{P}_G$ predicts future movement of the stand-on agent with Algorithm 11 on p. 88.
$\bar{P}_G$ determines recursively future event time instants with condition (7.4).

3: At $t = t_k$:
$\bar{P}_G$ invokes communication by sending the request $\boldsymbol{r}_G(t_k)$ (7.6).
4: At $t = t_k$:
$\bar{P}_S$ sends data $\boldsymbol{S}_S(t_k)$ (6.2).
5: At $t = t_k$:
$\bar{P}_G$ checks:
If $\boldsymbol{p}_S(t_k) = \boldsymbol{w}_S(t_k) \wedge$ No change of $\boldsymbol{w}_S(t) \wedge$ No change of $\boldsymbol{w}_G(t)$:
  If t_{EG1} is defined and $t_{EG1} < t_{k+1}$: go to step 7.
  Otherwise: Set $t_k = t_{k+1}$ and go to step 4.
Otherwise:
  Set $e_{count} = 0$ and go to step 6.
6: At $t = t_k$:
If condition (7.15) is fulfilled and $e_{count} = 0$:
  $\bar{P}_G$ generates event e_{G2}.
  $\bar{P}_G$ plans reactive trajectory $\boldsymbol{w}_{G,r}(t)$ with Algorithm 3 on p. 57 and Algorithm 13 on p. 157.
  $\bar{P}_G$ determines return time instant t_{ret} with (7.14).
  Set $t_k = t_{k+1}$, $e_{count} = 1$ and go to step 2.
If condition (7.11) is fulfilled and $e_{count} = 0$:
  $\bar{P}_G$ determines start time instant t_{EG1} of the proactive trajectory $\boldsymbol{w}_{G,p}(t)$ with (7.12).
  $\bar{P}_G$ determines return time instant t_{ret} with (7.14).
  Set $e_{count} = 1$.
  If $t_{k+1} < t_{EG1}$:
    Set $t_k = t_{k+1}$ and go to step 4.
  Otherwise:
    Go to step 7.
7: At $t = t_{EG1}$:
$\bar{P}_G$ generates event e_{G1}.
$\bar{P}_G$ plans proactive trajectory $\boldsymbol{w}_{G,p}(t)$ with Algorithm 3 and Algorithm 13.
Set $t_k = t_{k+1}$ and go to step 2.

Result: Movement of the agents $\bar{P}_G$ and $\bar{P}_S$ fulfilling control aim (A1).

The result of the event-based control method for a general movement of the agents is stated in Theorem 7.5.

Theorem 7.5 (Event-based control of agents). *Consider a stand-on agent and a give-way agent fulfilling Assumptions* 1.1, 1.2 *and* 1.4 *on p.* 16, *which are connected by a communication network satisfying Assumption* 6.1 *on p.* 127. *Furthermore, at time* $t = 0$ *the distance between the agents fulfils the inequalities*

$$\underline{s} + 2\,\bar{e}_G < s(0) < \bar{s} - 2\,\bar{e}_G, \tag{7.36}$$

for a given bound on the safety distance $\underline{s}$ *and the maximum separation* $\bar{s}$. *The fulfilment of the control aim* (A1) *is ensured if the following conditions hold:*

- *The give-way agent uses the set* $\mathcal{P}_S(t, t_k)$ *which is determined with* (7.1) *to estimate the future movement of the stand-on agent.*
- *The give-way agent generates events with conditions* (7.4), (7.11) *and* (7.15) *with the threshold* $\bar{e}_G$ *given by* (7.7).
- *The stand-on agent sends its data* $\boldsymbol{S}_S(t_k)$ *at any event time instant* t_{k+j}, $(j = 0, \ldots, N)$ *or after receiving a request* $\boldsymbol{r}_G(t_k)$ (7.6).
- *The stand-on agent and the give-way agent plan trajectories using Bézier curves.*
- *The stand-on agent and the give-way agent execute Algorithm* 14.

Proof. See Appendix B.9. ∎

The theorem states the main result of this chapter. Control aim (A1) is always guaranteed when the agents use Algorithm 14 with the components of the event-based control units derived in Chapters 6 and 7.

Circular movement of the agents. In case the two agents move on circular trajectories, they execute Algorithm 14 with small modifications to fulfil control aim (A1) where the requirements (1.1) and (1.2) are replaced by (1.3) and (1.7). The modifications of the algorithm are the following:

- Replace Algorithm 11 on p. 88 by Algorithm 12 on p. 92.
- Replace Algorithm 3 on p. 57 by Algorithm 5 on p. 68 and Algorithm 10 on p. 80.
- Replace conditions (7.4), (7.11) and (7.15) by (7.17), (7.21) and (7.22).
- Use the event thresholds $\bar{e}_\Phi$ and $\bar{e}_z$ (7.19) instead of $\bar{e}_G$ (7.7).

The result of the event-based control method for a circular movement of the agents is given in Corollary 7.1.

Corollary 7.1 (Event-based control of agents for a circular movement). *Consider a stand-on agent and a give-way agent fulfilling Assumptions* 1.1, 1.2 *and* 1.4 *on p.* 16, *which are connected by a communication network satisfying Assumption* 6.1 *on p.* 127 *and move on circular trajectories in different heights. Furthermore, at time $t = 0$ the phase difference and the height difference between the agents fulfil the inequalities*

$$\begin{aligned} \underline{s} + 2\,\bar{e}_{\Phi} &< \Phi_{\text{diff}}(0) < \bar{s} - 2\,\bar{e}_{\Phi} \\ \bar{s} - 2\,\bar{e}_{z} &< z_{\text{diff}}(0) \le \bar{z} \end{aligned} \tag{7.37}$$

for given bounds $\underline{s}$ and $\bar{s}$ on the phase difference and a desired height difference $\bar{z}$. The fulfilment of the control aim (A1) *for a circular movement of the agents is ensured if the following conditions hold:*

- *The give-way agent uses the set $\mathcal{P}_{S,c}(t, t_k)$ which is determined with* (7.2) *to estimate the future phases and heights of the stand-on agent.*
- *The give-way agent generates events with conditions* (7.17), (7.21) *and* (7.22) *with thresholds $\bar{e}_{\Phi}$ and $\bar{e}_{z}$ given by* (7.19).
- *The stand-on agent sends its data $\boldsymbol{S}_S(t_k)$ at any event time instant t_{k+j}, $(j = 0, \ldots, N)$ or after receiving a request $\boldsymbol{r}_G(t_k)$* (7.6).
- *The stand-on agent and the give-way agent plan trajectories using Bézier curves.*
- *The stand-on agent and the give-way agent execute Algorithm* 14 *with the modifications stated above.*

Proof. The proof is consistent to the proof of Theorem 7.5 where the control aim is initially satisfied by fulfilling Assumptions 1.2, 1.4 and 6.1 and ineq. (7.37). The proof is composed of the proofs of the parts of the control method. ∎

The corollary states that the event-based control units derived in Chapters 6 and 7 ensure the fulfilment of control aim (A1) for a circular movement. To this aim the control units execute the modified Algorithm 14 on p. 168.

Part IV

Event-based control over an unreliable communication network

Communication over an unreliable network

8

This chapter states the consequences on the event-based control of the agents and the satisfiability of the control aims arising from the use of an unreliable communication network (Section 8.1). The extensions of the event-based control units A_S and A_G to cope with the effects of an unreliable network are introduced in Section 8.2.

8.1 Consequences of unreliable communication links

8.1.1 *Assumptions*

The network for the data exchange between the moving agents considered in the remainder of Part IV of this thesis is not assumed to be ideal but it induces transmission delays and packet losses. Hence, Assumption 6.1 on p. 127 is replaced by the much less rigorous Assumption 8.1. The communication network fulfils this assumption in the remainder of Part IV.

Assumption 8.1. The mean packet loss probability of the channel is bounded from above by $p_{e,max}$. The statistics of the wireless channel are constant as long as the difference of the current distance $d(t)$ between the agents and the distance $d(\tilde{t}_i)$ at the last time instant $\tilde{t}_i$ at which the statistics were estimated does not exceed $\bar{d}$:

$$\bar{d} \geq |d(\tilde{t}_i) - d(t)|, \quad i = 1, 2, \ldots$$

Remark. The statement of Assumption 8.1 is shown in Fig. 8.1, where the difference $|d(\tilde{t}_i) - d(t)|$ of the distances between the agents is shown over time.

At the time instant $\tilde{t}_i$ the difference is zero because the agents have not yet moved on. With an increasing time, the distance between the agents changes depending on their movements. When the difference $|d(\tilde{t}_i) - d(t)|$ violates the threshold $\bar{d}$ the properties of the wireless channel can no longer be considered constant and $d(\tilde{t}_i) \stackrel{!}{=} d(\tilde{t}_{i+1})$ is set. As it can be seen in the figure, the distance between two time instants $\tilde{t}_i, \tilde{t}_{i+1}$ $(i = 1, .., N)$ depends on the movement of the agents.

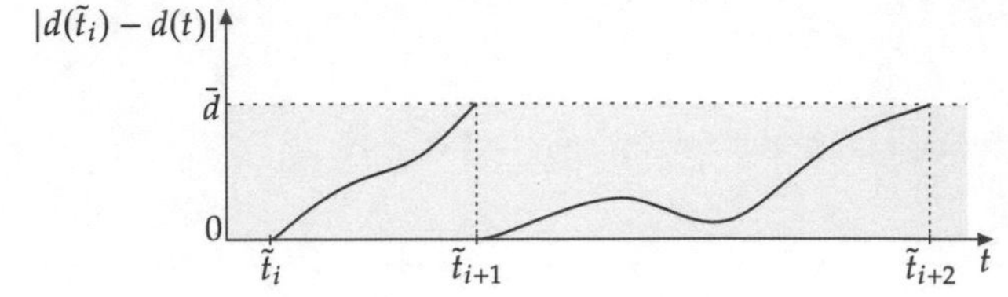

Fig. 8.1: Illustration of Assumption 8.1.

Furthermore, it is now assumed that computation times are caused by the agents during the execution of the control algorithms. Hence, Assumption 6.2 on p. 127 is replaced by the less rigorous Assumption 8.2 and the constraint (C3) of the problem statement in Section 1.4 is taken into account.

Assumption 8.2. The computation power onboard the moving agents is limited. This means that the execution of the control algorithms by the agents causes computational delays.

The further Assumptions 1.1 – 1.4 made in Chapter 1 on p. 16 regarding the dynamics of the quadrotors still apply. The extension of the event-based control units is also evaluated by an application to the quadrotors.

8.1.2 *Communication delays*

Description of the effect of communication delays. When an agent sends its data over an ideal communication network at an event time instant t_k, the information is received by the other agent instantaneously at the same time instant, as depicted in Fig. 8.2 (left). The data transmission over a network that induces transmission delays causes that information that is sent at t_k is first received after an unknown time delay τ_k at the time instant $t_k + \tau_k$ (Fig. 8.2 (right)). The time delay is composed of the transmission delay τ_{n} of the network and the computation times τ_{c} of the agents.

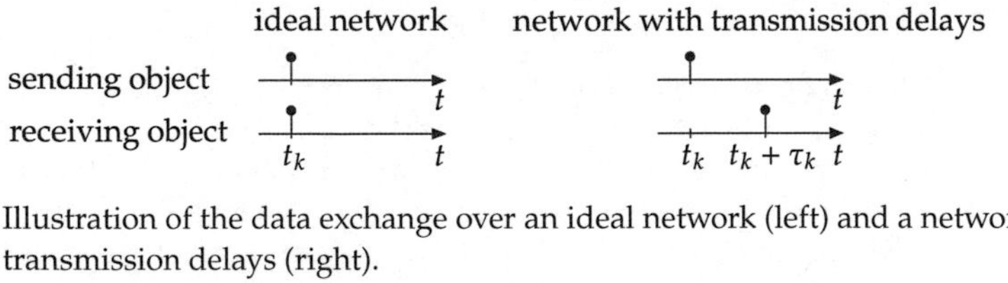

Fig. 8.2: Illustration of the data exchange over an ideal network (left) and a network with transmission delays (right).

The delayed reception of the information causes that these data are outdated by the current time delay τ_k, while the sending agent continues moving during this time span.

Hence, the agent is not at the communicated position at the event time instant t_k but in an area around this position, as illustrated in Fig. 8.4 (right).

Consequences on the event-based control units. The delayed data exchange over the network requires three extensions of the event-based control units to be able to fulfil the control aim (A1):

- The current time delay τ_k is unknown to the agents. Hence, a delay estimator must be utilised to generate an estimate $\tilde{\tau}_{\max}(d(\tilde{t}_i))$ of the time delay that is used for the event generation. The value $\tilde{\tau}_{\max}(d(\tilde{t}_i))$ is only a statistical estimate of the time delay. Hence, the current time delay τ_k can be smaller or larger.
- The event-based control unit $A_{\rm G}$ of the give-way agent requires new information immediately at an event time instant to satisfy control aim (A1). Since this can no longer be guaranteed due to the time delay, communication must be triggered prior to the event time instant t_k. The event generator has to be modified in a way that communication is triggered at a time instant $t_{{\rm c},k} = t_k - \tilde{\tau}_{\max}(d(\tilde{t}_i))$. The time instant is prior to t_k by the estimated time delay, as illustrated in Fig. 8.3. This causes the data to be received at the latest at the event time instant if it is not lost.

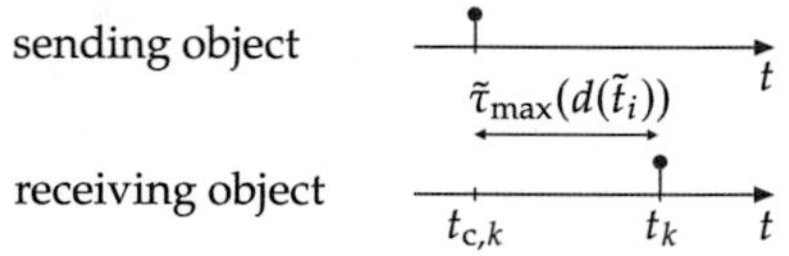

Fig. 8.3: Invocation of communication prior to an event time instant.

- The prediction of the future movement of a nearby agent and the determination of the current distance between the agents was based on the communicated information (Fig. 8.4 (left)). As the data is outdated by the time delay and the neighbouring agent moves on, the prediction of the agent's future movement and the determination of the distance between the agents has to be executed with respect to the uncertainty about the current position. The agent is not located at the communicated position but remains in a set $\mathcal{P}_{\rm S,d}(t_{{\rm c},k}, t_{{\rm c},k}, \tau_k)$ around this position (Fig. 8.4 (right)).

Consequences on the satisfiability of the control aims. Without considering the time delay in the event-based control units the satisfaction of control aim (A1) cannot be guaranteed, because required information is not available at an event time instant. However, control aims (A2) and (A3) are still satisfiable. As it is described in the next chapter, the extension of the control units leads to the fulfilment of control aim (A1) even in the presence of time delays.

Furthermore, with the extended control units the control aim (A1) for a circular movement is satisfiable.

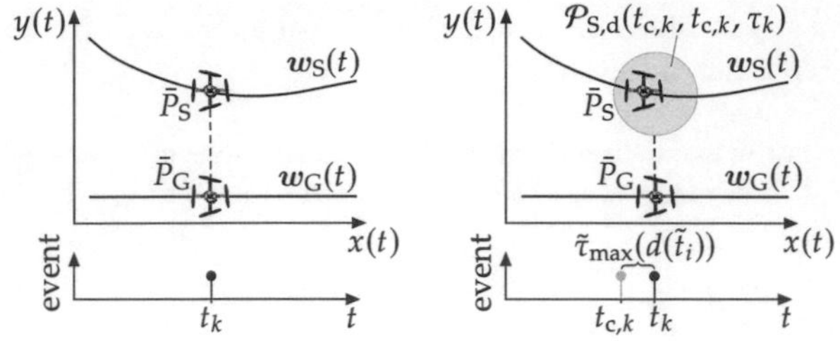

Fig. 8.4: Event-based control method with instantaneous communication (left) and delayed communication (right).

8.1.3 *Packet losses*

Description of the effect of packet losses. A packet loss means that sent data does not arrive at the receiver and the information is lost. A packet loss can occur for two reasons:

- A packet is indeed lost, for example because the received signal power is below the sensitivity of the receiver.
- A packet is received by an agent, but the reception was so late that the data are outdated and are considered to be lost. For the consideration of the time delay of the data transmission, the statistical estimate $\tilde{\tau}_{\max}(d(\tilde{t}_i))$ is used in the event generation. However, the actual time delay may be larger ($\tau_k > \tilde{\tau}_{\max}(d(\tilde{t}_i))$). When the information is received by an agent after the time span $\tilde{\tau}_{\max}(d(\tilde{t}_i))$, this data are too outdated to be used. Therefore, packets are considered to be lost if $\tau_k > \tilde{\tau}_{\max}(d(\tilde{t}_i))$ applies, as illustrated in Fig. 8.5.

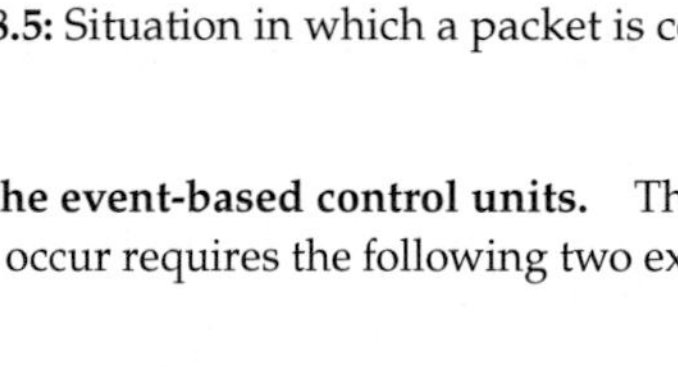

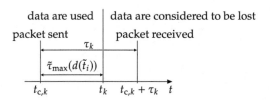

Fig. 8.5: Situation in which a packet is considered to be lost.

Consequences on the event-based control units. The data exchange over a network in which packet losses occur requires the following two extensions of the event-based control units:

- The event generator needs to be extended so that it is able to detect whether the packet has been received by the time instant t_k. If a packet is lost, it has to request new data by invoking communication. To this aim the communication scheme is extended by sending acknowledgment (ACK) messages.

- In the time span until the newly requested data are received at a time instant $t_{r,k}$, there is no information about the position of the neighbouring agent. Therefore, it can move unnoticed in such a way that the control aim (A1) is violated as shown in Fig. 8.6 (right). The trajectory planning unit has to be extended so that a trajectory is planned to guarantee at least the collision avoidance, despite the lack of information.

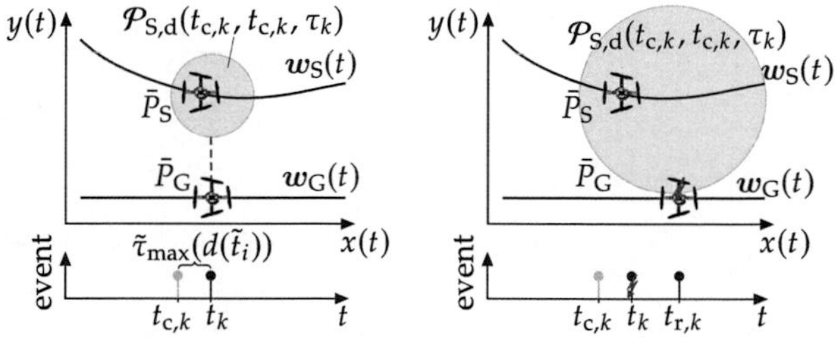

Fig. 8.6: Event-based control method with delayed communication (left) and communication with packet losses (right).

Consequences on the satisfiability of the control aims. Without considering packet losses in the event-based control units the fulfilment of control aim (A1) cannot be guaranteed due to the lack of required information. The extended control units described in the next chapter are only able to guarantee one requirement (1.1) or (1.2) of control aim (A1). As the collision avoidance is of superior importance compared to keeping a maximum separation between the agents, the control units are extended so as to guarantee requirement (1.1) of control aim (A1). Requirement (1.2) of control aim (A1) can only be fulfilled with a certain probability with the extended control units.

As both the lower bound (1.4) and the upper bound (1.5) must be fulfilled in order to satisfy the requirement (1.3) of control aim (A1) for a circular movement of the agents, the extended event-based control units are not able to guarantee this control aim, but can fulfil it only with a certain probability.

8.2 Extension of the event-based control units

8.2.1 *Event-based control unit A_S of the stand-on agent for an unreliable network*

The structure of the extended event-based unit A_S of the stand-on agent for an unreliable network is shown in Fig. 8.7. As it can fulfil its local control aim (S1) without the use of a communication network, it still has to execute the same tasks described in Section 6.2. Hence, the structure of the unit A_S remains unchanged. A delay estimator D_S is added to the unit, because both agents can be assigned with the tasks 'stand-on' and 'give-way' depending on the respective situation.

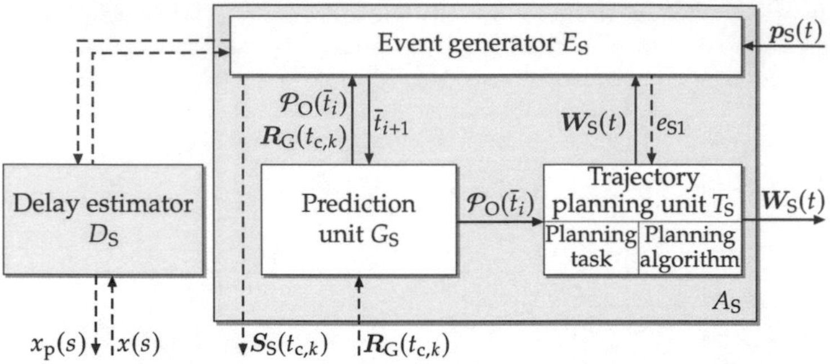

Fig. 8.7: Structure of the event-based control unit A_S with the delay estimator D_S of the stand-on agent.

The stand-on agent receives now an extended request $\boldsymbol{R}_G(t_{c,k})$ given by (9.23) on p. 195 from the give-way agent that contains the future event time instants t_{k+j}, $(j = 1, \ldots, N)$ and the estimated time delays $\tilde{\tau}_{max}(d(t_{k+j}))$ at these time instants. Using the request the stand-on agent determines the communication time instants $t_{c,k+j}$ at which it sends its data automatically to ensure the data to be received in time by the give-way agent if no packet loss occurs.

As the data $\boldsymbol{S}_S(t_{c,k})$ can get lost, the event generator E_S expects an ACK message from the give-way agent at least after the time delay $2\,\tilde{\tau}_{max}(d(t_k))$ after sending its data. Otherwise it assumes the data to be lost. If it does not receive an ACK message it sends the new data $\boldsymbol{S}_S(t_{c,k} + 2\,\tilde{\tau}_{max}(d(t_k)))$ at the time instant $t = t_{c,k} + 2\,\tilde{\tau}_{max}(d(t_k))$.

Remark. The extended request includes the future event time instants as well as the estimated time delays at these time instants instead of including only the communication time instants. The reason for this is that the stand-on agent requires the estimate $\tilde{\tau}_{max}(d(t_k))$ to send the ACK message, as described in Section 9.7.

8.2.2 *Event-based control unit A_G of the give-way agent for an unreliable network*

The structure of the extended event-based unit A_G of the give-way agent is depicted in Fig. 8.8. As the give-way agent is responsible to fulfil the control aim (A1), its control unit has to execute new tasks to fulfil the control aim despite the uncertainties arising from the communication network. The extensions of the control unit A_G are described in more detail in the next chapter.

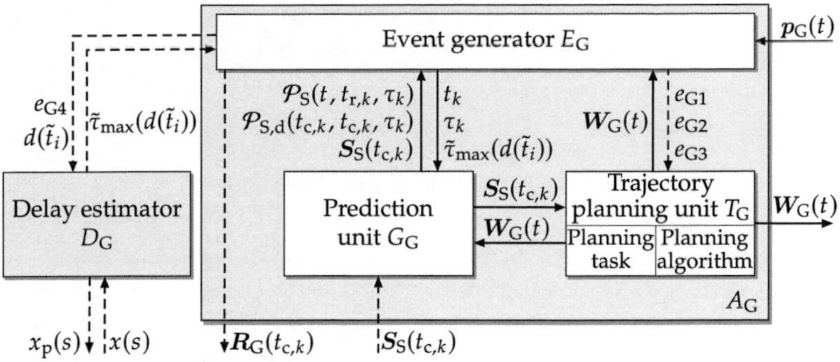

Fig. 8.8: Structure of the event-based control unit A_G with the delay estimator D_G of the give-way agent.

The prediction unit G_G generates the inclusion $\mathcal{P}(t, t_{r,k}, \tau_k)$ of the future positions of the stand-on agent with respect to the time delay τ_k. Furthermore, it generates the set $\mathcal{P}(t_{c,k}, t_{c,k}, \tau_k)$ that includes all possible positions of the stand-on agent during the time delay τ_k. This set is used by the event generator for the determination of the distance between the agents at an event time instant t_k.

The event generator E_G takes the time delay into account by the generation of the communication event e_{G0} and the reactive event e_{G1}. It invokes communication by sending the extended request $\boldsymbol{R}_G(t_{c,k})$ (9.23). Additionally, it uses two further events. With the *packet loss event* e_{G3} it triggers the planning of an *avoidance trajectory* $\boldsymbol{w}_{G,a}(t)$ by the trajectory planning unit in order to guarantee the collision-free movement of the agents. The *estimation event* e_{G4} invokes a new estimate $\tilde{\tau}_{max}(d(\tilde{t}_i))$ of the time delay whenever the current estimate gets too inaccurate.

The trajectory planning unit T_G is extended by a planning task for the avoidance trajectory $\boldsymbol{w}_{G,a}(t)$. This trajectory is planned so as to guarantee the collision-free movement despite the appearance of a packet loss.

A delay estimator is utilised that estimates the time delay of the data transmission between the agents. It generates an estimate whenever it is requested by the event e_{G4} from the event generator. For the estimate it uses the Markov model introduced in Section 5.3, which represents the channel properties depending on the distance $d(\tilde{t}_i)$ between the agents provided by the event generator. The parameters of the Markov model are initially estimated by sending the sequence of packets $x_p(s)$ into the network and evaluating the observation sequence $x(s)$, as stated in Section 5.3.4.

For a distinction between the event generation to invoke communication and to change the trajectory of the give-way agent and the event generation to trigger a new estimate of the time delay, two different notations for the event time instants are used corresponding to two different timelines, as illustrated in Fig. 8.9.

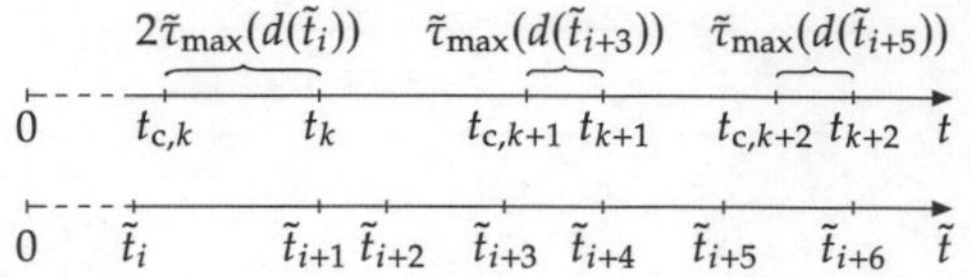

Fig. 8.9: Timelines of the event generation.

1. The event time instants at which the give-way agent requires new information of the stand-on agent or its trajectory has been changed due to the movement of the stand-on agent are denoted by t_{k+j}, $(j = 0, 1, \ldots, N)$. The communication time instants at which communication is invoked are denoted by $t_{c,k+j}$, $(j = 0, 1, \ldots, N)$.

2. As the time delay depends on the movement of the two agents, the event time instants to invoke a new estimate of the time delay are separated from the event time instants to trigger communication. The event time instants are denoted by $\tilde{t}_{i+l}$, $(l = 0, 1, \ldots, N)$. Whenever communication occurs at time instant t_k, the time delay is newly estimated as well so that $t_k = \tilde{t}_i$ holds. As the time delay can change rapidly with the movement of the agents between two consecutive event time instants t_k, t_{k+1}, several event time instants $\tilde{t}_i$ can occur between them, as shown exemplarily in Fig. 8.9.

In the next chapter the components of the event-based control unit A_G of the give-way agent are stated in detail.

Additional components of the event-based control unit

9

This chapter describes the extensions of the event-based control unit A_G of the give-way agent for the use of a communication network that induces transmission delays and packet losses. In Section 9.2 the delay estimator D_G is described for the estimation of the time delay. The extensions of the prediction unit G_G, the event generator E_G and the trajectory planning unit T_G are given in Sections 9.3 – 9.5. Furthermore, the event generation method is analysed in Section 9.6 and the adapted communication scheme is stated in Section 9.7. The event-based control method is summarised by an algorithm in Section 9.8.

9.1 Overview

This chapter describes which component has to be added to the event-based control unit of the give-way agent and how the existing components need to be changed in order to fulfil the requirements of control aim (G1) by using an unreliable network. As the give-way agent is responsible to fulfil control aim (G1), in this chapter the control unit A_G with the structure shown in Fig. 8.8 on p. 181 is considered. The control unit should solve the Problems 1.1 – 1.4. To this aim the method stated in Chapter 7 is combined with a delay estimator that generates an estimate of the transmission delay by using the communication model introduced in Section 5.3. The agents are connected by an unreliable communication network that induces transmission delays and packet losses according to Assumption 8.1 on p. 175. In addition, the computational delays need to be considered for the control according to Assumption 8.2 on p. 176.

The main contribution of this chapter is the combination of the method derived in Chapter 7 with a delay estimator. The result is a modification of Algorithm 14 on p. 218, which execution leads to the satisfaction of control aim (G1) even in the presence of time delays and packet losses. This main result is given in Theorem 9.5 on p. 219. Furthermore, it is shown that between two consecutive events a minimum time span is ensured to exclude Zeno behaviour even by using an unreliable network. In Theorem 9.3 on p. 217 this result is given for a general movement of the agents and in Theorem 9.4 on p. 218 it is stated for a circular movement. The prediction method is modified to cope with delayed data as stated in Theorem 9.1 on p. 191 for a general movement and in Theorem 9.2 on p.

193 for a circular movement. Important results for the event generation are summarised in the Lemmas 9.1 on p. 199 and 9.2 on p. 206. In the following sections the components are stated in detail.

9.2 Delay estimator D_G

9.2.1 *Task of the delay estimator*

The delay estimator determines a statistical estimate of the mean time delay

$$\tau_k = \tau_c + \tau_{n,k}$$

of a signal transfer to solve Problem 1.1. The time delay consists of the computation times τ_c of the agents and the mean transmission delay $\tau_{n,k}$ induced by the communication network. The estimation result is given by $\tilde{\tau}_{max}(d(\tilde{t}_i))$ and passed to the event generator. For the generation of the estimate the delay estimator contains the Markov model, introduced in Section 5.3.2. The model is initialised at time instant $t = 0$ by a channel estimation, described in Section 5.3.4. The delay estimator uses the maximum distance

$$d(\tilde{t}_i) = \text{dist}_{\max}\left(w_G(\tilde{t}_i), \mathcal{P}_S\left(\tilde{t}_i, t_{r,k}, \tau_k\right)\right), \tag{9.1}$$

which is determined with (4.19) on p. 94 and provided by the event generator. As the transmission delay of the network changes with a change of the distance between the agents, a new estimate is only triggered if the estimation event e_{G4} is generated by the event generator. In this case, the difference $|d(\tilde{t}_i) - d(t)|$ of the maximum distance $d(\tilde{t}_i)$ between the agents at the last event time instant $\tilde{t}_i$ and the current maximum distance $d(t)$ between the agents exceeds an event threshold and the present estimate has become too uncertain. The parameters of the Rician fading model, which depend on the distance between the agents, are newly estimated using the distance $d(\tilde{t}_i)$. Hence, the result of the estimate (9.13) on p. 189 holds in the time interval $[\tilde{t}_i, \tilde{t}_{i+1}]$. The use of the maximum possible distance between the agents leads to an estimate of the properties of the worst possible channel, which states a lower bound of the quality of the channel. If the agents are closer to one another, the channel quality increases on average. Hence, the delay estimator generates a conservative estimate of the mean time delay.

The computation times τ_c of the agents stay nearly constant for a broad range of parameters. Hence, it is sufficient to estimate the computational delay once at time instant $t = 0$. The delay estimator generates an estimate of the upper bound $\tau_{c,max}$ of the computation times.

As the result of the estimate (9.13) is a statistical value, the current time delay of an information transfer can be shorter or larger. As it is not required to get an exact estimate of the channel properties, but it is sufficient to keep the uncertainty of the estimate below a threshold, an event-based estimation approach is suitable.

Remark. The delay estimator in this section is designed for UAV applications since the control method is evaluated by using UAVs in Chapters 11 and 12. By changing the parametrisation of the Markov model and using a different radio propagation model, the delay estimator can also be used for other types of agents in different applications.

9.2.2 *Estimation of the time delay*

Estimation of the computation times of the agents. The computation times of the agents include the execution times of the control algorithms and the times required to transfer the data on board between the components of the agents as sensors, actuators, controllers etc. via fieldbus systems and the AD/DA conversion delays. Hence, the overall computation delay is given by

$$\tau_{\mathrm{c}} = \tau_{\mathrm{sc}} + \tau_{\mathrm{ct}} + \tau_{\mathrm{ca}} \tag{9.2}$$

where τ_{ct} is the delay caused by the execution of the control algorithms by the controller. The values τ_{sc} and τ_{ca} are the sensor-to-controller delay and the controller-to-actuator delay, respectively. In (9.2) the processing delay of the sensor has been included in τ_{sc}, while the processing delay of the actuator is included in τ_{ca}. As the proposed method is applied to quadrotors, which have embedded microcontrollers with a limited CPU time to execute the control algorithms, the execution times can be large and have to be considered by the event-based control unit.

For an estimate of the computation times methods from the real time programming are used. In order to derive an upper bound $\tau_{\mathrm{c,max}}$ of the computation times the delay estimator considers the Worst Case Execution Time (WCET) for each component and each task [60]. The WCET gives the longest possible duration of the execution of a task. It is determined by an end-to-end analysis. The analysis takes into account the structure of the algorithms to be executed, the input data, the compiler and the architecture and clock frequency of the executing processors. It is sufficient to estimate the computation time delay once at time instant $t = 0$, because the WCET stays constant in every interval $[\tilde{t}_i, \tilde{t}_{i+1}]$. As an embedded control system is used, the delays caused by sample and hold also have to be taken into account [60].

Result. The result of the estimation process is an upper bound $\tau_{\mathrm{c,max}}$ on the computation times of the agents given by (9.2). The value stays constant and holds for the entire time course $t \in [t_{\mathrm{start}}, t_{\mathrm{end}}]$ of the movement of the agents.

Estimation of the transmission delay. For the generation of an estimate of the transmission delay $\tau_{\mathrm{n},k}$ of a signal transfer, the mean packet loss probability of the wireless channel needs to be considered. For a preferably reliable communication link between the agents it is claimed that the mean packet loss probability $p_{\mathrm{e}}(d(\tilde{t}_i))$ that is derived with the Markov model in (5.15) on p. 119 does not exceed a required upper bound $p_{\mathrm{e,max}}$ (Assumption 8.1 on p. 175):

$$p_{\mathrm{e}}(d(\tilde{t}_i)) \overset{!}{\leq} p_{\mathrm{e,max}}. \tag{9.3}$$

The mean packet loss probability depends with (5.15) on the packet loss probabilities $p_{\mathrm{g}}(d(\tilde{t}_i))$ and p_{b} of the state 'good' and the state 'bad' of the Markov model, respectively. As $p_{\mathrm{b}} = 1$ holds (5.12), the requirement (9.3) results in a requirement on the packet loss probability $p_{\mathrm{g}}(d(\tilde{t}_i))$ as

$$p_{\mathrm{g}}(d(\tilde{t}_i)) \overset{!}{\leq} p_{\mathrm{g,max}}. \tag{9.4}$$

The packet loss probability $p_{\mathrm{g}}(d(\tilde{t}_i))$ of the state 'good' of the Markov model is supposed not to exceed an upper bound $p_{\mathrm{g,max}}$. Substituting $p_{\mathrm{e}}(d(\tilde{t}_i))$ by requirement $p_{\mathrm{e,max}}$ and inserting in (5.15) leads with (5.12) and the substitution of $p_{\mathrm{g}}(d(\tilde{t}_i))$ by $p_{\mathrm{g,max}}$ to

$$p_{\mathrm{e,max}} = p_{\mathrm{g,max}}\, \pi_{\mathrm{g}} + \pi_{\mathrm{b}}. \tag{9.5}$$

Solving eqn. (9.5) for $p_{\mathrm{g,max}}$ results in the upper bound of the packet loss probability of the state 'good':

$$p_{\mathrm{g,max}} = \frac{p_{\mathrm{e,max}} - \pi_{\mathrm{b}}}{\pi_{\mathrm{g}}}. \tag{9.6}$$

In the following the statistical estimate $\tau_{\mathrm{n,max}}(d(\tilde{t}_i))$ of the transmission delay is determined with respect to the requirement (9.3). Inserting (9.4) in (5.14) as

$$p_{\mathrm{g,max}} = 1 - Q_1\left(\frac{\nu(d(\tilde{t}_i))}{\sigma(d(\tilde{t}_i))}, \frac{\sqrt{2\, S_{\mathrm{min}}(d(\tilde{t}_i))}}{\sigma(d(\tilde{t}_i))}\right) \tag{9.7}$$

and solving for $S_{\mathrm{min}}(d(\tilde{t}_i))$ by using eqn. (5.10) gives the minimum received signal power $S_{\mathrm{min}}(d(\tilde{t}_i))$, which is only deceeded with the required mean packet loss probability $p_{\mathrm{g,max}}$ given by (9.6). In (9.7) the receiver sensitivity S_{rs} is replaced by $S_{\mathrm{min}}(d(\tilde{t}_i))$, because the probability $p_{\mathrm{g,max}}$ depends on the minimum received signal power. $S_{\mathrm{min}}(d(\tilde{t}_i))$ is derived iteratively. Starting from the value S_{rs} that results in the lowest possible packet loss probability $p_{\mathrm{g}}(d(\tilde{t}_i))$ the signal power is increased until the right side of (9.7) equals $p_{\mathrm{g,max}}$.

Hence, if a signal is received with at least the power $S_{\mathrm{min}}(d(\tilde{t}_i))$, the requirement (9.3) is satisfied. Furthermore, as it is stated in Section 5.1, the smallest signal power $S_{\mathrm{min}}(d(\tilde{t}_i))$ has to equal at least the receiver sensitivity S_{rs}:

$$S_{\mathrm{min}}(d(\tilde{t}_i)) \geq S_{\mathrm{rs}}. \tag{9.8}$$

Otherwise no signal and hence no packets can be received, which causes requirement (9.3) to be violated. This fact shows that $p_{\mathrm{e,max}}$ has to be chosen appropriately, because the curve of the Rice distribution and thus the location of the value $S_{\mathrm{min}}(d(\tilde{t}_i))$ varies with the distance between the agents. If $p_{\mathrm{e,max}}$ is chosen too small, the resulting signal power $S_{\mathrm{min}}(d(\tilde{t}_i))$ is smaller than the receiver sensitivity S_{rs} and no communication is possible.

For the determination of the maximum data rate $R_{\text{max}}(d(\tilde{t}_i))$ the channel is assumed to be affected by additive white Gaussian noise (AWGN) with zero mean and variance N. Moreover, the average noise power level does not change over time. Hence, it is sufficient to estimate the noise variance once at the beginning of the movement of the agents. The maximum data rate $R_{\text{max}}(d(\tilde{t}_i))$ that is supported by the channel with probability $1 - p_{\text{e,max}}$ of a successful data transfer is determined with (5.1) by

$$R_{\text{max}}(d(\tilde{t}_i)) = B \cdot \log\left(1 + \frac{S_{\text{min}}(d(\tilde{t}_i))}{N}\right), \tag{9.9}$$

where B corresponds to the bandwidth of the channel and N denotes the noise variance. When transmitting data with $R_{\text{max}}(d(\tilde{t}_i))$, the data gets only lost with the required probability $p_{\text{e,max}}$. If data is transmitted with a higher data rate, the information is received with a higher packet loss probability. The statistical estimate of the transmission delay is obtained with (5.7) as

$$\tau_{\text{n,max}}(d(\tilde{t}_i)) = \frac{M}{R_{\text{max}}(d(\tilde{t}_i))} \tag{9.10}$$

in which M denotes the number of bits that have to be transmitted. The estimate is newly determined whenever the distance between the agents changes significantly and the channel statistics change accordingly. In this case, the Rice distribution is updated with the new maximum distance $d(\tilde{t}_i)$ to derive the current packet loss probability $p_{\text{g}}(d(\tilde{t}_i))$ of the Markov model.

Result. The result of the estimation process is a statistical estimate $\tau_{\text{n,max}}(d(\tilde{t}_i))$ of the mean transmission delay given by (9.10). Only a statistical estimate results by the Markov model, because the stationary distribution of the Markov model is used, which states the properties of the channel just on average.

Discussion. For the estimate, the maximum distance $d(\tilde{t}_i)$ (9.1) between the agents provided by the event generator is utilised to get an estimate of the worst possible channel between the agents. Hence, the estimate of the transmission delay is variable and overestimates the mean transmission delay in the general case, because only the worst possible situation is taken into account in the calculation. As data transmissions over shorter distances require less time on average, the estimate of the delay estimator can also be used for smaller distances between the agents. This means that if a packet takes a longer time to transmit over a shorter distance than on average, the information can be received, because the event generator waits longer for the information due to the larger estimate of the time delay.

Output of the delay estimator D_G. The estimate of the time delay is given by

$$\tau_{\text{max}}(d(\tilde{t}_i)) = \tau_{\text{c,max}} + \tau_{\text{n,max}}(d(\tilde{t}_i)). \tag{9.11}$$

It is composed of the estimate of the computation times given by (9.2) and the estimate of the transmission delay derived by (9.10). It describes only the statistical mean of the transmission delay for the maximum possible distance between the agents in a time interval. Hence, the actual time delay τ_k of an information transfer can be smaller ($\tau_k < \tau_{\max}(d(\tilde{t}_i))$) or larger ($\tau_k > \tau_{\max}(d(\tilde{t}_i))$) than the estimated time delay although the longest possible distance between the agents was considered in the calculations. If $\tau_k < \tau_{\max}(d(\tilde{t}_i))$ holds, the control aims can always be satisfied by an event generation of the event generator, because the necessary information is received in time. In contrast, if $\tau_k > \tau_{\max}(d(\tilde{t}_i))$ holds, it is not possible to invoke events so as to guarantee control aim (G1) based only on this estimate, because the information is received too late or never by the give-way agent. During the lack of information the stand-on agent could move in a way to violate the control aim. In order to guarantee collision avoidance, in Section 9.5 a method is introduced to cope with this uncertainty.

In order to reduce the probability that $\tau_k > \tau_{\max}(d(\tilde{t}_i))$ holds, the following considerations are made.

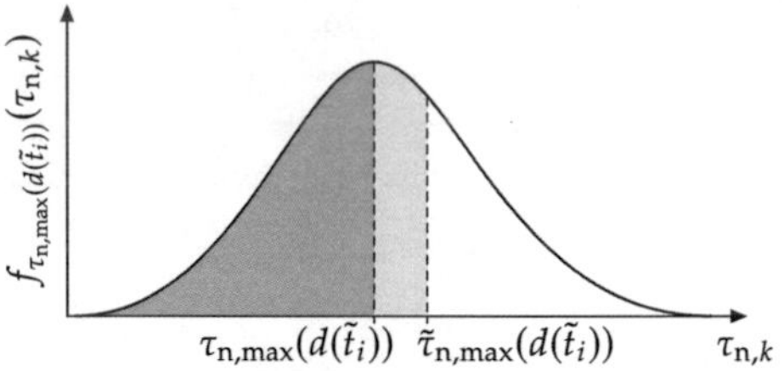

Fig. 9.1: Probability density function of $\tau_{n,k}$.

In Fig. 9.1 the probability density function of the transmission delay $\tau_{n,k}$ is shown. It can be seen that the probability that $\tau_{n,k}$ is smaller than $\tau_{n,\max}(d(\tilde{t}_i))$ (dark grey area in Fig. 9.1) is equal to the probability that $\tau_{n,k}$ is larger than $\tau_{n,\max}(d(\tilde{t}_i))$ (light grey area and white are in the figure). This is reasonable, because the estimate (9.10) is the mean of the transmission delay. This causes that necessary information at a time instant will be often considered to be lost although it will be received by the agent later. As a result, at these time instants the maximum separation between the agents (requirement (1.2) of control aim (A1)) cannot be guaranteed any more but only satisfied with a certain probability according to Section 8.1. In order to reduce the probability that $\tau_{n,k} > \tau_{n,\max}(d(\tilde{t}_i))$ holds, the delay estimator does not use the estimate (9.10). It uses an estimate that corresponds to an adjustment of the mean of the transmission delay by the factor δ_e as

$$\tilde{\tau}_{n,\max}(d(\tilde{t}_i)) = \delta_e \, \tau_{n,\max}(d(\tilde{t}_i)) \tag{9.12}$$

with $\delta_e > 1$. In Fig. 9.1 it can be seen that the probability of receiving data in time increases (dark grey area and light grey area in the figure).

An appropriate trade-off has to be found to determine the factor δ_e. On the one hand, the estimate (9.10) increases, which causes a higher number of communication events. On the other hand, the probability that packets will be received after the time span $\tau_{n,max}(d(\tilde{t}_i))$ reduces so that the number of packet losses detected by the event-based control unit reduces. The determination of δ_e is stated in Section 9.6.

Result. The result of the estimation process is the estimated time delay given by

$$\tilde{\tau}_{max}(d(\tilde{t}_i)) = \tau_{c,max} + \tilde{\tau}_{n,max}(d(\tilde{t}_i)). \tag{9.13}$$

It is composed of the estimates (9.2) and (9.12). It is newly derived after the event e_{G4} is generated and passed to the event generator.

9.2.3 *Algorithm of the event-based delay estimation*

The event-based operating principle of the delay estimator to derive an estimate of the time delay of a signal transfer is summarised in Algorithm 15. The parts of the algorithm are executed by both the event generator and the delay estimator. The event generator supervises the distance between the agents continuously and generates an event at time instant $\tilde{t}_i$ if the threshold $\bar{d}$ is violated, as stated in Section 9.4.6. Furthermore, it determines the distance $d(\tilde{t}_i)$. The delay estimator generates a new estimate of the current time delay at the discrete time instants t_k and $\tilde{t}_i$.

Algorithm 15 Event-based delay estimation

Given: Event threshold $\bar{d}$.
Event time instants t_{k+j}, $(j = 0, \ldots, N)$.
P_0 at distance d_0.
Upper bound on the mean packet loss probability $p_{e,max}$ (9.3).

1: If $t = 0$:
Generate estimation event e_{G4}.
Determine upper bound of the computation time $\tau_{c,max}$ with (9.2).
Perform channel estimation with (5.21), (5.22) and obtain parameters (5.24), (5.26).
Initialise Markov model with (5.27) using (5.23), (5.26).

2: If $t = t_{k+j}$, $(j = 0, 1, \ldots) \vee |d(\tilde{t}_i) - d(t)| = \bar{d}$:
Generate estimation event e_{G4}.
Set $\tilde{t}_i = t$ and determine $d(\tilde{t}_i)$ with (9.1).
Determine mean received signal power $P(d(\tilde{t}_i))$ with (5.2) using the distance $d(\tilde{t}_i)$.
Determine packet loss probability $p_g(d(\tilde{t}_i))$ with (5.4) and substitution of (5.8).
Update Markov model with $p_g(d(\tilde{t}_i))$.
Determine maximum data rate $R_{max}(d(\tilde{t}_i))$ for the data transmission with (9.9).
Determine statistical estimate of the transmission delay $\tau_{n,max}(d(\tilde{t}_i))$ with (9.10).

Determine time delay $\tau_{\max}(d(\tilde{t}_i))$ with (9.11).
Determine adjusted time delay $\tilde{\tau}_{\max}(d(\tilde{t}_i))$ with (9.13) and transfer it to the event generator.

Result: Discrete online statistical estimate of the time delay $\tilde{\tau}_{\max}(d(\tilde{t}_i))$ of the communicated signals.

Remark. The computation times of the agents are only estimated at time $t = 0$ because this delay remains constant over time. Hence, an estimate performed more often is not necessary. The Markov model is also initialised with the results of the channel estimation at time $t = 0$. At runtime the parameters of the Rice distribution incorporated in the Markov model are updated based on the current distance between the agents. A different packet loss probability $p_{\mathrm{g}}(d(\tilde{t}_i))$ results, which influences the transmission delay $\tau_{\mathrm{n,max}}(d(\tilde{t}_{i+1}))$.

9.3 Prediction unit G_{G}

9.3.1 *Task of the prediction unit*

The prediction unit has to execute four tasks to generate a feasible estimate of the future movement of the stand-on agent despite the time delay of the information transfer:

1. It receives the communicated data $\boldsymbol{S}_{\mathrm{S}}(t_{\mathrm{c},k})$ of the stand-on agent.
2. It determines the current time delay τ_k of the received information. The time instant $t_{\mathrm{c},k}$ at which the stand-on agent sent its data is communicated in the element $s_{\mathrm{S},2,1}$ of $\boldsymbol{S}_{\mathrm{S}}(t_{\mathrm{c},k})$ according to (6.2) on p. 129. The time delay is determined by
$$\tau_k = t_{\mathrm{r},k} - t_{\mathrm{c},k} \tag{9.14}$$
with $t_{\mathrm{r},k}$ as the time instant of reception. Furthermore, if the prediction unit does not receive the data until the event time instant $t_k = t_{\mathrm{c},k} + \tilde{\tau}_{\max}(d(\tilde{t}_i))$ it informs the event generator about the packet loss.
3. The prediction unit generates the set $\mathcal{P}_{\mathrm{S,d}}(t_{\mathrm{c},k}, t_{\mathrm{c},k}, \tau_k)$, which includes all possible positions of the stand-on agent that it could reach during the time delay τ_k.
4. It generates an inclusion $\mathcal{P}_{\mathrm{S}}(t, t_{\mathrm{r},k}, \tau_k)$ of the future movement of the stand-on agent. Due to the uncertainty about the communicated data resulting from the time delay, the set based on the position $\boldsymbol{p}_{\mathrm{S}}(t_{\mathrm{c},k})$ and speed $v_{\mathrm{S}}(t_{\mathrm{c},k})$ of the stand-on agent at the communication time instant $t_{\mathrm{c},k}$ has to incorporate the delay τ_k.

9.3.2 *Prediction for a general movement*

The time delay of a data transmission causes that the received information of the stand-on agent is outdated by the delay τ_k. As the stand-on agent continues moving, it is not located at the communicated position $p(t_{c,k})$ but in an area around this position. The ellipsoidal set $\mathcal{P}_{S,d}(t_{c,k}, t_{c,k}, \tau_k)$ includes this area and is used for the event generation. It is determined using Theorem 9.1 with the substitutions $t = t_{c,k}$ and $t_{r,k} = t_{c,k}$. The theorem extends Theorem 4.1 on p. 87 by incorporating the time delay τ_k.

Remark. The set $\mathcal{P}_{S,d}(t_{c,k}, t_{c,k}, \tau_k)$ can also be determined using Theorem 4.1 with the substitutions $t = t_{c,k} + \tau_k$ and $t_k = t_{c,k}$.

For an inclusion of the future movement of the stand-on agent the uncertainty about the communicated data needs to be considered. The set of future positions has to be generated from each position $p_S(t_{c,k} + \tau_k)$ inside the set $\mathcal{P}_{S,d}(t_{c,k}, t_{c,k}, \tau_k)$, $(p_S(t_{c,k} + \tau_k) \in \mathcal{P}_{S,d}(t_{c,k}, t_{c,k}, \tau_k))$, separately. The union of all these sets results in the inclusion $\mathcal{P}_S(t, t_{r,k}, \tau_k)$ of the future movement. However, this approach requires a considerable amount of computation resources.

Hence, a simpler approach is used for the inclusion of the future movement of the stand-on agent. The idea is to include the movement during the delay τ_k in the set $\mathcal{P}_S(t, t_{r,k}, \tau_k)$. To this aim the set is based on the position $p_S(t_{c,k})$ and speed $v_S(t_{c,k})$ of the stand-on agent at the communication time instant and incorporates the time delay τ_k. For the generation of the set $\mathcal{P}_S(t, t_{r,k}, \tau_k)$ Theorem 9.1 is used.

Theorem 9.1 (Prediction of the movement of an agent with delayed information). *Consider an agent, that has the position* $\boldsymbol{p}(t_{c,k})$ *and the speed* $\boldsymbol{v}(t_{c,k})$ *at the communication time instant* $t_{c,k}$ *and moves on trajectories given by Bézier curves* (3.19)*, that satisfy the limitations of Assumptions* 1.2 *and* 1.4 *on p.* 17*. Assume that the information received at time instant* $t_{r,k} = t_{c,k} + \tau_k$ *is outdated by the time delay* τ_k*. Then, the future positions* $\boldsymbol{p}(t)$*,* $(t \geq t_{c,k} + \tau_k)$ *of the agent are included in the set*

$$\mathcal{P}(t, t_{r,k}, \tau_k) = \left\{ \boldsymbol{p}(t) \in \mathbb{R}^3 : \frac{(x(t) - \bar{x}(t, t_{r,k}, \tau_k))^2}{r_x^2(t, t_{r,k}, \tau_k)} + \frac{(y(t) - \bar{y}(t, t_{r,k}, \tau_k))^2}{r_y^2(t, t_{r,k}, \tau_k)} + \frac{(z(t) - \bar{z}(t, t_{r,k}, \tau_k))^2}{r_z^2(t, t_{r,k}, \tau_k)} - 1 \leq 0 \right\} \tag{9.15}$$

with

$$\begin{aligned}
\bar{x}(t, t_{r,k}, \tau_k) &= x(t_{c,k}) + v_x(t_{c,k}) \cdot e \cdot (t - t_{r,k} + \tau_k) \\
\bar{y}(t, t_{r,k}, \tau_k) &= y(t_{c,k}) + v_y(t_{c,k}) \cdot e \cdot (t - t_{r,k} + \tau_k) \\
\bar{z}(t, t_{r,k}, \tau_k) &= z(t_{c,k}) + v_z(t_{c,k}) \cdot e \cdot (t - t_{r,k} + \tau_k) \\
r_x(t, t_{r,k}, \tau_k) &= r_y(t, t_{r,k}, \tau_k) = f \cdot v_{max} \cdot (t - t_{r,k} + \tau_k) \\
r_z(t, t_{r,k}, \tau_k) &= v_{max} \cdot (t - t_{r,k} + \tau_k),
\end{aligned} \tag{9.16}$$

and $e = 1 - f$*, where* f *is given by* (4.6).

Proof. See Appendix B.10. ∎

Remark. The coefficients e and f are again scaling factors to reduce the conservatism of the prediction method. As for the determination of an inclusion of the positions of the stand-on agent at future time instants t_{k+j} the current time delay τ_k is not available, τ_k is replaced by $\tilde{\tau}_{\max}(d(t_{k+j}))$ in (9.15). The value $\tilde{\tau}_{\max}(d(t_{k+j}))$ is the estimated time delay at time instants t_{k+j} provided by the delay estimator that is expected to occur based on the distances $d(t_{k+j})$ between the trajectories of the agents.

The theorem is an extension of Theorem 4.1 to cope with delayed data in a simple way. It is an interesting result since the set (9.15) is an extension of the set (4.4) by taking the time delay τ_k into account in the generation of the set (9.15). The consideration of τ_k causes the radii and the centre of the ellipsoid to be adapted by τ_k. Generating the set (9.15) solves Problem 1.2 in the presence of time delays. For the derivation of the sets $\mathcal{P}_{\mathrm{S,d}}(t_{\mathrm{c},k}, t_{\mathrm{c},k}, \tau_k)$ and $\mathcal{P}_{\mathrm{S}}(t, t_{\mathrm{r},k}, \tau_k)$ the prediction unit executes Algorithm 11 on p. 88 where the position $\boldsymbol{p}_{\mathrm{S}}(t_{\mathrm{c},k})$ and the speed $v_{\mathrm{S}}(t_{\mathrm{c},k})$ is used. In addition, eqns. (4.4) and (4.5) are replaced by (9.15) and (9.16).

Example 9.1. The estimation method of the future positions of the stand-on agent is illustrated in Fig. 9.2 in the 2D space. The prediction method stated in Theorem 9.1 is compared with the method given in Theorem 4.1 for the ideal network. Here, first the movement of the agent during the time delay τ_k is estimated and second the future movement of the agent is included in $\mathcal{P}_{\mathrm{S}}(t, t_{\mathrm{r},k}, \tau_k)$ using the generated set $\mathcal{P}_{\mathrm{S,d}}(t_{\mathrm{c},k}, t_{\mathrm{c},k}, \tau_k)$. The figure shows the situation at time $t = 4\,\mathrm{s}$, where the agent started its movement at $t = 0$ at the position marked as the black dot. The time delay is determined to be $\tau_k = 0.5\,\mathrm{s}$. During the time delay the moving stand-on agent is able to reach a position marked as the grey area, which corresponds to the set $\mathcal{P}_{\mathrm{S,d}}(t_{\mathrm{c},k}, t_{\mathrm{c},k}, \tau_k)$. Hence, for each position inside the grey area ($\boldsymbol{p}_{\mathrm{S}}(t_{\mathrm{c},k} + \tau_k) \in \mathcal{P}_{\mathrm{S,d}}(t_{\mathrm{c},k}, t_{\mathrm{c},k}, \tau_k)$) the prediction of the future movement of the agent needs to be executed, which causes a high computation effort.

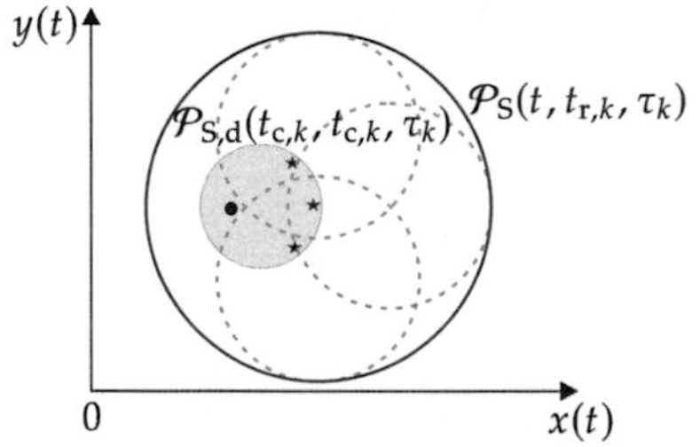

Fig. 9.2: Illustration of the position estimation method.

The black stars in the figure mark exemplarily three positions from which the prediction of the future movement is started. The prediction using Theorem 4.1 results exemplarily in three sets indicated by the green dotted circles. In contrast, using the method stated in Theorem 9.1 the prediction is only carried out once starting from the communicated position $\boldsymbol{p}_{\mathrm{S}}(t_{\mathrm{c},k})$ of the stand-on agent, marked in Fig. 9.2 by the black dot. The resulting set $\mathcal{P}_{\mathrm{S}}(t, t_{\mathrm{r},k}, \tau_k)$ is indicated by the red solid circle. As it can be seen, the sets generated with the method from Theorem 4.1 and from Theorem 9.1 overlap exactly and lead to the same prediction result. □

9.3.3 *Prediction for a circular movement*

With the same considerations as for the general movement of the agents, the prediction method for an inclusion of the future circular movement of the stand-on agent can be adapted. Again, the method is based on the position $p_S(t_{c,k})$ at the communication time instant. With this data the phase $\Phi_S(t_{c,k})$ and the height $z_S(t_{c,k})$ of the stand-on agent is determined. Furthermore, the generated set of future phases and heights of the stand-on agent incorporates the time delay τ_k.

The prediction result is given in Theorem 9.2. The set (9.17) is an extension of the set (4.11) by an incorporation of the time delay. The generation of (9.17) solves Problem 1.2 for a circular movement in the presence of time delays.

Theorem 9.2 (Prediction of the circular movement of an agent with delayed information). *Consider an agent, that has the phase $\Phi(t_{c,k})$ and the height $z(t_{c,k})$ at time $t_{c,k}$ and moves on trajectories given by Bézier curves* (2.11) *satisfying the limitations of Assumption* 1.3 *on p.* 17. *Assume that the information received at time instant $t_{r,k} = t_{c,k} + \tau_k$ is outdated by the time delay τ_k. Then the future phases $\Phi(t)$ and heights $z(t)$, $(t \geq t_{c,k} + \tau_k)$ of the agent are included in the set*

$$\mathcal{P}_c(t, t_{r,k}, \tau_k) = \left\{ p(t) \in \mathbb{R}^3 : \ \Phi(t) \in \Delta\Phi(t, t_{r,k}, \tau_k), \ z(t) \in \Delta z(t, t_{r,k}, \tau_k) \right\} \quad (9.17)$$

where

$$\Delta\Phi(t, t_{r,k}, \tau_k) = [\Phi_{\min}(t, t_{r,k}, \tau_k), \Phi_{\max}(t, t_{r,k}, \tau_k)] \quad (9.18)$$

$$\Delta z(t, t_{r,k}, \tau_k) = [z_{\min}(t, t_{r,k}, \tau_k), z_{\max}(t, t_{r,k}, \tau_k)] \quad (9.19)$$

with

$$\Phi_{\min}(t, t_{r,k}, \tau_k) = \operatorname{mod}\left(\Phi_S(t_{c,k}) + \tfrac{v_{\min}}{2\pi r} \cdot (t - t_{r,k} + \tau_k) \cdot 360^\circ, 360^\circ\right) \quad (9.20)$$

$$\Phi_{\max}(t, t_{r,k}, \tau_k) = \operatorname{mod}\left(\Phi_S(t_{c,k}) + \tfrac{v_{\max}}{2\pi r} \cdot (t - t_{r,k} + \tau_k) \cdot 360^\circ, 360^\circ\right) \quad (9.21)$$

$$z_{\min}(t, t_{r,k}, \tau_k) = \min\left(z_1(t, t_{r,k}, \tau_k), z_2(t, t_{r,k}, \tau_k)\right) \quad (9.22)$$

$$z_{\max}(t, t_{r,k}, \tau_k) = \max\left(z_1(t, t_{r,k}, \tau_k), z_2(t, t_{r,k}, \tau_k)\right)$$

and

$$z_1(t, t_{r,k}, \tau_k) = |z_S(t_{c,k}) + v_{\max} \cdot (t - t_{r,k} + \tau_k)|$$

$$z_2(t, t_{r,k}, \tau_k) = |z_S(t_{c,k}) - v_{\max} \cdot (t - t_{r,k} + \tau_k)|\,.$$

r states the radius of the circle.

Proof. See Appendix B.11. ∎

The prediction unit executes Algorithm 12 on p. 92 to generate the sets $\mathcal{P}_{\mathrm{S,d,c}}(t_{\mathrm{c},k}, t_{\mathrm{c},k}, \tau_k)$ and $\mathcal{P}_{\mathrm{S,c}}(t, t_{\mathrm{r},k}, \tau_k)$. The algorithm is based on the position $p_{\mathrm{S}}(t_{\mathrm{c},k})$ at the communication time instant $t_{\mathrm{c},k}$ and eqns. (4.11), (4.12) and (4.13) are replaced by (9.17), (9.18) and (9.19).

9.4 Event generator E_{G}

9.4.1 *Overview of the general event generation*

The event generator still acts as a supervisor and is responsible for invoking communication and for triggering a change of the trajectory $w_{\mathrm{G}}(t)$ of the give-way agent to keep control aim (G1) fulfilled. Furthermore, it triggers the sending of ACK messages whenever it received data from the stand-on agent. In order to cope with the uncertainty of the communicated data resulting from the time delay τ_k the conditions for the generation of the events e_{G0}, e_{G1} and e_{G2} have to be adjusted compared to the event generation for the ideal communication network stated in Chapter 7. In addition, two new events are introduced to handle the time delays and packet losses.

Sending of an ACK message. The event generator sends an acknowledgement (ACK) message to the stand-on agent to inform it about the successful information transfer whenever new information $\boldsymbol{S}_{\mathrm{S}}(t_{\mathrm{c},k})$ has been received:

$$\text{reception of } \boldsymbol{S}_{\mathrm{S}}(t_{\mathrm{c},k}) \Rightarrow \text{ sending of an ACK message.}$$

Furthermore, if the give-way agent does not receive an ACK message of the stand-on agent after the time span $2\,\tilde{\tau}_{\max}(d(t_k))$ after sending a request $\boldsymbol{R}_{\mathrm{G}}(t_{\mathrm{c},k})$ (9.23) it considers the request to be lost and sends it again.

Event generation in the presence of transmission delays. For the generation of the events the event generator monitors the following distances between the agents:

1. The distance $\mathrm{dist}(w_{\mathrm{G}}(t), \mathcal{P}_{\mathrm{S}}(t, t_{\mathrm{r},k}, \tilde{\tau}_{\max}(d(t_k))))$ between the trajectory of the give-way agent and the set of predicted positions of the stand-on agent given by (4.19) on p. 94.

2. The distance $\mathrm{dist}(w_{\mathrm{G}}(t), w_{\mathrm{S}}(t))$ between the trajectories of the agents determined with (4.25) on p. 97.

3. The distance $\mathrm{dist}(p_{\mathrm{G}}(t_k), \mathcal{P}_{\mathrm{S,d}}(t_{\mathrm{c},k}, t_{\mathrm{c},k}, \tilde{\tau}_{\max}(d(t_k))))$ between the current position of the give-way agent and the positions of the stand-on agent during the delay $\tilde{\tau}_{\max}(d(t_k))$, which is derived by (4.19).

For the determination of the distance in point 1. the set $\mathcal{P}_S(t, t_k)$ generated with (4.4) used in condition (7.4) is replaced by the set $\mathcal{P}_S(t, t_{r,k}, \tilde{\tau}_{max}(d(t_k)))$ derived with (9.15) that is used in condition (9.24) to take the time delay into account. For the same reason for the determination of the distance in point 3. the current position $p_S(t_k)$ of the stand-on agent, which is used in condition (7.15) is replaced by the set of possible positions $\mathcal{P}_{S,d}(t_{c,k}, t_{c,k}, \tilde{\tau}_{max}(d(t_k)))$ during the time delay that is used in condition (9.29). The determination of the distance in point 2. remains unchanged as the uncertainty about the future trajectory of the stand-on agent does not change by considering the time delay.

The situations that lead to the generation of the events e_{G0}, e_{G1} and e_{G2} and the reactions after triggering the events are the same as for the use of an ideal communication network and are given in Section 7.3. The only difference is that the event generator sends an extended request

$$\boldsymbol{R}_G(t_{c,k}) = \begin{pmatrix} t_{c,k} & t_{c,k+1} & \dots \\ \tilde{\tau}_{max}(d(t_k)) & \tilde{\tau}_{max}(d(t_{k+1})) & \dots \end{pmatrix}^T \tag{9.23}$$

compared to (7.6) to the stand-on agent. It contains the future communication time instants $t_{c,k+j}$, $(j = 0, \dots, N)$ and the statistical estimate $\tilde{\tau}_{max}(d(t_{k+j}))$, $(j = 0, \dots, N)$ (9.13) of the mean time delay that occurs at the future event time instants. $\tilde{\tau}_{max}(d(t_{k+j}))$ is provided by the delay estimator based on the maximum possible distances $d(t_{k+j})$ between the agents at time instants t_{k+j} that are determined by the event generator with (4.25).

The following sections focus on the adaptation of the event conditions for the events e_{G0}, e_{G1} and e_{G2} in the presence of time delays. Furthermore, the *packet loss event* e_{G3} and the *estimation event* e_{G4} are introduced. The situations in which these events need to be generated and the reactions after triggering the events are given. In particular it is described how the event generator uses the statistical estimate of the time delay to generate events deterministically so as always to satisfy the collision avoidance between the agents. The event generation for a network that induces transmission delays and packet losses is summarised in Tab. 9.1.

Table 9.1: Summary of the event generation for a general movement for an unreliable communication network.

Situation	Event	Reaction
Position of agent $\bar{P}_S$ becomes too uncertain.	e_{G0}	$\bar{P}_G$ sends $\boldsymbol{R}_G(t_{c,k})$, $\bar{P}_S$ responds by sending $\boldsymbol{S}_S(t_{c,k})$ if $\boldsymbol{w}_S(t)$ has changed. $\bar{P}_S$ sends $\boldsymbol{S}_S(t_{c,k})$ autonomously if $\boldsymbol{w}_S(t)$ is unchanged.
$\bar{P}_S$ or $\bar{P}_G$ changed their trajectories and a violation of control aim (G1) is detected in the future.	e_{G1}	$\bar{P}_G$ plans a proactive trajectory $\boldsymbol{w}_{G,p}(t)$.

Table 9.1: Summary of the event generation for a general movement for an unreliable communication network. (continued)

Situation	Event	Reaction
$\bar{P}_S$ changed its trajectory without notification and a violation of control aim (G1) is imminent.	e_{G2}	$\bar{P}_G$ plans a reactive trajectory $w_{G,r}(t)$.
Information from the agent $\bar{P}_S$ gets lost.	e_{G3}	$\bar{P}_G$ plans an avoidance trajectory $w_{G,a}(t)$ introduced in Section 9.5.
Uncertainty about the channel properties becomes too large.	e_{G4}	$\bar{P}_G$ generates a new statistical estimate $\tilde{\tau}_{max}(d(\tilde{t}_i))$ of the mean time delay with Algorithm 15 on p. 189.

9.4.2 *Communication event e_{G0}*

Deterministic event generation. For the generation of the communication event it is first assumed that the time delay τ_k equals the estimate $\tilde{\tau}_{max}(d(\tilde{t}_i))$ of the time delay, $(\tau_k = \tilde{\tau}_{max}(d(\tilde{t}_i)))$. This means that the time delay is uniquely known and the event can be deterministically generated. In the last paragraph of this section the handling of the statistical uncertainty is described when the current delay τ_k does not equal the statistical time delay $(\tau_k \neq \tilde{\tau}_{max}(d(\tilde{t}_i)))$. Hence, with the method stated in this section a deterministic event generation based on a statistical estimate is possible. The condition (9.24) stated below for event generations with respect to the time delay is again only evaluated if one of the following situations occur:

- the give-way agent has received the initial information $\boldsymbol{S}_S(0)$ of the stand-on agent.
- the stand-on agent has changed its trajectory.
- the give-way agent has changed its trajectory.

Otherwise there is no need to determine new event time instants, because the agents move in a way for which the already determined event time instants are valid. The event condition is again evaluated recursively for the time interval $t_{c,k} \leq t \leq t_{end}$ included in

$\mathcal{S}_S(t_{c,k})$. Communication events are generated if the following condition is satisfied at time instants $t = t_{c,k+j}$:

$$\text{Event } e_{G0}: \begin{cases} \quad \text{dist}_{\min}(\boldsymbol{w}_G(t), \mathcal{P}_S(t, t_{r,k+1}, \tilde{\tau}_{\max}(d(t_{k+1})))) \leq \underline{s} + \bar{e}_G + \bar{e}_{2d}(d(t_{k+1})) \\ \vee\, \text{dist}_{\max}(\boldsymbol{w}_G(t), \mathcal{P}_S(t, t_{r,k+1}, \tilde{\tau}_{\max}(d(t_{k+1})))) \geq \bar{s} - \bar{e}_G - \bar{e}_{2d}(d(t_{k+1})) & (I) \\ \wedge \quad \text{dist}_{\min}(\boldsymbol{w}_G(t), \mathcal{P}_S(t, t_{r,k+j}, \tilde{\tau}_{\max}(d(t_{k+j})))) \leq \underline{s} + \bar{e}_G + \bar{e}_{d}(d(t_{k+j})) \\ \vee\, \text{dist}_{\max}(\boldsymbol{w}_G(t), \mathcal{P}_S(t, t_{r,k+j}, \tilde{\tau}_{\max}(d(t_{k+j})))) \geq \bar{s} - \bar{e}_G - \bar{e}_{d}(d(t_{k+j})) & (II) \\ \wedge \quad t - t_{k+j} \geq t_{\text{com}} & (III) \end{cases} \tag{9.24}$$

for $t_{c,k} \leq t \leq t_{\text{end}}$, $(j = 2, \ldots, N)$. After the generation of a communication event at time instant $t_{c,k+1}$, the set $\mathcal{P}_S(t, t_{r,k+1}, \tilde{\tau}_{\max}(d(t_{k+1})))$ is reset and newly generated to determine the next communication time instant. The set considers the estimates $\tilde{\tau}_{\max}(d(t_{k+j}))$, $(j = 1, \ldots, N)$ of the future time delays at time instants t_{k+j}. The parameters $\underline{s}$ and $\bar{s}$ correspond to the requirements of control aim (G1). The threshold $\bar{e}_G$ and the parameter t_{com} are derived in Section 7.3.2 and 7.5.2, respectively and are applied unchanged. The parameters $\bar{e}_{2d}(d(t_{k+1}))$ and $\bar{e}_d(d(t_{k+j}))$, $(j = 2, \ldots, N)$ are thresholds to consider the time delay. They are given in (9.27) and (9.28) on p. 199. Hence, for taking the time delays into account the event condition (7.4) for the ideal network is extended by the parameters $\bar{e}_{2d}(d(t_{k+1}))$ and $\bar{e}_d(d(t_{k+j}))$.

Remark. In the time delay-free case ($\tilde{\tau}_{\max}(d(\tilde{t}_i)) = 0$) it holds $\bar{e}_{2d}(d(t_{k+1})) = \bar{e}_d(d(t_{k+j})) = 0$. This means, condition (9.24) is equal to condition (7.4), which shows that the condition is only extended to handle the time delays.

The decision with which part ((I), (II) or (III)) of condition (9.24) the events are triggered depends on the distance between the agents. Events are generated

$$\text{with}: \begin{cases} \text{part } (I) \quad \text{for} & t = 0 \vee t = t_{c,k+1} \\ & \vee\, \text{dist}(\boldsymbol{w}_G(t), \mathcal{P}_S(t, t_{r,k+1}, \tilde{\tau}_{\max}(d(t_{k+1})))) \in [\underline{s} + \bar{e}_G, \bar{s} - \bar{e}_G] \\ \text{part } (II) \quad \text{for} & t = t_{c,k+j},\ j = 2, \ldots, N \\ & \vee\, \text{dist}(\boldsymbol{w}_G(t), \mathcal{P}_S(t, t_{r,k+j}, \tilde{\tau}_{\max}(d(t_{k+j})))) \in [\underline{s} + \bar{e}_G, \bar{s} - \bar{e}_G] \\ \text{part } (III) \text{ for} & \text{dist}(\boldsymbol{w}_G(t), \mathcal{P}_S(t, t_{r,k+j}, \tilde{\tau}_{\max}(d(t_{k+j})))) \in]\underline{s}, \underline{s} + \bar{e}_G[\\ & \vee\, \text{dist}(\boldsymbol{w}_G(t), \mathcal{P}_S(t, t_{r,k+j}, \tilde{\tau}_{\max}(d(t_{k+j})))) \in]\bar{s} - \bar{e}_G, \bar{s}[\,. \end{cases}$$

Evaluation of the conditions of (9.24). Conditions (I) and (II) are evaluated with the method presented in Section 4.3.1 by adapting the radii of the ball. The third condition is evaluated using a clock, as described in Section 7.3.2.

Communication time instant. Part (I) or part (III) of the condition (9.24) cause the first communication time instant for the event e_{G0} to be generated at

$$\text{Time}: \begin{cases} \quad t_{c,k+1} = \arg\min_{t>t_k} \left\{\text{dist}(\boldsymbol{w}_G(t), \mathcal{P}_S(t, t_{r,k+1}, \tilde{\tau}_{\max}(d(t_{k+1})))) \le \underline{s} + \bar{e}_G + \bar{e}_{2d}(d(t_{k+1}))\right\} \\ \vee\, t_{c,k+1} = \arg\min_{t>t_k} \{\text{dist}(\boldsymbol{w}_G(t), \mathcal{P}_S(t, t_{r,k+1}, \tilde{\tau}_{\max}(d(t_{k+1})))) \ge \bar{s} - \bar{e}_G - \bar{e}_{2d}(d(t_{k+1}))\} \\ \wedge\, t_{c,k+1} = \arg\min_{t>t_k} \{t - t_{k+1} \ge t_{\text{com}}\}\,. \end{cases} \tag{9.25}$$

Every further communication time instant $t_{c,k+j}$, $(j = 2, \ldots, N)$ for e_{G0} is generated if part (II) or part (III) of (9.24) are fulfilled at

$$\text{Time}: \begin{cases} \quad t_{c,k+j} = \arg\min_{t>t_{k+j}} \left\{\text{dist}(\boldsymbol{w}_G(t), \mathcal{P}_S(t, t_{r,k+j}, \tilde{\tau}_{\max}(d(t_{k+j})))) \le \underline{s} + \bar{e}_G + \bar{e}_d(d(t_{k+j}))\right\} \\ \vee\, t_{c,k+j} = \arg\min_{t>t_{k+j}} \left\{\text{dist}(\boldsymbol{w}_G(t), \mathcal{P}_S(t, t_{r,k+j}, \tilde{\tau}_{\max}(d(t_{k+j})))) \ge \bar{s} - \bar{e}_G - \bar{e}_d(d(t_{k+j}))\right\} \\ \wedge\, t_{c,k+j} = \arg\min_{t>t_{k+j}} \left\{t - t_{k+j} \ge t_{\text{com}}\right\}\,. \end{cases} \tag{9.26}$$

The reason for this structure of the event condition is the communication flow with respect to the time delays. At the time instant $t_{c,k+1}$ the give-way agent has to send a new request $\boldsymbol{R}_G(t_{c,k+1})$. Subsequently the stand-on agent responds by sending its data $\boldsymbol{S}_S(t_{c,k+1})$. Hence, the give-way agent receives the data after twice the time delay. As the stand-on agent sends its data autonomously at every further communication time instant, the agent $\bar{P}_G$ receives the data only delayed by τ_k. This means that the first communication time instant needs to be generated earlier before the event time instant t_k. The different parts of the condition in which $\bar{e}_{2d}(d(t_{k+1})) = 2\,\bar{e}_d(d(t_{k+1}))$ holds, as stated in the next paragraph ensure this requirement.

Internal event generation. In addition to the generation of the event e_{G0} at the communication time instants $t_{c,k+j}$, the event generator has to determine the time instants t_{k+j} at which new data is required. The event time instants t_{k+j} need to be known, because the communication time instants are determined under the assumption that the information of the stand-on agent is received at event time instants t_{k+j} if the current delays τ_{k+j} equal the estimated mean time delays $\tilde{\tau}_{\max}(d(t_{k+j}))$, $(\tau_{k+j} = \tilde{\tau}_{\max}(d(t_{k+j})))$. If $\tau_{k+j} \neq \tilde{\tau}_{\max}(d(t_{k+j}))$ holds, the considerations stated in the last paragraph of this section have to be taken into account. The internal event is generated with condition (7.4) where the sets $\mathcal{P}_S(t, t_{r,k+j}, \tilde{\tau}_{\max}(d(t_{k+j})))$ (9.15) are used. The corresponding time instants of the event generation are given by (7.5).

Remark. The communication time instants (9.25) and (9.26) can also be determined with the time instants (7.5) of the internal event as

$$\begin{aligned} t_{c,k+1} &= t_{k+1} - 2\,\tilde{\tau}_{\max}(d(t_{k+1})) \\ t_{c,k+j} &= t_{k+j} - \tilde{\tau}_{\max}(d(t_{k+j})), \quad j = 2, \ldots, N. \end{aligned}$$

As a result, communication between the agents is invoked. The give-way agent sends the request (9.23) to the stand-on agent, so that it is able to send its data autonomously at the following communication time instants.

Derivation of the delay thresholds. The delay thresholds $\bar{e}_{2d}(d(t_{k+1}))$ and $\bar{e}_d(d(t_{k+j}))$ are determined in a way so that the movement of the stand-on agent during the time delays of data packets are taken into account. When the data are received by the give-way agent, then the agents are separated at least by the distances $\underline{s} + \bar{e}_G$ or $\bar{s} - \bar{e}_G$. The thresholds are variable over time, because they depend on the variable estimated time delays $\tilde{\tau}_{max}(d(t_{k+j}))$, $(j = 1, \ldots, N)$ at the time instants t_{k+j} provided by the delay estimator. The thresholds are derived in Lemma 9.1, where $t_{r,k+j}$ are the time instants at which the give-way agent receives the data of the stand-on agent.

Lemma 9.1 (Event thresholds $\bar{e}_{2d}(d(t_{k+1})), \bar{e}_d(d(t_{k+j}))$). *Consider a stand-on agent and a give-way agent fulfilling Assumptions* 1.1, 1.2, 1.4 *on p.* 17 *and* 8.2 *on p.* 176 *and that are connected over a communication network fulfilling Assumption* 8.1 *on p.* 175. *It is assumed that communication is invoked with condition* (9.24) *and no packet losses occur. The separations* $s(t_{r,k+j})$, $(j = 1, \ldots, N)$ *between the agents fulfil*

$$s(t_{r,k+j}) \geq \underline{s} + \bar{e}_G$$
$$s(t_{r,k+j}) \leq \bar{s} - \bar{e}_G$$

at time instant $t_{r,k+j}$ *if the delay thresholds are given by*

$$\bar{e}_{2d}(d(t_{k+1})) = v_{S,max} \cdot 2 \cdot \tilde{\tau}_{max}(d(t_{k+1})) \tag{9.27}$$
$$\bar{e}_d(d(t_{k+j})) = v_{S,max} \cdot \tilde{\tau}_{max}(d(t_{k+j})), \quad j = 2, \ldots, N. \tag{9.28}$$

Proof. See Appendix B.12. ■

The lemma states that the effects resulting from time delays are completely considered in (9.27) and (9.28). Hence, the event threshold (7.7) of Theorem 7.1 on p. 140 still holds. The utilisation of (7.7) together with (9.27) and (9.28) leads to a solution of Problem 1.3 in the presence of time delays.

Illustration of the event generation. The method of the determination of the communication time instants is depicted in Fig. 9.3, in which the trajectories of the stand-on agent and the give-way agent and the generation of the set $\mathcal{P}_S(t, t_{r,k+j}, \tilde{\tau}_{max}(d(t_{k+j})))$ are shown in the 2D space. For simplicity, communication is only invoked with the lower thresholds $\text{dist}_{min}(w_G(t), \mathcal{P}_S(t, t_{r,k+j}, \tilde{\tau}_{max}(d(t_{k+j}))))$ of (9.24). The bound $\underline{s} + \bar{e}_G$ is marked as a dotted line, the bound $\underline{s} + \bar{e}_G + \bar{e}_d(d(t_{k+j}))$ is marked as a dashed line. In the left part of the figure the bound $\underline{s} + \bar{e}_{\mathrm{G}} + \bar{e}_{2\mathrm{d}}(d(t_{k+1}))$ is represented as a dashed line for the first event. For an evaluation of the next communication time instants $t_{\mathrm{c},k+j}$, $(j = 1, \ldots, N)$ the give-way agent assumes the stand-on agent to follow its communicated trajectory. The condition (9.24) is evaluated recursively and the set $\mathcal{P}_{\mathrm{S}}(t, t_{\mathrm{r},k+j}, \tilde{\tau}_{\max}(d(t_{k+j})))$ is generated recursively as well. The movement of the stand-on agent during the time delays $\tilde{\tau}_{\max}(d(t_{k+j}))$ is included in the sets $\mathcal{P}_{\mathrm{S,d}}(t_{\mathrm{c},k+j}, t_{\mathrm{c},k+j}, \tilde{\tau}_{\max}(d(t_{k+j})))$, $(j = 1, \ldots, N)$. They are marked as the black shaded areas in Fig. 9.3 under the assumption that the stand-on agent follows its trajectory. For the examination of the condition, the positions $w_{\mathrm{G}}(t_{\mathrm{c},k+j})$ of the give-way agent on its trajectory at the time instants $t_{\mathrm{c},k+j}$ are used. The communication time instants are depicted as the grey beams. The black beams indicate the time instants of the internal event at which the information of the stand-on agent needs to be received by the give-way agent.

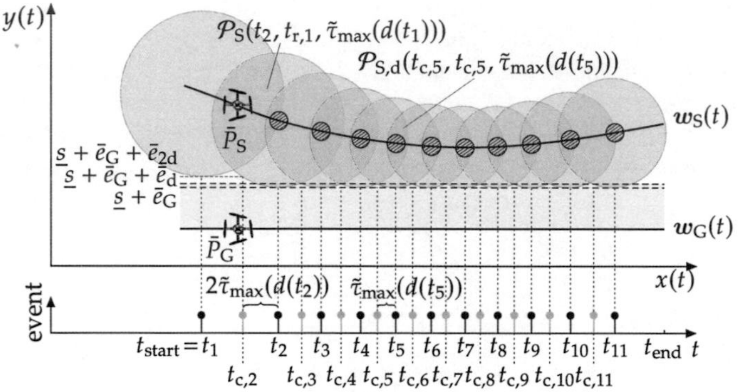

Fig. 9.3: Illustration of the determination of the communication time instants.

Handling of the statistical uncertainty. The estimate of the mean time delay $\tilde{\tau}_{\max}(d(\tilde{t}_i))$ obtained by the delay estimator with (9.13) is only a statistical value. Usually, the current time delay τ_k differs from $\tilde{\tau}_{\max}(d(\tilde{t}_i))$. Two cases have to be distinguished:

1. τ_k is smaller than $\tilde{\tau}_{\max}(d(\tilde{t}_i))$, $(\tau_k < \tilde{\tau}_{\max}(d(\tilde{t}_i)))$: In this case, the data is received before the event time instant $t = t_k$ and $t_{\mathrm{c},k} + \tau_k < t_k$ holds. The event generation described in the preceding paragraph is still applicable, because the required data is received even earlier than necessary.

2. τ_k is larger than $\tilde{\tau}_{\max}(d(\tilde{t}_i))$, $(\tau_k > \tilde{\tau}_{\max}(d(\tilde{t}_i)))$: If the data has not been received by the give-way agent within the time delay $\tilde{\tau}_{\max}(d(\tilde{t}_i))$, the communicated data of the stand-on agent is considered to be lost. During the lack of information the stand-on agent moves on and it could change its trajectory unnoticed so that a violation of control aim (G1) threatens. Hence, the give-way agent reacts on a packet loss with a

change of its trajectory so as to guarantee the collision avoidance (1.1) even though the give-way agent has no exact information about the movement of the stand-on agent. Due to its superior importance, collision avoidance has always to be fulfilled accepting that requirement (1.2) could be violated by a change of the trajectory. In case of a packet loss it is not possible to guarantee requirement (1.2). The event generation and the corresponding trajectory planning method are stated in Sections 9.4.5 and 9.5.

9.4.3 *Proactive event* e_{G1}

The proactive event e_{G1} is generated in the case of time delays in the same situation as stated in Section 7.3.3 when no time delays occur. The event condition (7.11) can be used unchanged, because the current trajectory of the stand-on agent is communicated at the event time instant. As a reaction a proactive trajectory $w_{G,p}(t)$ is planned at the time instant given by (7.12), as stated in Section 7.4.1.

9.4.4 *Reactive event* e_{G2}

The situation in which the reactive event e_{G2} in case of time delays is triggered is the same as without the occurrence of time delays, given in Section 7.3.4.

For the generation of the event the uncertainty about the position of the stand-on agent at an event time instant t_k has to be taken into account. The possible positions of the stand-on agent are included in the set $\mathcal{P}_{S,d}(t_{c,k}, t_{c,k}, \tau_k)$. As data were received at the time instant $t_{r,k}$, the current time delay τ_k can be determined with (9.14). The event is generated at a time instant t_k with the condition

$$\text{Reactive event } e_{G2}: \begin{cases} \quad \text{dist}_{\min}(p_G(t_k), \mathcal{P}_{S,d}(t_{c,k}, t_{c,k}, \tau_k)) \leq \underline{s} + \bar{e}_G & (I) \\ \vee \ \text{dist}_{\max}(p_G(t_k), \mathcal{P}_{S,d}(t_{c,k}, t_{c,k}, \tau_k)) \geq \bar{s} - \bar{e}_G & (II) \end{cases} \tag{9.29}$$

where the distance between the current position of the give-way agent and the predicted positions of the stand-on agent during the time delay τ_k is determined with (4.19) on p. 94.

As a reaction a reactive trajectory $w_{G,r}(t)$ is determined at the time instant given by (7.16), derived in Section 7.4.1.

9.4.5 *Packet loss event* e_{G3}

Situation. Data from the stand-on agent is not received by the give-way agent within the time delay $\tilde{\tau}_{\max}(d(\tilde{t}_i))$, so that $\tau_k > \tilde{\tau}_{\max}(d(\tilde{t}_i))$ holds. Hence, the data are considered to be lost.

Event generation. The event e_{G3} is triggered if the condition

$$\text{Packet loss event } e_{G3}: \quad \tau_k = t - t_{c,k} > \tilde{\tau}_{\max}(d(\tilde{t}_i)), \tag{9.30}$$

is satisfied at the time instant

$$t_{EG3} = t_{c,k} + \tilde{\tau}_{\max}(d(\tilde{t}_i)), \tag{9.31}$$

Reaction. As a result an avoidance trajectory $w_{G,a}(t)$ is generated by the trajectory planning unit, stated in Section 9.5. The trajectory changes the movement of the give-way agent so as to guarantee the collision avoidance.

9.4.6 *Estimation event e_{G4}*

Situation. The uncertainty about the properties of the channel has become too large due to the relative movement of the agents.

Event generation. The event generator continuously determines the maximum distance $d(t)$ between the agents with (4.19) using the set $\mathcal{P}_S(t, t_{r,k}, \tau_k))$. An event is triggered if the condition

$$\text{Estimation event } e_{G4}: \quad |d(\tilde{t}_i) - d(t)| = \bar{d} \tag{9.32}$$

is satisfied at a time instant

$$t_{EG4} = \tilde{t}_{i+1} = \min_{t > \tilde{t}_i} \left\{ |d(\tilde{t}_i) - d(t)| = \bar{d} \right\}.$$

The value $d(\tilde{t}_i)$ states the maximum distance between the agents at the last estimation time instant $\tilde{t}_i$. Furthermore, for the implementation of the method the current distance $d(t)$ is saved as the reference distance $d(\tilde{t}_i)$, $(d(\tilde{t}_i) = d(t))$ for the next event generation. The threshold $\bar{d}$ is chosen appropriately so that the channel properties can be assumed to be constant according to Assumption 8.1 on p. 175.

Reaction. After the event generation a new statistical estimate $\tilde{\tau}_{\max}(d(\tilde{t}_i))$ of the time delay is invoked, which is determined by the delay estimator.

9.4.7 *Specific event generation for a circular movement*

This section gives the conditions for the event generation for a circular movement of the agents in the presence of time delays. According to the event generation for a general movement of the agents, the conditions for the events e_{G0} and e_{G2} need to be modified, while the condition for event e_{G1} remains unchanged. As control aim (G1) for a circular movement cannot be guaranteed in the presence of packet losses, the give-way agent does not plan an avoidance trajectory. Hence, the packet loss event e_{G3} is not triggered. The estimation event e_{G4} can be applied unchanged for the circular movement. In the following paragraphs the modified event conditions for a circular movement in the presence of time delays are stated, which are summarised in Tab. 9.2.

Table 9.2: Summary of the event generation for a circular movement for an unreliable communication network.

Situation	Event	Reaction
Phase and height of agent $\bar{P}_S$ become too uncertain.	e_{G0}	$\bar{P}_G$ sends $\boldsymbol{r}_G(t_k)$, $\bar{P}_S$ responds by sending $\boldsymbol{S}_S(t_k)$ if $\boldsymbol{w}_S(t)$ has changed. $\bar{P}_S$ sends $\boldsymbol{S}_S(t_k)$ autonomously if $\boldsymbol{w}_S(t)$ is unchanged.
$\bar{P}_S$ or $\bar{P}_G$ changed their trajectories and a violation of control aim (G1) for a circular movement is detected in the future.	e_{G1}	$\bar{P}_G$ plans a proactive trajectory $\boldsymbol{w}_{G,p}(t)$.
$\bar{P}_S$ changed its trajectory without notification and a violation of control aim (G1) for a circular movement is imminent.	e_{G2}	$\bar{P}_G$ plans a reactive trajectory $\boldsymbol{w}_{G,r}(t)$.
Uncertainty about the channel properties becomes too large.	e_{G4}	$\bar{P}_G$ generates a new statistical estimate $\tilde{\tau}_{max}(d(\tilde{t}_i))$ of the mean time delay with Algorithm 15 on p. 189.

Communication event e_{G0}**.** The event condition (9.33) given below is only evaluated if

- the give-way agent receives the initial data $\boldsymbol{S}_S(0)$.
- the stand-on agent changed its trajectory.
- the give-way agent changed its trajectory.

Events are triggered if the condition

$$\text{Event } e_{\mathrm{G0}}: \begin{cases} \left.\begin{array}{l} \Phi_{\mathrm{diff,min}}(t, t_{\mathrm{r},k+1}, \tilde{\tau}_{\max}(d(t_{k+1}))) \leq \underline{s} + \bar{e}_{\Phi} + \bar{e}_{\mathrm{2dc}}(d(t_{k+1})) \\ \vee\, \Phi_{\mathrm{diff,max}}(t, t_{\mathrm{r},k+1}, \tilde{\tau}_{\max}(d(t_{k+1}))) \geq \bar{s} - \bar{e}_{\Phi} - \bar{e}_{\mathrm{2dc}}(d(t_{k+1})) \end{array}\right. & (I) \\ \vee\;\; z_{\mathrm{diff,min}}(t, t_{\mathrm{r},k+1}, \tilde{\tau}_{\max}(d(t_{k+1}))) \leq \bar{z} - \bar{e}_{\mathrm{z}} - \bar{e}_{\mathrm{2dz}}(d(t_{k+1})) & (II) \\ \left.\begin{array}{l} \Phi_{\mathrm{diff,min}}(t, t_{\mathrm{r},k+j}, \tilde{\tau}_{\max}(d(t_{k+j}))) \leq \underline{s} + \bar{e}_{\Phi} + \bar{e}_{\mathrm{dc}}(d(t_{k+j})) \\ \wedge\;\; \vee\, \Phi_{\mathrm{diff,max}}(t, t_{\mathrm{r},k+j}, \tilde{\tau}_{\max}(d(t_{k+j}))) \geq \bar{s} - \bar{e}_{\Phi} - \bar{e}_{\mathrm{dc}}(d(t_{k+j})) \end{array}\right. & (III) \\ \vee\;\; z_{\mathrm{diff,min}}(t, t_{\mathrm{r},k+j}, \tilde{\tau}_{\max}(d(t_{k+j}))) \leq \bar{z} - \bar{e}_{\mathrm{z}} - \bar{e}_{\mathrm{dz}}(d(t_{k+j})) & (IV) \\ \wedge \qquad t - t_{k+j} \geq t_{\mathrm{com}} & (V) \end{cases} \tag{9.33}$$

for $t_{\mathrm{c},k} \leq t \leq t_{\mathrm{end}}$, $(j = 2, \ldots, N)$ is fulfilled at time instants $t_{\mathrm{c},k+j}$. The phase differences $\Phi_{\mathrm{diff,min}}(t, t_{\mathrm{r},k+j}, \tilde{\tau}_{\max}(d(t_{k+j})))$, $\Phi_{\mathrm{diff,max}}(t, t_{\mathrm{r},k+j}, \tilde{\tau}_{\max}(d(t_{k+j})))$ and the height difference $z_{\mathrm{diff,min}}(t, t_{\mathrm{r},k+j}, \tilde{\tau}_{\max}(d(t_{k+1})))$ are determined with (4.30) and (4.31) using the sets (9.20), (9.21) and (9.22). The value $\tilde{\tau}_{\max}(d(t_{k+j}))$ is again the estimated time delay at time instants t_{k+j}, which are expected to occur due to the distances $d(t_{k+j})$ between the trajectories of the agents. The parameters $\underline{s}$ and $\bar{s}$ correspond to the requirements of control aim (G1) for a circular movement. The thresholds $\bar{e}_{\Phi}$ and $\bar{e}_{\mathrm{z}}$ are given in Section 7.3.6 and the parameter t_{com} is derived in Section 7.5.3. The thresholds $\bar{e}_{\mathrm{2dc}}(d(t_{k+1}))$, $\bar{e}_{\mathrm{dc}}(d(t_{k+j}))$, $\bar{e}_{\mathrm{2dz}}(d(t_{k+1}))$ and $\bar{e}_{\mathrm{dz}}(d(t_{k+j}))$ consider the time delay and are given by (9.34) – (9.37).

The decision with which part ((I), (II), (III), (IV) or (V)) of condition (9.33) the events are triggered is as follows:

$$\text{Event generated with}: \begin{cases} \text{part } (I) \text{ for} & \begin{array}{l} t = 0 \vee t = t_{\mathrm{c},k+1} \\ \vee\, \Phi_{\mathrm{diff}}(t) \in \left[\underline{s} + \bar{e}_{\Phi}, \bar{s} - \bar{e}_{\Phi}\right] \end{array} \\ \text{part } (II) \text{ for} & \begin{array}{l} t = 0 \vee t = t_{\mathrm{c},k+1} \\ \vee\, z_{\mathrm{diff}}(t) \in [\bar{z} - \bar{e}_{\mathrm{z}}, \bar{z}] \end{array} \\ \text{part } (III) \text{ for} & \begin{array}{l} t = t_{\mathrm{c},k+j},\ j = 2, \ldots, N \\ \vee\, \Phi_{\mathrm{diff}}(t) \in \left[\underline{s} + \bar{e}_{\Phi}, \bar{s} - \bar{e}_{\Phi}\right] \end{array} \\ \text{part } (IV) \text{ for} & \begin{array}{l} t = t_{\mathrm{c},k+j},\ j = 2, \ldots, N \\ \vee\, z_{\mathrm{diff}}(t) \in [\bar{z} - \bar{e}_{\mathrm{z}}, \bar{z}] \end{array} \\ \text{part } (V) \text{ for} & \begin{array}{l} \Phi_{\mathrm{diff}}(t) \in \left]\underline{s}, \underline{s} + \bar{e}_{\Phi}\right[\\ \vee\, \Phi_{\mathrm{diff}}(t) \in \left]\bar{s} - \bar{e}_{\Phi}, \bar{s}\right[\\ \vee\;\; z_{\mathrm{diff}}(t) \in [0, \bar{z} - \bar{e}_{\mathrm{z}}[\,. \end{array} \end{cases}$$

The parts (I), (II) or (V) of the condition (9.33) cause the first communication time instant at

$$\text{Time}: \begin{cases} t_{c,k+1} = \arg\min_{t>t_k} \{\Phi_{\text{diff,min}}(t, t_{r,k+1}, \tilde{\tau}_{\max}(d(t_{k+1}))) \leq \underline{s} + \bar{e}_\Phi + \bar{e}_{2dc}(d(t_{k+1}))\}\,, \\ \vee\, t_{c,k+1} = \arg\min_{t>t_k} \{\Phi_{\text{diff,max}}(t, t_{r,k+1}, \tilde{\tau}_{\max}(d(t_{k+1}))) \geq \bar{s} - \bar{e}_\Phi - \bar{e}_{2dc}(d(t_{k+1}))\}\,, \\ \vee\, t_{c,k+1} = \arg\min_{t>t_k} \{z_{\text{diff,min}}(t, t_{r,k+1}, \tilde{\tau}_{\max}(d(t_{k+1}))) \leq \bar{z} - \bar{e}_z - \bar{e}_{2dz}(d(t_{k+1})))\}\,, \\ \wedge\, t_{c,k+1} = \arg\min_{t>t_k} \{t - t_{k+1} \geq t_{\text{com}}\}\,. \end{cases}$$

Every further communication time instant $t_{c,k+j}$, $(j = 2, \ldots, N)$ is generated if parts (III), (IV) or (V) of (9.33) are fulfilled at

$$\text{Time}: \begin{cases} t_{c,k+j} = \arg\min_{t>t_{k+j}} \left\{\Phi_{\text{diff,min}}(t, t_{r,k+j}, \tilde{\tau}_{\max}(d(t_{k+j}))) \leq \underline{s} + \bar{e}_\Phi + \bar{e}_{dc}(d(t_{k+j}))\right\}\,, \\ \vee\, t_{c,k+j} = \arg\min_{t>t_{k+j}} \left\{\Phi_{\text{diff,max}}(t, t_{r,k+j}, \tilde{\tau}_{\max}(d(t_{k+j}))) \geq \bar{s} - \bar{e}_\Phi - \bar{e}_{dc}(d(t_{k+j}))\right\}\,, \\ \vee\, t_{c,k+j} = \arg\min_{t>t_{k+j}} \left\{z_{\text{diff,min}}(t, t_{r,k+j}, \tilde{\tau}_{\max}(d(t_{k+1}))) \leq \bar{z} - \bar{e}_z - \bar{e}_{dz}(d(t_{k+j}))\right\}\,, \\ \wedge\, t_{c,k+j} = \arg\min_{t>t_{k+j}} \left\{t - t_{k+j} \geq t_{\text{com}}\right\}\,. \end{cases}$$

The internal event to determine the event time instants t_{k+j} is generated with condition (7.17) at the time instant (7.18).

The delay thresholds are derived in Lemma 9.2 with the same considerations as for the general movement of the agents. With the thresholds given in the lemma, Problem 1.3 for a circular movement can be solved in the presence of time delays.

Lemma 9.2 (Event thresholds $\bar{e}_{2dc}(d(t_{k+1}))$, $\bar{e}_{2dz}(d(t_{k+1}))$, $\bar{e}_{dc}(d(t_{k+j}))$, $\bar{e}_{dz}(d(t_{k+j}))$**).** *Consider a stand-on agent and a give-way agent fulfilling Assumptions* 1.1, 1.2, 1.4 *on p.* 17 *and* 8.2 *on p.* 176 *that are moving on circular trajectories. The agents are connected by a communication network satisfying Assumption* 8.1 *on p.* 175. *It is assumed that communication is invoked with condition* (9.33) *and no packet losses occur. The phase difference* $\Phi_{\text{diff}}(t_{\text{r},k+j})$ *between the agents fulfils*

$$\Phi_{\text{diff}}(t_{\text{r},k+j}) \in [\underline{s}, \bar{s}]$$

with $\underline{s}$, $\bar{s}$ *given by* (1.4), (1.5) *on p.* 11 *and the height difference satisfies*

$$z_{\text{diff}}(t_{\text{r},k+j}) = \bar{z}$$

at time instants $t_{\text{r},k+j}$ *if the delay thresholds are given by*

$$\bar{e}_{2dc}(d(t_{k+1})) = \frac{v_{\text{S,max}}}{r} \cdot 2 \cdot \tilde{\tau}_{\max}(d(t_{k+1})) \tag{9.34}$$

$$\bar{e}_{2dz}(d(t_{k+1})) = v_{\text{S,max}} \cdot 2 \cdot \tilde{\tau}_{\max}(d(t_{k+1})) \tag{9.35}$$

$$\bar{e}_{dc}(d(t_{k+j})) = \frac{v_{\text{S,max}}}{r} \cdot \tilde{\tau}_{\max}(d(t_{k+j})), \quad j = 2, \ldots, N \tag{9.36}$$

$$\bar{e}_{dz}(d(t_{k+j})) = v_{\text{S,max}} \cdot \tilde{\tau}_{\max}(d(t_{k+j})), \quad j = 2, \ldots, N \tag{9.37}$$

where r *is the radius of the circle.*

Proof. See Appendix B.13. ■

Reactive event e_{G2}**.** For the generation of the reactive event the uncertainty of the phase and the height of the stand-on agent has to be considered. The possible phases and heights of the agent are included in the set $\mathcal{P}_{\text{S,d,c}}(t_{\text{c},k}, t_{\text{c},k}, \tau_k)$ determined with (9.17). The event is generated if the condition

$$\text{Reactive event } e_{\text{G2}}: \begin{cases} \quad \Phi_{\text{diff,min}}(t_{\text{c},k}, t_{\text{c},k}, \tau_k) \leq \underline{s} + \bar{e}_{\Phi} & (I) \\ \vee \quad \Phi_{\text{diff,max}}(t_{\text{c},k}, t_{\text{c},k}, \tau_k) \geq \bar{s} - \bar{e}_{\Phi} & (II) \\ \vee \quad z_{\text{diff,min}}(t_{\text{c},k}, t_{\text{c},k}, \tau_k) \leq \bar{z} - \bar{e}_{\text{z}} & (III) \end{cases} \tag{9.38}$$

is fulfilled at a time instant t_k. The phase difference and the height difference are determined with (4.30) and (4.31) on p. 99.

9.5 Trajectory planning unit T_G

The tasks of the trajectory planning unit remain unchanged compared to the tasks of the unit in the delay-free case, stated in Section 7.4 in the presence of time delays. The trajectory planning unit still plans the trajectories with the method, given in Chapter 3. Furthermore, the planning tasks of the proactive trajectory $w_{G,p}(t)$ and the reactive trajectory $w_{G,r}(t)$ after the generation of the events e_{G1} and e_{G2} are the same as if no time delays occur. Hence, in this section only the planning task for the avoidance trajectory $w_{G,a}(t)$ triggered by event e_{G3} is presented. The trajectory is generated in one phase and solves Problem 1.4 in the presence of packet losses.

As the control aim (G1) for a circular movement cannot be guaranteed in the case of packet losses, only the planning task for the avoidance trajectory for the general movement of the agents is given in this section.

Planning task for the avoidance trajectory $w_{G,a}(t)$. Packet losses of the communication channel cause a lack of information of the give-way agent until the next data is received. In the meantime the stand-on agent could move unnoticed in a way to violate requirements (1.1) or (1.2) of control aim (G1). As the collision avoidance is of superior importance and should always be guaranteed, the give-way agent precautionally plans an avoidance trajectory, which exploits the dynamical limitations of the agent to change the direction of its movement so that no collision is possible. At the same time it is not possible to plan a trajectory to guarantee requirement (1.2) of control aim (G1). Control aim (G1) requires the distance $s(t) = ||p_G(t) - p_S(t)||$ between the agents to be in the interval

$$\underline{s} \leq s(t) \leq \bar{s}.$$

Hence, it is clear that a change of the trajectory of the give-way agent to keep the lower bound $\underline{s}$ for collision avoidance satisfied leads to a more probable violation of the upper bound $\bar{s}$, which is illustrated in the last paragraph of this section. The planning task for the trajectory $w_{G,a}(t)$ is given by

Given:

- Conditions on the start point $W_{G,a}(t_s)$ given by (7.23) and the end point $W_{G,a}(t_e)$ given by (9.39).
- Transition time interval t_t given by (9.42).
- Inclusion $\mathcal{P}_S(t, t_{r,k}, \tau_k)$ (9.15) of the movement of the stand-on agent.
- Parameter $\underline{s}$ of control aim (G1).

Find: Trajectory $w_{G,a}(t)$ that guarantees requirement (1.1) of control aim (G1) in the presence of packet losses.

Boundary conditions:
- System and actuator limitations given by (1.8), (1.9).
- Maximum speed (1.10).

1) Start conditions of the avoidance trajectory. For a smooth transition between the current trajectory of the give-way agent and the avoidance trajectory the start conditions $\boldsymbol{W}_{\mathrm{G,a}}(t_{\mathrm{s}})$ of the trajectory are given by (7.23). The start time instant is given by $t_{\mathrm{s}} = t_{\mathrm{EG3}}$, where t_{EG3} is determined by (9.31).

2) End conditions of the avoidance trajectory. Due to the packet loss the information about the stand-on agent is not available to the give-way agent. Hence, the inclusion $\mathcal{P}_{\mathrm{S}}(t, t_{\mathrm{r},k}, \tau_k)$ of the movement of the stand-on agent is used and the worst-case scenario is assumed that the stand-on agent moves directly towards the give-way agent with its maximum speed. This considerations results in the end conditions of the avoidance trajectory given by

$$\boldsymbol{W}_{\mathrm{a}}(t_{\mathrm{e}}) = \begin{pmatrix} \hat{x}_{\mathrm{S}}(t_{\mathrm{s}}) + \varepsilon_{\mathrm{x}}\, d_{\mathrm{S}} & \hat{y}_{\mathrm{S}}(t_{\mathrm{s}}) + \varepsilon_{\mathrm{y}}\, d_{\mathrm{S}} & \hat{z}_{\mathrm{S}}(t_{\mathrm{s}}) + \varepsilon_{\mathrm{z}}\, d_{\mathrm{S}} & 0 \\ \varepsilon_{\mathrm{x}}\, ||\boldsymbol{v}_{\mathrm{S,max}}|| & \varepsilon_{\mathrm{y}}\, ||\boldsymbol{v}_{\mathrm{S,max}}|| & \varepsilon_{\mathrm{z}}\, ||\boldsymbol{v}_{\mathrm{S,max}}|| & 0 \\ \ddot{w}_{\mathrm{G,x}}(t_{\mathrm{e}}) & \ddot{w}_{\mathrm{G,y}}(t_{\mathrm{e}}) & \ddot{w}_{\mathrm{G,z}}(t_{\mathrm{e}}) & 0 \\ w^{(3)}_{\mathrm{G,x}}(t_{\mathrm{e}}) & w^{(3)}_{\mathrm{G,y}}(t_{\mathrm{e}}) & w^{(3)}_{\mathrm{G,z}}(t_{\mathrm{e}}) & 0 \\ w^{(4)}_{\mathrm{G,x}}(t_{\mathrm{e}}) & w^{(4)}_{\mathrm{G,y}}(t_{\mathrm{e}}) & w^{(4)}_{\mathrm{G,z}}(t_{\mathrm{e}}) & 0 \end{pmatrix}. \tag{9.39}$$

The position $\hat{\boldsymbol{p}}_{\mathrm{S}}(t_{\mathrm{s}}) = (\hat{x}_{\mathrm{S}}(t_{\mathrm{s}}) \;\; \hat{y}_{\mathrm{S}}(t_{\mathrm{s}}) \;\; \hat{z}_{\mathrm{S}}(t_{\mathrm{s}}))^{\mathrm{T}}$, where

$$\hat{\boldsymbol{p}}_{\mathrm{S}}(t_{\mathrm{s}}) = \min_{\boldsymbol{p}_{\mathrm{S}}(t_{\mathrm{s}}) \in \mathcal{P}_{\mathrm{S}}(t, t_k, \tau_k)} (||\boldsymbol{p}_{\mathrm{G}}(t_{\mathrm{s}}) - \boldsymbol{p}_{\mathrm{S}}(t_{\mathrm{s}})||) \tag{9.40}$$

denotes the closest possible position of the stand-on agent in the set $\mathcal{P}_{\mathrm{S}}(t, t_{\mathrm{r},k}, \tau_k)$ to the give-way agent at the start time instant t_{s}. The end time instant is given by $t_{\mathrm{e}} = t_{\mathrm{s}} + t_{\mathrm{t}}$ with t_{t} given by (9.42) on p. 209. The give-way agent is forced to increase the distance to the stand-on agent by

$$d_{\mathrm{S}} = ||\boldsymbol{v}_{\mathrm{S,max}} \cdot 2 \cdot \tilde{\tau}_{\max}(d(\tilde{t}_i))|| \tag{9.41}$$

and to move on in the same direction and with the same speed as the stand-on agent. The scaling factor $\boldsymbol{\varepsilon} = (\varepsilon_{\mathrm{x}} \;\; \varepsilon_{\mathrm{y}} \;\; \varepsilon_{\mathrm{z}})^{\mathrm{T}}$ is used to force the give-way agent to adjust the distance to the stand-on agent in the right direction so as to move away from the stand-on agent. The further entries in (9.39) make the give-way agent to move uniformly with the speed $||\boldsymbol{v}_{\mathrm{S,max}}||$.

After a packet loss the newly sent information from the stand-on agent is available to the give-way agent after the delay $2\,\tilde{\tau}_{\max}(d(\tilde{t}_i))$ if no second packet loss occurs. During this time span the stand-on agent is able to cover a maximum distance given by (9.41). An adjustment of the position of the give-way agent about the distance d_{S} in relation to the closest possible position of the stand-on agent leads at least to the same distance between the agents after the time delay $2\,\tilde{\tau}_{max}(d(\tilde{t}_i))$. Hence, the possible collision is avoided.

If data of the stand-on agent are received after the avoidance manoeuvre a proactive trajectory is planned by the give-way agent to keep control aim (G1) satisfied. Otherwise, the event e_{G3} is generated again and the give-way agent adjusts the distance to the stand-on agent again.

3) Transition time interval t_t of the avoidance trajectory. It is desired that the give-way agent adjusts the distance to the stand-on agent within the estimated time delay $2\,\tilde{\tau}_{max}(d(\tilde{t}_i))$. Hence, the transition time interval of the avoidance trajectory is given by

$$t_t = 2 \cdot \tilde{\tau}_{max}(d(\tilde{t}_i)). \tag{9.42}$$

Illustration of the planning method. The trajectory planning method is exemplarily shown in Fig. 9.4 for a movement in the 2D space. The packet sent at time instant $t = t_{c,k}$ gets lost as indicated by the flash. At time instant $t = t_k$ the event e_{G3} is generated and the give-way agent increases the distance to the stand-on agent about d_S (9.41) to avoid the possible collision with the stand-on agent. At time instant $t = t_{k+1}$ the newly sent information of the stand-on agent is received. As no collision threatens, the give-way agent returns to its initially planned trajectory.

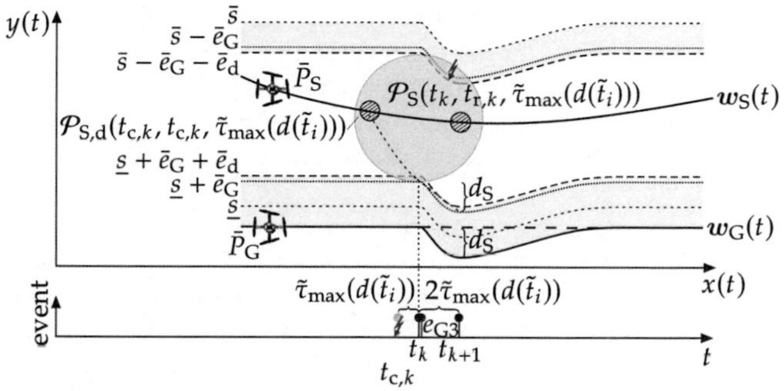

Fig. 9.4: Illustration of the reaction of the give-way agent after a packet loss.

In the upper part of the figure the upper bound $\bar{s}$ of control aim (G1) and the corresponding thresholds for an event generation are shown in dependence upon the trajectory of the give-way agent. It can be seen that before event e_{G3} is generated the thresholds are not violated. Due to the change of the trajectory of the give-way agent the set $\mathcal{P}_S(t, t_{r,k}, \tilde{\tau}_{max}(d(\tilde{t}_i)))$ of possible positions of the stand-on agent intersects with the thresholds indicated by the red flash in the upper part of the figure. Hence, a violation of requirement (1.2) of control aim (G1) is now possible. This behaviour indicates that only one bound $\underline{s}$ or $\bar{s}$ can be guaranteed at the same time. The reaction of the agent to maintain the fulfilment of one bound could lead to the violation of the other bound.

Furthermore, based on this exemplary situation it can be seen that although a packet got lost the stand-on agent moves in a way that no collision threatens and the planning of the avoidance trajectory of the give-way agent was not necessary. Due to the lack of information the give-way agent has no knowledge about the current movement of the stand-on agent and has to assume the worst-case scenario.

Remark. The trajectories for a circular movement are still planned using the planning tasks, given in Section 7.4. As the control aim (G1) for a circular movement cannot be guaranteed if packet losses occur, an avoidance trajectory is not planned in this scenario.

9.6 Analysis of the event generation in the presence of time delays and packet losses

Aim of the analysis. This section evaluates the suitability of the estimate (9.13) on p. 189 of the time delay of a data transmission. Generating an appropriate estimate is important, because it is used for the invocation of communication. A poor estimate causes either communication to be invoked unnecessarily often or the data received at a time instant cannot be used, because it is too old. It is analysed by a simulation how often the current time delay τ_k of an information transfer is larger than the statistical estimate $\tilde{\tau}_{\max}(d(\tilde{t}_i))$ of the mean time delay, $(\tau_k > \tilde{\tau}_{\max}(d(\tilde{t}_i)))$, which implies that the data are not lost but are received late by the give-way agent. As a result the event e_{G3} is invoked unnecessarily, because the received data has become too uncertain in the meantime to be used for the control method.

The wireless communication channel is modelled according to Section 5.3, where the agents send their information with a transmission power P_0 as packets with size M over the channel with a bandwidth B. The data can be received with the sensitivity S_{rs}. The parameters of the channel are summarised in Tab. 9.4. The two-state Markov model has the transition matrix

$$\boldsymbol{G}_{\mathrm{M}} = \begin{pmatrix} p_{\mathrm{gg}} & p_{\mathrm{gb}} \\ p_{\mathrm{bg}} & p_{\mathrm{bb}} \end{pmatrix}$$

where the transition probabilities are given by

$$p_{\mathrm{gg}} = 0.995, \quad p_{\mathrm{bb}} = 0.5, \quad p_{\mathrm{gb}} = 0.005, \quad p_{\mathrm{bg}} = 0.5$$

to represent the correlation of the packet losses. The stationary distribution is determined with (5.18) to be

$$\pi_{\mathrm{g}} = 0.99, \quad \pi_{\mathrm{b}} = 0.01.$$

Remark. For the analysis the bandwidth $B = 35\,\text{kbps}$ is used. This is a small bandwidth for communication systems, which results in large transmission times $\tau_{n,k}$. However, this bandwidth is used in the analysis, because it provides a good illustration of the effect of the method. The variation of the value of the estimated time delay depending on the distance between the agents is greater and thus more illustrative for the interpretation. For a larger bandwidth the results of the analysis still hold, but the transmission times are generally reduced and the variation of the values of the estimated time delay for different distances $d(\tilde{t}_i)$ is smaller. Furthermore, the transmission power P_0 influences the results, because with a higher transmission power data can be sent with a lower packet loss probability.

Table 9.4: Parameters of the simulation analysis.

Description	Value
Transmission power level P_0	2 W
Reference distance d_0	1 m
Path loss exponent α	2
Packet size M	128 bytes
Bandwidth B	35 kbps
Rician factor $\mathcal{K}(d(\tilde{t}_i))$	10
Receiver sensitivity S_{rs}	−40 dBm = 100 nW
Bound on the mean packet loss probability $p_{\text{e,max}}$	0.03

A Worst Case Execution Time (WCET) analysis of the runtimes of the control algorithms of the agents gives that the computation times are bounded by

$$\tau_{\text{c,max}} = 33\,\text{ms}.$$

A data transmission under the condition that no packets get lost causes large time delays, because the data rate with which the information is transmitted may not exceed the capacity of the channel. In this case signals are received with a power $S(d(\tilde{t}_i))$ that equals the receiver sensitivity $S(d(\tilde{t}_i)) = S_{\text{rs}}$. In order to reduce the conservatism of the estimate, which is overestimating the actual time delay, it is accepted that communication takes place with the mean packet loss probability $p_{\text{e,max}}$ from Assumption 8.1 on p. 175. In this case the signals are received with a power $S(d(\tilde{t}_i)) = S_{\min}(d(\tilde{t}_i))$, where $S_{\min}(d(\tilde{t}_i))$ is derived in (9.8) on p. 186. This fact results in smaller values for the estimate $\tilde{\tau}_{\max}(d(\tilde{t}_i))$ but might also cause the unnecessary invocation of event e_{G3}, because a packet is not lost but just received after the time span $\tilde{\tau}_{\max}(d(\tilde{t}_i))$. According to Section 8.1 this packet is considered to be lost, because the information is not useful any more. To reduce the frequency of communication the event thresholds in Section 9.4 are determined under the assumption that at an event time instant t_k the data are received. If the information are received later, the stand-on agent could approach the give-way agent unnoticed and a violation of control aim (G1) threatens.

In the following paragraphs the analysis is split in three parts. First, the results are stated when signals are received with a power $S(d(\tilde{t}_i)) = S_{rs}$. Second, the results are given when signals are received with a power $S(d(\tilde{t}_i)) = S_{min}(d(\tilde{t}_i))$ and packets get lost with probability $p_{e,max}$. Third, the results are compared and analysed.

Results for $S(d(\tilde{t}_i)) = S_{rs}$. The results of the simulation are stated in Tab. 9.5 for eight different distances $d(\tilde{t}_i)$ between the agents.

Table 9.5: Results of the simulation for $S(d(\tilde{t}_i)) = S_{rs}$.

$d(\tilde{t}_i)$/m	$\tau_{max}(d(\tilde{t}_i))$/ms	$p_e(d(\tilde{t}_i))$	$p_g(d(\tilde{t}_i))$
10	333	0.01	0
50	333	0.01	0
100	333	0.01	0
1000	333	0.01	0
1250	333	0.012	0.002
1500	333	0.018	0.008
1750	333	0.03	0.02
2000	333	0.06	0.05

The time delay $\tau_{max}(d(\tilde{t}_i))$ is determined with (9.11) using the values stated in Tab. 9.4. The mean packet loss probability $p_e(d(\tilde{t}_i))$ is obtained with (5.15), the mean packet loss probability of the good state $p_g(d(\tilde{t}_i))$ is determined with (5.14). It can be seen that the time delay $\tau_{max}(d(\tilde{t}_i))$ stays constant for all distances $d(\tilde{t}_i)$, because for the received signal power $S(d(\tilde{t}_i)) = S_{rs}$ holds. This causes due to eqn. (9.9) a constant data rate $R_{max}(d(\tilde{t}_i))$, which results with (9.10) in a constant transmission delay $\tau_{n,max}(d(\tilde{t}_i))$. In contrast, with an increasing distance $d(\tilde{t}_i)$ the packet loss probability of the state 'good' $p_g(d(\tilde{t}_i))$ increases and hence the mean packet loss probability $p_e(d(\tilde{t}_i))$ increases as well. At low distances $d(\tilde{t}_i) < 1000$ m no packet losses result from the state 'good' of the Markov model. Packets get only lost if the Markov model is in the state 'bad'. At the distance $d(\tilde{t}_i) = 1750$ m the mean packet loss probability equals the bound on the mean packet loss probability,

$(p_e(d(\tilde{t}_i)) - p_{e,max} = 0.03)$. Hence, for every distance $d(\tilde{t}_i) > 1750\,\text{m}$ the required mean packet loss probability cannot be satisfied any more, because the agents are moving out of range of their communication modules as it can be seen in the last row of Tab. 9.5.

The estimation of the time delay by assuming a communication where the signals are received with power $S(d(\tilde{t}_i)) = S_{rs}$ causes no unnecessary invocation of event e_{G3}. Every detected packet loss due to $\tau_k > \tau_{n,max}(d(\tilde{t}_i))$ is indeed a packet loss. If data are not received within the time span $\tau_{n,max}(d(\tilde{t}_i))$, its signal power $S(d(\tilde{t}_i))$ has to be below the receiver sensitivity so that no information transfer is possible. The drawback of this estimation technique is a large estimated time delay $\tau_{n,max}(d(\tilde{t}_i))$ compared with the estimated time delay that uses the assumption that signals are received with a higher signal power as stated in the next paragraph. A large time delay $\tau_{max}(d(\tilde{t}_i))$ causes larger thresholds $\bar{e}_{2d}(d(t_{k+1}))$ and $\bar{e}_d(d(t_{k+j}))$ so that communication is invoked more frequently.

Results for $S(d(\tilde{t}_i)) = S_{min}(d(\tilde{t}_i)) > S_{rs}$**.** To overcome the drawback of large estimated time delays it is accepted that communication takes place with the mean packet loss probability $p_{e,max}$. This results in shorter time delays $\tau_{max}(d(\tilde{t}_i))$, but might also cause the invocation of event e_{G3} unnecessarily, because the packet is not lost but it is just received after the time span $\tau_{max}(d(\tilde{t}_i))$. The results of the simulation are given in Tab. 9.6 for six different distances $d(\tilde{t}_i)$ between the agents.

Table 9.6: Results of the simulation for $S(d(\tilde{t}_i)) = S_{min}(d(\tilde{t}_i)) > S_{rs}$.

$d(\tilde{t}_i)$ /m	S_{min} /nW	$\tau_{max}(d(\tilde{t}_i))$ /ms	$p_e(d(\tilde{t}_i))$ $= p_{e,max}$	$p_g(d(\tilde{t}_i))$ $= p_{g,max}$	$\tilde{\tau}_{max}(d(\tilde{t}_i))$ /ms	$\tilde{p}_e(d(\tilde{t}_i))$	$\tilde{p}_g(d(\tilde{t}_i))$
10	$3 \cdot 10^6$	218	0.03	0.02	227	0.01	0
50	$1.2 \cdot 10^5$	241	0.03	0.02	251	0.01	0
100	$3 \cdot 10^4$	253	0.03	0.02	264	0.01	0
1000	300	311	0.03	0.02	325	0.012	0.002
1250	190	319	0.03	0.02	333	0.012	0.002
1500	130	325	0.03	0.02	340	0.012	0.002

The required packet loss probability of the state 'good' $p_{g,max}$ is determined with (9.6) using $p_{e,max}$ given in Tab. 9.4. With $p_{g,max}$ and eqn. (9.7) the signal power $S_{min}(d(\tilde{t}_i))$ is obtained iteratively with the method described on p. 186. When signals are communicated with this power, they are received by the give-way agent with the mean packet loss probability $p_e(d(\tilde{t}_i)) = p_{e,max}$. It can be seen in the third column of Tab. 9.6 that the estimated time delay $\tau_{max}(d(\tilde{t}_i))$ is significantly smaller compared to the estimate using the receiver sensitivity S_{rs} (column 2 of Tab. 9.5). It increases with an increasing distance between the agents. The reason for this is that a larger received signal power $S(d(\tilde{t}_i))$ results in a higher maximum data rate $R_{max}(d(\tilde{t}_i))$ (9.9) that is supported by the channel with probability $1 - p_{e,max}$. With eqn. (9.10) a lower transmission delay $\tau_{n,max}(d(\tilde{t}_i))$ results. Hence, with this estimation technique communication is invoked less frequently.

In contrast data are communicated with a higher packet loss probability than possible compared to the preceding paragraph (third column in Tab. 9.5 to fourth column in Tab. 9.6). This means that data are considered to be lost even though the packet is just received too late by the give-way agent. The packet loss event e_{G3} is invoked unnecessarily at about 2 % of the communication time instants as it can be seen by comparing the packet loss probability $p_g(d(\tilde{t}_i))$ of the state 'good' of the fourth column in Tab. 9.5 with the probability $p_g(d(\tilde{t}_i))$ of the fifth column in Tab. 9.6. Several simulations with each 10000 communication time instants with different distances between the agents have shown that the effective packet loss probability $p_e(d(\tilde{t}_i))$ varies between 1.9 % and 4.2 %. Hence, the variance of the packet loss probability is quite low.

In order to reduce the packet loss probability $p_e(d(\tilde{t}_i))$ but not to have to consider such long time delays as in the preceding paragraph, eqn. (9.13) is used to adjust the estimated time delay. The adjustment factor is chosen to be

$$\delta_e = 1.05,$$

which means that the estimated time delay is increased by 5 %. The results of the adjustment are shown in the three right-hand columns of Tab. 9.6. It can be seen that the packet loss probability $\tilde{p}_e(d(\tilde{t}_i))$ is reduced significantly and equals the packet loss probability $p_e(d(\tilde{t}_i))$ derived in the preceding paragraph. The estimated time delay is still significantly smaller compared to the preceding paragraph. With larger distances between the agents the packet loss probability increases slightly. At a distance of $d(\tilde{t}_i) = 1500\,\mathrm{m}$, $\tilde{p}_e(d(\tilde{t}_i))$ is still smaller compared to $p_{e,max}$, but the time delay $\tilde{\tau}_{max}(d(\tilde{t}_i))$ is larger than $\tau_{max}(d(\tilde{t}_i))$ derived in the preceding paragraph. Hence, the unnecessary invocation of e_{G3} could be nearly completely prevented. Again several simulations with each 10000 communication time instants were made, where only a few packets got lost.

Evaluation of the results. The results of the preceding paragraphs show that the estimation method for the time delay $\tilde{\tau}_{max}(d(\tilde{t}_i))$ is suitable to reduce the unnecessary invocation of event e_{G3}. The estimate is still less conservative compared to the estimate, which considers that no packets get lost. An adjustment of the time delay $\tau_{max}(d(\tilde{t}_i))$ by only 5 % leads to the fact that packets, which are not lost but just received late by the give-way agent can still be utilised for the event generation.

9.7 Extended communication method

The communication flow of the extended event-based control method is shown in the flow chart in Fig. 9.5. The time instants when communication is invoked are written at the top of an execution step. The time delays are indicated at the arrows. Compared to the communication flow described in Section 7.5.1 the sending of the ACK messages and the time delays are considered. The communication is still split into an initialisation phase and an execution phase. Initially at time instant $t = 0$ the give-way agent sends a request $\boldsymbol{R}_{\mathrm{G}}(0)$ to the stand-on agent to obtain its initial data including the initial trajectory. The give-way agent confirms the receipt of the data by sending an ACK message.

In the execution phase, communication is as follows:

- The give-way agent determines based on the initial information the future communication time instants $t_{\mathrm{c},k+j}$, $(j = 1, ..., N)$ at which communication must be invoked and sends them with a request $\boldsymbol{R}_{\mathrm{G}}(t_{\mathrm{c},k})$ (9.23) to the stand-on agent at the communication time instant $t_{\mathrm{c},k}$. If it does not respond within the time span $2\,\tilde{\tau}_{\max}(d(t_k))$, the request is considered to be lost and is sent again.

- After the reception of a request the stand-on agent sends its current data $\boldsymbol{S}_{\mathrm{S}}(t_{\mathrm{c},k})$ automatically at the time instants $t_{\mathrm{c},k+j}$ included in the request and waits for an ACK message from the give-way agent.

- The give-way agent waits the time delay $\tilde{\tau}_{\max}(d(t_{k+j}))$ after the time instants $t_{\mathrm{c},k+j}$ for the information of the stand-on agent. If it does not receive the data within this time span, the data is considered to be lost. If the agents approach one another the current estimated time delay $\tilde{\tau}_{\max}(d(\tilde{t}_i))$ is smaller compared to the time delay $\tilde{\tau}_{\max}(d(t_{k+j}))$, which is determined for the future event time instants t_{k+j}, $(\tilde{\tau}_{\max}(d(\tilde{t}_i)) < \tilde{\tau}_{\max}(d(t_{k+j})))$. In this case the give-way agent does not wait the time span $\tilde{\tau}_{\max}(d(t_{k+j}))$ for the information of the stand-on agent, but only the time span $\tilde{\tau}_{\max}(d(\tilde{t}_i))$. After this shorter time span the data is considered to be lost and the give-way agent sends a request $\boldsymbol{R}_{\mathrm{G}}(t_{\mathrm{c},k})$ to receive the data from the stand-on agent earlier.

- When the give-way agent does not send an ACK message, the stand-on agent sends its information again after the time span $2\,\tilde{\tau}_{\max}(d(t_{k+j}))$ after the communication time instant $t_{\mathrm{c},k+j}$. If it receives the request $\boldsymbol{R}_{\mathrm{G}}(t_{\mathrm{c},k})$, it sends its data directly.

- After reception of the data at a time instant $t_{\mathrm{r},k}$ the give-way agent checks whether the stand-on agent is on its communicated trajectory as $\boldsymbol{p}_{\mathrm{S}}(t_{\mathrm{r},k}) = \boldsymbol{w}_{\mathrm{S}}(t_{\mathrm{r},k})$ and if it did not change its future trajectory. If both is the case, the give-way agent sends an ACK message and waits for the information of the stand-on agent at the next event time instant t_{k+1}. On the other hand, if the give-way agent detects that the stand-on agent deviated from its trajectory as $\boldsymbol{p}_{\mathrm{S}}(t_{\mathrm{r},k}) \neq \boldsymbol{w}_{\mathrm{S}}(t_{\mathrm{r},k})$ or it changed its future trajectory

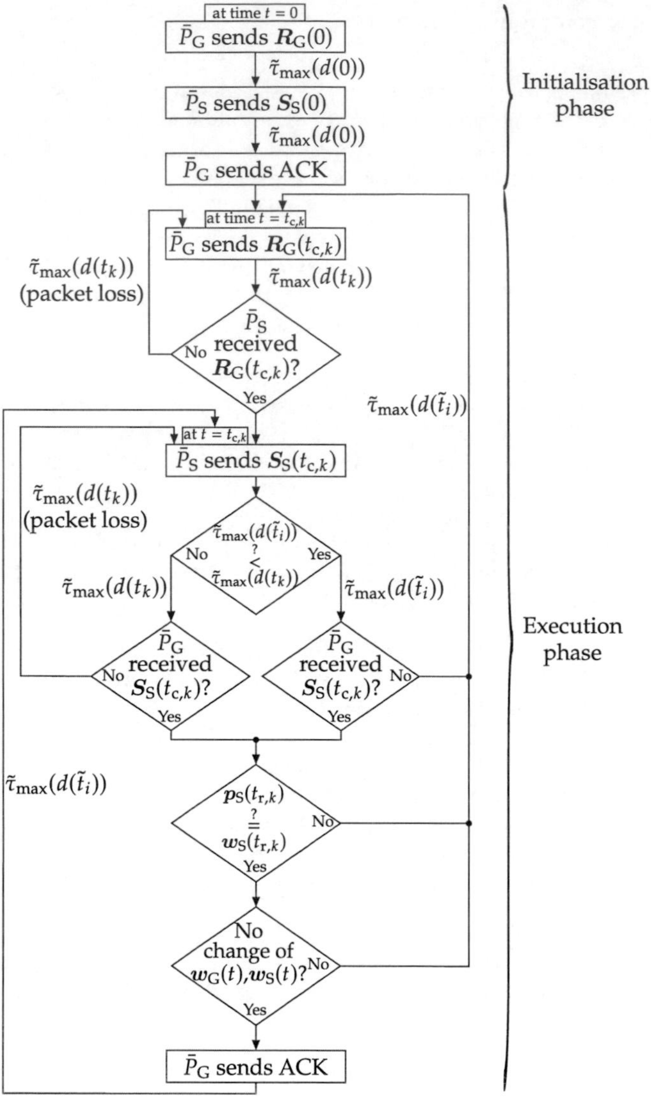

Fig. 9.5: Communication flow of the extended event-based control method.

$\boldsymbol{w}_\mathrm{S}(t), (t > t_{\mathrm{r},k})$ or the give-way agent changed its future trajectory $\boldsymbol{w}_\mathrm{G}(t), (t > t_{\mathrm{r},k})$, the give-way agent determines new future communication time instants $t_{\mathrm{c},k+j}$ and sends them as a request $\boldsymbol{R}_\mathrm{G}(t_{\mathrm{c},k+1})$ to the stand-on agent.

- The stand-on agent responds by sending its data $\boldsymbol{S}_\mathrm{S}(t_{\mathrm{c},k+1})$ at the next communication time instant.

Communication frequency for a general movement. In the presence of time delays the communication frequency is increased due to the changed communication condition (9.24) compared to the delay-free communication. The minimum time span $t_{\mathrm{min,com}}$ between two events e_{G0} is stated in Theorem 9.3. It is an important result, because it states that even in the presence of time delays and packet losses the Zeno behaviour is excluded.

Theorem 9.3 (Minimum time span between two events e_{G0} using an unreliable network). *Consider a stand-on agent and a give-way agent that satisfy Assumptions* 1.1, 1.2, 1.4 *on p.* 17 *and* 8.2 *on p.* 176 *and that are connected over a communication network fulfilling Assumption* 8.1 *on p.* 175, *which induces transmission delays and packet losses. The minimum time span between two consecutive communication events e_{G0} is bounded from below by*

$$t_{\mathrm{min,com}} \geq \min\left(t_{\mathrm{sep}}, t_{\mathrm{com}}\right)$$

where

$$t_{\mathrm{sep}} \geq \frac{\bar{e}_\mathrm{G} - \bar{e}_{\mathrm{2d}}(d(t_{k+1}))}{v_{\mathrm{SG,max}}} \tag{9.43}$$

and t_{com} is given by (7.33).

Proof. The proof is similar to the proof of Theorem 7.3 on p. 165. The difference results from the changed event condition (9.24). Due to this condition communication is invoked if the agents have covered the relative distance $\bar{e}_\mathrm{G} - \bar{e}_{\mathrm{2d}}(d(t_{k+1}))$ to one another after the last event time instant. Hence, the minimum time span between two consecutive events e_{G0} is given by (9.43), which proves the theorem. ∎

Communication frequency for a circular movement. For a circular movement of the agents the communication frequency is also increased in the presence of time delays compared to the delay-free communication. This results from the changed communication condition (9.33). The minimum time span $t_{\mathrm{c,min,com}}$ between two events e_{G0} is given in Theorem 9.4. Similar to the result for a general movement, the theorem states that Zeno behaviour is excluded also for a circular movement in the presence of time delays and packet losses.

Theorem 9.4 (Minimum time span between two events e_{G0} using an unreliable network for a circular movement). *Consider a stand-on agent and a give-way agent that satisfy Assumptions* 1.1, 1.2, 1.4 *on p.* 17 *and* 8.2 *on p.* 176 *and move on circular trajectories. The agents communicate over a network that fulfils Assumption* 8.1 *on p.* 175, *which induces transmission delays and packet losses. The minimum time span between two consecutive communication events e_{G0} is bounded from below by*

$$t_{c,min,com} \geq \min\left(t_{c,sep}, t_{com}\right)$$

where $t_{c,sep}$ is given by

$$t_{c,sep} = \min\left(t_{\Phi,sep}, t_{z,sep}\right) \tag{9.44}$$

with

$$t_{\Phi,sep} \geq \frac{(\bar{e}_{\Phi} - \bar{e}_{2dc}(d(t_{k+1})))\, 2\, \pi\, r}{v_{SG,max}}, \quad t_{z,sep} \geq \frac{\bar{e}_{z} - \bar{e}_{2dz}(d(t_{k+1}))}{v_{SG,z,max}}.$$

and t_{com} is specified in (7.35).

Proof. The proof is similar to the proof of Theorem 7.4 on p. 168. The difference results from the changed event condition (9.33). Due to this condition communication is invoked if the agents have covered the relative phase $\bar{e}_{\Phi} - \bar{e}_{2dc}(d(t_{k+1}))$ or the relative height $\bar{e}_{z} - \bar{e}_{2dz}(d(t_{k+1}))$ to one another after the last event time instant. Hence, the minimum time span between two consecutive events e_{G0} is given by (9.44), which proves the theorem. ∎

9.8 Algorithms for the event-based control of mobile agents for an unreliable network

General movement of the agents. Algorithm 14 on p. 168 summarises the event-based control method for a general movement of the agents and the use of an ideal communication network. This algorithm can be used with small modifications when the agents are connected by a network that induces time delays and packet losses. The execution of the algorithm guarantees requirement (1.1) of (G1) and satisfies requirement (1.2) of (G1) with a probability defined in (9.46). The modifications are as follows:

- Replace conditions (7.4) and (7.15) by (9.24) and (9.29).
- Add condition (9.32), which causes the execution of Algorithm 15 on p. 189.
- Add condition (9.30), which causes the planning of an avoidance trajectory at a time instant given by (9.31) with the start condition (7.23) and the end condition (9.39) and the transition time interval (9.42).

The result of the extended event-based control method for a general movement of the agents is stated in Theorem 9.5.

Theorem 9.5 (Event-based control of agents over an unreliable network). *Consider a stand-on agent and a give-way agent fulfilling Assumptions* 1.1, 1.2, 1.4 *on p.* 17 *and* 8.2 *on p.* 176*, which are connected by a communication network satisfying Assumption* 8.1 *on p.* 175 *that induces transmission delays and packet losses. Furthermore, at time* $t = 0$ *the distance between the agents fulfils the inequalities*

$$\underline{s} + 2\,\bar{e}_{\mathrm{G}} < s(0) < \bar{s} - 2\,\bar{e}_{\mathrm{G}}, \tag{9.45}$$

for a given bound on the safety distance $\underline{s}$ *and the maximum separation* $\bar{s}$*. Requirement* (1.1) *of control aim* (A1) *is ensured and requirement* (1.2) *of control aim* (A1) *is fulfilled with probability*

$$p_{\mathrm{A1}} \geq 1 - p_{\mathrm{e,max}} \tag{9.46}$$

if the following holds

- *The give-way agent uses the sets* $\mathcal{P}_{\mathrm{S}}(t, t_{\mathrm{r},k}, \tilde{\tau}_{\max}(d(t_k)))$ *and* $\mathcal{P}_{\mathrm{S,d}}(t_{\mathrm{c},k}, t_{\mathrm{c},k}, \tilde{\tau}_{\max}(d(t_k)))$ *which are determined with* (9.15) *to estimate the future movement of the stand-on agent.*
- *The give-way agent generates events with conditions* (9.24), (7.11), (9.29), (9.30) *and* (9.32) *with the threshold* $\bar{e}_{\mathrm{G}}$ *given by* (7.7) *and the delay thresholds* $\bar{e}_{\mathrm{2d}}(d(t_{k+1}))$, $\bar{e}_{\mathrm{d}}(d(t_{k+j}))$ *given by* (9.27) *and* (9.28).
- *The stand-on agent sends its data* $\boldsymbol{S}_{\mathrm{S}}(t_{\mathrm{c},k})$ *at any communication time instant* $t_{\mathrm{c},k+j}$, $(j = 0, \ldots, N)$ *or after receiving a request* $\boldsymbol{R}_{\mathrm{G}}(t_{\mathrm{c},k})$ (9.23).
- *The stand-on agent and the give-way agent plan trajectories using Bézier curves.*
- *The stand-on agent and the give-way agent execute Algorithm* 14 *on p.* 168 *with the modifications stated above.*

Proof. See Appendix B.14. ∎

The theorem states the main result of the chapter. The important point is that the execution of Algorithm 14 with the modifications stated above by the agents leads to a guaranteed collision avoidance even in the presence of time delays and packet losses. Due to the packet losses, the maximum separation between the agents can only be satisfied with the probability given by (9.46).

Circular movement of the agents. Algorithm 14 can further be used with small modifications to guarantee control aim (G1) for a circular movement in the presence of time delays.

In the presence of packet losses it is not possible to guarantee the control aim, but it can only be satisfied with the probability given by (9.48). The modifications are the following:

- Replace Algorithm 11 on p. 88 by Algorithm 12 on p. 92.
- Replace Algorithm 3 on p. 57 by Algorithm 5 on p. 68 and Algorithm 10 on p. 80.
- Replace conditions (7.4), (7.11) and (7.15) by (9.33), (7.21) and (9.38).
- Add condition (9.32), which causes the execution of Algorithm 15 on p. 189.
- Use the event thresholds $\bar{e}_\Phi$ and $\bar{e}_z$ (7.19) instead of $\bar{e}_G$ (7.7).

The result of the extended event-based control method for a circular movement of the agents is stated in Corollary 9.1.

Corollary 9.1 (Event-based control of agents over an unreliable network for a circular movement). *Consider a stand-on agent and a give-way agent fulfilling Assumptions* 1.1, 1.2, 1.4 *on p.* 17 *and* 8.2 *on p.* 176 *that are moving on circular trajectories. The agents are connected by a communication network satisfying Assumption* 8.1 *on p.* 175, *which induces transmission delays and packet losses. Furthermore, at time* $t = 0$ *the phase difference and the height difference between the agents fulfils the inequalities*

$$\begin{aligned} \underline{s} + 2\,\bar{e}_\Phi < \Phi_{\mathrm{diff}}(0) < \bar{s} - 2\,\bar{e}_\Phi \\ \bar{s} - 2\,\bar{e}_z < z_{\mathrm{diff}}(0) < \bar{z} \end{aligned} \tag{9.47}$$

for given bounds $\underline{s}$ *and* $\bar{s}$ *on the phase difference and a desired height difference* $\bar{z}$. *Control aim (A1) for a circular movement of the agents is fulfilled with probability*

$$p_{\mathrm{A1,c}} \geq 1 - p_{\mathrm{e,max}} \tag{9.48}$$

if the following holds

- *The give-way agent uses the sets* $\mathcal{P}_{\mathrm{S,c}}(t, t_{\mathrm{r},k}, \tilde{\tau}_{\max}(d(t_k)))$ *and* $\mathcal{P}_{\mathrm{S,d,c}}(t_{\mathrm{c},k}, t_{\mathrm{c},k}, \tilde{\tau}_{\max}(d(t_k)))$ *which are determined with* (9.17) *to estimate the future phases and heights of the stand-on agent.*
- *The give-way agent generates events with conditions* (9.33), (7.21), (9.38) *and* (9.32). *It uses the thresholds* $\bar{e}_\Phi$ *and* $\bar{e}_z$ *given by* (7.19) *and the delay thresholds* $\bar{e}_{\mathrm{2dc}}(d(t_{k+1}))$, $\bar{e}_{\mathrm{2dz}}(d(t_{k+1}))$, $\bar{e}_{\mathrm{dc}}(d(t_{k+j}))$ *and* $\bar{e}_{\mathrm{dz}}(d(t_{k+j}))$ *given by* (9.34), (9.35), (9.36) *and* (9.37).
- *The stand-on agent sends its data* $\boldsymbol{S}_{\mathrm{S}}(t_{\mathrm{c},k})$ *at any communication time instant* $t_{\mathrm{c},k+j}$, $(j = 0, \ldots, N)$ *or after receiving a request* $\boldsymbol{R}_{\mathrm{G}}(t_{\mathrm{c},k})$ (9.23).
- *The stand-on agent and the give-way agent plan trajectories using Bézier curves.*
- *The stand-on agent and the give-way agent execute Algorithm* 14 *on p.* 168 *with the modifications stated above.*

Proof. The proof is similar to the proof of Theorem 9.5 where the control aim is initially satisfied by fulfilling Assumptions 1.2, 1.4 and 8.1 and eqn. (9.47). The difference is that the method does not react on packet losses and control aim (A1) for a circular movement can only be satisfied with probability (9.48) according to the last part of the proof of Theorem 9.5. The proof is composed of the proofs of the parts of the extended control method. ∎

The corollary states that the event-based control method is able to fulfil control aim (A1) for a circular movement with the probability given by (9.48). Due to the packet losses it is not possible any more to ensure the satisfaction of the control aim.

Part V

Simulation studies and Experiments

Demonstration example: Cooperative quadrotors

10

The quadrotor as a demonstration example is introduced in this chapter. Section 10.1 presents the hardware of the used quadrotors. Their dynamics are stated in Section 10.2, where a model of the UAV is derived. Section 10.3 presents the local control loop of the quadrotor, which is a flatness-based two-degrees-of-freedom controller that stabilises the system and makes it follow the trajectories. In Section 10.4 the event-based control loop of two quadrotors is given.

10.1 Hardware of the quadrotors

Motivation. In this thesis the event-based control method is applied to two quadrotors as they are well suited for testing control engineering concepts due to their complex dynamics. Based on their broad movement possibilities (e.g. vertical take-offs and hovering flights), these UAVs are suitable for a wide range of applications in which the cooperation of two agents can be tested. As the quadrotor dynamics are less restricted compared to other agents (e.g. cars), a method that is suitable for the application to quadrotors can also be applied to other agents.

Structure of the quadrotor. The quadrotors are constructed using commercial off-the-shelf components. Two quadrotors of the type 'Crazyflie 2.1' from the company 'Bitcraze AB' are used, which is shown in Fig. 10.1. The quadrotors consist of a symmetrical frame with four brush motors mounted on the outriggers at a diagonal distance of 6.5 cm, which propulsive powers act downwards. The motors are controlled by PWM signals. In addition, a self-developed rotor guard made by a 3D printer is mounted on which infrared marker balls are attached. With these balls the quadrotors can be tracked by a camera system at the experimental test bed as described in Section 12.2.

A lithium-polymer (LiPo) accumulator with a capacity of 250 mAh and a maximum charging voltage of 4.2 V is used to power the motors. The quadrotor together with the accumulator and the rotor guard has a mass of 32 g. All parameters of the quadrotor are summarised in Tab. 10.1.

The system architecture on board the UAV is shown in Fig. 10.2. The quadrotor possesses two microcontrollers and an inertial measurement unit (IMU) that are stated in detail

Fig. 10.1: Quadrotor of type 'Crazyflie 2.1'.

in the next paragraphs. The components of the circuit board are connected by an I^2C interface.

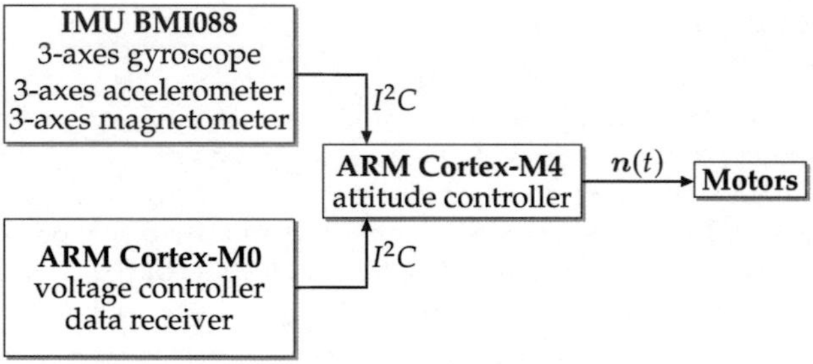

Fig. 10.2: Block diagram of the on-board system architecture.

Sensors of the quadrotor. An inertial measurement unit of type 'BMI088' is used to measure the attitude on board the quadrotors. It consists of the following three types of sensors: a 3-axes gyroscope, a 3-axes accelerometer and a 3-axes magnetometer.

The gyroscope reacts to rotational movements around the three body axes of the quadrotor. Three types of rotational movements exists: roll, pitch and yaw, which are illustrated in Fig. 10.3. Yaw is the rotation around the M_z-axis, while roll describes the rotation around the M_x-axis and pitch is the rotation around the M_y-axis. The gyroscopes measure the angular speeds around these three body axes so that the current attitude of the sensor in relation to its initial attitude can be determined.

In order to obtain the absolute attitude the accelerometers are used, which measure the accelerations acting on the sensor. With the measurements, the vectorial direction of gravity in the body-fixed coordinate system M can be determined. However, the measurements of the accelerometers are affected by a high measurement noise.

The magnetometers measure the magnetic flux density. They are used as a magnetic compass and utilise the magnetic field of the earth to determine the north direction to obtain the absolute attitude for the yaw axis.

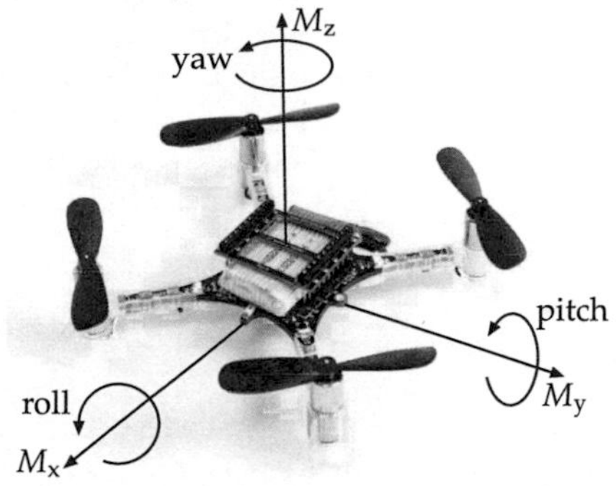

Fig. 10.3: Rotational axes of the quadrotor.

Microcontrollers. The quadrotor possesses two microcontrollers. The controller of type 'ARM Cortex-M0 nRF51822' has two tasks. First, it ensures the voltage supply of the circuit board by the LiPo accumulator. Second, it receives commands from a ground station sent over a 'Crazyradio PA' antenna introduced in Section 12.2. The controller processes the received data and forwards them to the main 32-bit microcontroller of type 'ARM Cortex-M4 STM32F405' on which the firmware of the Crazyflie runs. This controller receives the data from the IMU and executes the attitude controller, which generates PWM signals. The microcontroller contains electronic speed controllers (ESC), which control the rotor speeds of the motors using the PWM signals.

10.2 Quadrotor dynamics

10.2.1 *Nonlinear model*

In this section a model for the quadrotor is derived. The dynamics of a multirotor UAV are described by considering the UAV in two different coordinate systems, as shown in Fig. 10.4 (left). First, it is a three dimensional earth-fixed coordinate frame I and second, it is a three dimensional body-fixed coordinate frame M. Hence, the UAV can be modelled by a translational and a rotational subsystem as depicted in Fig. 10.5. The rotational subsystem models the attitude of the UAV, which is the rotation of the body-fixed frame M against the earth-fixed frame I given by the Euler angles $\phi(t)$, $\vartheta(t)$, $\psi(t)$ (Fig. 10.4 (right)). The description of the attitude is based on the DIN 9300 aerospace standard with the yaw angle $\psi(t) \in [-\pi, \pi[$, the pitch angle $\vartheta(t) \in]-\frac{\pi}{2}, \frac{\pi}{2}[$ and the roll angle $\phi(t) \in [-\pi, \pi[$. The rotation angles are defined in the body-fixed frame M, which means that changing the rotation sequence causes a change of the determined attitude in the earth-fixed frame I. According to DIN 9300 the rotation follows the yaw-pitch-roll order. The attitude results from torques that are generated by differences in the rotor speeds $n_i(t)$, $(i = 1, \ldots, 4)$.

The translational subsystem describes the $(x(t), y(t))$ position and the altitude $z(t)$ of the centre of gravity (COG) of the UAV in I. It is assumed that the COG is in the origin of M and the quadrotor is modelled as a point mass. The position and the altitude are the result of the attitude of the quadrotor and the thrust

$$F_z(t) = 4\pi^2 c_{\mathrm{th}} \left(\hat{u}_1(t) + \hat{u}_2(t) + \hat{u}_3(t) + \hat{u}_4(t)\right) \tag{10.1}$$

with $\hat{u}_i(t) = n_i(t)$, which acts along the M_z axis of the body-fixed frame. The effectiveness of the rotors is represented by the thrust constant c_{th} in (10.1).

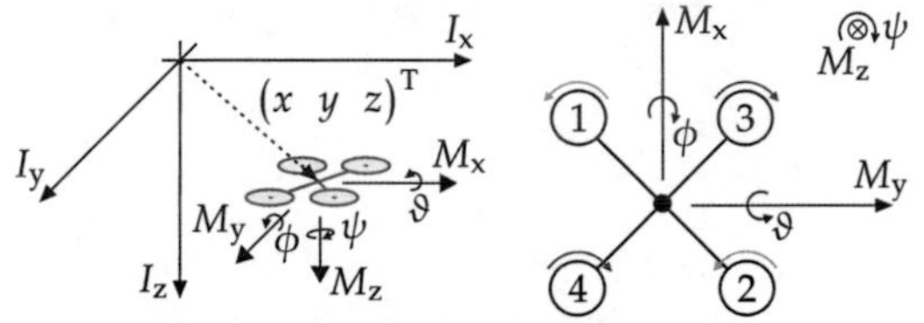

Fig. 10.4: Coordinates of the quadrotor.

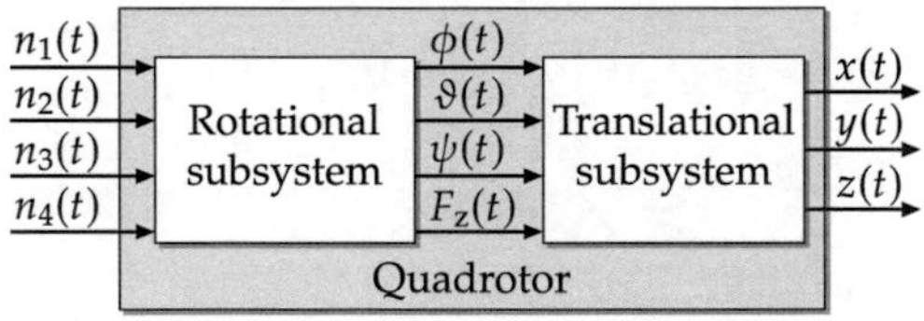

Fig. 10.5: Block diagram of the quadrotor.

Rotational subsystem. With the state

$$\boldsymbol{x}_{\mathrm{r}}(t) = \begin{pmatrix} \dot{\phi}(t) & \dot{\vartheta}(t) & \dot{\psi}(t) & \phi(t) & \vartheta(t) & \psi(t) \end{pmatrix}^{\mathrm{T}}$$

containing the Euler angles and the angular speeds $\dot{\phi}(t)$, $\dot{\vartheta}(t)$, $\dot{\psi}(t)$ the nonlinear state space model is given by

$$\dot{\boldsymbol{x}}_{\mathrm{r}}(t) = \boldsymbol{f}_{\mathrm{r}}(\boldsymbol{x}_{\mathrm{r}}(t), \hat{\boldsymbol{u}}(t)) = \begin{pmatrix} \frac{(J_y - J_z)\dot{\vartheta}\dot{\psi}}{J_x} + \frac{4\pi^2 c_{\mathrm{th}} \alpha l(-\hat{u}_1+\hat{u}_2+\hat{u}_3-\hat{u}_4)}{J_x} - \frac{J_r \Omega \dot{\vartheta}}{J_x} - \frac{c_\phi \dot{\phi}}{J_x} \\ \frac{(J_z - J_x)\dot{\phi}\dot{\psi}}{J_y} + \frac{4\pi^2 c_{\mathrm{th}} \alpha l(\hat{u}_1+\hat{u}_2-\hat{u}_3-\hat{u}_4)}{J_y} - \frac{J_r \Omega \dot{\phi}}{J_y} - \frac{c_\vartheta \dot{\vartheta}}{J_y} \\ \frac{(J_x - J_y)\dot{\phi}\dot{\vartheta}}{J_z} + \frac{4\pi^2 c_{\mathrm{th}} c_{\mathrm{yaw}}(\hat{u}_1-\hat{u}_2+\hat{u}_3-\hat{u}_4)}{J_z} - \frac{c_\psi \dot{\psi}}{J_z} \\ \dot{\phi} \\ \dot{\vartheta} \\ \dot{\psi} \end{pmatrix} \tag{10.2}$$

$$\boldsymbol{y}_{\mathrm{r}}(t) = (\boldsymbol{0}_{3\times 3}\ \boldsymbol{I}_3)\,\boldsymbol{x}_{\mathrm{r}}(t), \qquad \boldsymbol{x}_{\mathrm{r}}(0) = \boldsymbol{x}_{\mathrm{r}0}$$

with $\Omega(t) = 2\pi(\hat{u}_1(t) + \hat{u}_3(t) - \hat{u}_2(t) - \hat{u}_4(t))$ and $\alpha = \sin(45°)$. The dependencies on the time t are omitted for simplicity. The input

$$\hat{\boldsymbol{u}}(t) = \begin{pmatrix} n_1(t) & n_2(t) & n_3(t) & n_4(t) \end{pmatrix}^{\mathrm{T}}$$

refers to the rotor speeds. The constants are given in Tab. 10.1.

The first three rows of eqn. (10.2) state the angular accelerations of the quadrotor. These accelerations result of torques, which are composed of the precession (first term), the rotor moment (second term), the propeller gyro effect for $\ddot{\phi}(t)$ and $\ddot{\vartheta}(t)$ (third term) and the drag moment (last term). The schematic view of the UAV is depicted in Fig. 10.4 (right) and illustrates that the torque for $\ddot{\phi}(t)$ and $\ddot{\vartheta}(t)$ is a result of a lever between the rotors 1 and 4 and the rotors 2 and 3 or between the rotors 1 and 3 and the rotors 2 and 4. The yaw moment $\ddot{\psi}(t)$ appears from an angular momentum caused by a speed difference of the clockwise (CW) and the counter clockwise (CCW) rotating rotors. The last three rows of eqn. (10.2) state the integration from acceleration to speed.

Table 10.1: Parameters of the quadrotor.

Constant	Description	Value
l	Distance COG / rotor	$32.5 \cdot 10^{-3}$ m
m	Mass	$37 \cdot 10^{-3}$ kg
g	Gravitation constant	$9.81\ \mathrm{m} \cdot \mathrm{s}^{-2}$
n_{op}	Rotor speed at the operating point	$317\ \mathrm{s}^{-1}$
$\underline{n}$	Minimum rotor speed	$0\ \mathrm{s}^{-1}$

Table 10.1: Parameters of the quadrotor. (continued)

Constant	Description	Value
$\bar{n}$	Maximum rotor speed	$398\ \mathrm{s}^{-1}$
$\underline{\phi}, \bar{\phi}, \underline{\vartheta}, \bar{\vartheta}$	Angle limitations	$\pm 60°$
T_{M}	Time constant of the motors	$30 \cdot 10^{-3}\ \mathrm{s}$
c_{th}	Thrust constant	$2.287 \cdot 10^{-8}\ \mathrm{kg} \cdot m$
c_{yaw}	Yaw constant	$1.885 \cdot 10^{-5} \mathrm{kg} \cdot \mathrm{m}^2$
$c_{\mathrm{x,y}}$	x, y drag coefficients	$12 \cdot 10^{-3}\ \mathrm{kg} \cdot s^{-1}$
c_{z}	z drag coefficient	$14 \cdot 10^{-3}\ \mathrm{kg} \cdot s^{-1}$
$c_{\phi,\vartheta,\psi}$	Angle drag coefficients	$1 \cdot 10^{-6}\ \mathrm{kg} \cdot m^2 \cdot s^{-1}$
J_{x}	x axis moment of inertia	$1.957 \cdot 10^{-5}\ \mathrm{kg} \cdot \mathrm{m}^2$
J_{y}	y axis moment of inertia	$1.938 \cdot 10^{-5}\ \mathrm{kg} \cdot \mathrm{m}^2$
J_{z}	z axis moment of inertia	$3.770 \cdot 10^{-5} \mathrm{kg} \cdot \mathrm{m}^2$
J_{r}	z rotor moment of inertia	$65 \cdot 10^{-6} \mathrm{kg} \cdot \mathrm{m}^2$

Translational subsystem. Using the state

$$\boldsymbol{x}_{\mathrm{t}}(t) = \begin{pmatrix} \dot{x}(t) & \dot{y}(t) & \dot{z}(t) & x(t) & y(t) & z(t) \end{pmatrix}^{\mathrm{T}},$$

which is composed of the $(x(t), y(t))$ position, the altitude $z(t)$ and the speeds $\dot{x}(t)$, $\dot{y}(t)$, $\dot{z}(t)$ along the axes the nonlinear model is given by

$$\begin{aligned} \dot{\boldsymbol{x}}_{\mathrm{t}}(t) =& \boldsymbol{f}_{\mathrm{t}}\left(\boldsymbol{x}_{\mathrm{t}}(t), \boldsymbol{x}_{\mathrm{r}}(t), \boldsymbol{u}_{\mathrm{t}}(t)\right) = \\ & \begin{pmatrix} -\frac{F_{\mathrm{z}}}{m}\left(\cos(\phi)\sin(\vartheta)\cos(\psi) + \sin(\phi)\sin(\psi)\right) - \frac{c_{\mathrm{x}}}{m}\dot{x} \\ -\frac{F_{\mathrm{z}}}{m}\left(\cos(\phi)\sin(\vartheta)\sin(\psi) - \sin(\phi)\cos(\psi)\right) - \frac{c_{\mathrm{y}}}{m}\dot{y} \\ g - \frac{F_{\mathrm{z}}}{m}\left(\cos(\phi)\cos(\vartheta)\right) - \frac{c_{\mathrm{z}}}{m}\dot{z} \\ \dot{x} \\ \dot{y} \\ \dot{z} \end{pmatrix} \\ \boldsymbol{y}_{\mathrm{t}}(t) =& \left(\boldsymbol{O}_{3\times 3}\ \boldsymbol{I}_3\right)\boldsymbol{x}_{\mathrm{t}}(t), \qquad \boldsymbol{x}_{\mathrm{t}}(0) = \boldsymbol{x}_{\mathrm{t}0}. \end{aligned} \tag{10.3}$$

The time dependencies are again omitted for simplicity. The input $\boldsymbol{u}_{\mathrm{t}}(t) = \begin{pmatrix} \boldsymbol{y}_{\mathrm{r}}(t) & F_z(t) \end{pmatrix}^{\mathrm{T}}$ is composed of the Euler angles (output of the rotational subsystem) and the thrust given by (10.1). The model (10.3) contains the gravity g in the third term to describe the altitude and some drag forces. The constants are described in Tab. 10.1. The combined model has the state vector

$$\boldsymbol{x}(t) = \begin{pmatrix} \boldsymbol{x}_{\mathrm{r}}^{\mathrm{T}}(t) & \boldsymbol{x}_{\mathrm{t}}^{\mathrm{T}}(t) \end{pmatrix}^{\mathrm{T}}. \tag{10.4}$$

10.2.2 *Partially linearised model*

The rotational subsystem is linearised, the translational subsystem is just simplified, which leads again to a nonlinear model. Due to the lower complexity this model will be used in Section 10.3.3 to derive the feedforward controller of the quadrotor. The hover flight of the quadrotor is characterised by

$$\begin{aligned} \boldsymbol{x}_{\mathrm{r}}(t) &= \begin{pmatrix} \boldsymbol{0}_{1\times 5} & \bar{\psi} \end{pmatrix}^{\mathrm{T}}, & \dot{\boldsymbol{x}}_{\mathrm{r}}(t) &= \boldsymbol{0}_{6\times 1} \\ \boldsymbol{x}_{\mathrm{t}}(t) &= \begin{pmatrix} \boldsymbol{0}_{1\times 3} & \bar{x} & \bar{y} & \bar{z} \end{pmatrix}^{\mathrm{T}}, & \dot{\boldsymbol{x}}_{\mathrm{t}}(t) &= \boldsymbol{0}_{6\times 1} \end{aligned} \tag{10.5}$$

for the states and their derivatives. The bar denotes constant operating points. Due to small angular velocities around (10.5) the effects of precession and friction and the propeller gyro effect are negligible. With the state $\boldsymbol{x}_{\mathrm{r}}(t)$ and the input $u_i(t) = n_i^2(t)$, $(i = 1, \ldots, 4)$ the nonlinear model of the rotational subsystem (10.2) can be linearised as

$$\begin{aligned} \dot{\boldsymbol{x}}_{\mathrm{r}}(t) &= \begin{pmatrix} \boldsymbol{0}_{3\times 3} & \boldsymbol{0}_{3\times 3} \\ \boldsymbol{I}_3 & \boldsymbol{0}_{3\times 3} \end{pmatrix} \boldsymbol{x}_{\mathrm{r}}(t) + \begin{pmatrix} \boldsymbol{I}_3 \\ \boldsymbol{0}_{3\times 3} \end{pmatrix} \boldsymbol{B}_{\mathrm{r}} \boldsymbol{u}(t) \\ \boldsymbol{y}_{\mathrm{r}}(t) &= \begin{pmatrix} \boldsymbol{0}_{3\times 3} & \boldsymbol{I}_3 \end{pmatrix} \boldsymbol{x}_{\mathrm{r}}(t), \quad \boldsymbol{x}_{\mathrm{r}}(0) = \boldsymbol{0}_{6\times 1} \end{aligned} \tag{10.6}$$

with

$$\boldsymbol{B}_{\mathrm{r}} = \begin{pmatrix} -\alpha b_{\phi} & \alpha b_{\phi} & \alpha b_{\phi} & -\alpha b_{\phi} \\ \alpha b_{\vartheta} & \alpha b_{\vartheta} & -\alpha b_{\vartheta} & -\alpha b_{\vartheta} \\ b_{\psi} & -b_{\psi} & b_{\psi} & -b_{\psi} \end{pmatrix}$$

and the coefficients

$$b_{\phi} = \frac{4\pi^2 c_{\mathrm{th}} l}{J_{\mathrm{x}}}, \quad b_{\vartheta} = \frac{4\pi^2 c_{\mathrm{th}} l}{J_{\mathrm{y}}}, \quad b_{\psi} = \frac{4\pi^2 c_{\mathrm{th}} c_{\mathrm{yaw}}}{J_{\mathrm{z}}}.$$

Supposing the quadrotor to be in the hover flight, small angles $\phi(t)$ and $\vartheta(t)$ can be assumed. A linearisation around these angles leads to the simplified nonlinear, translational model

$$\begin{aligned}\dot{\boldsymbol{x}}_{\mathrm{t}}(t) =& \boldsymbol{f}_{\mathrm{t}}\left(\boldsymbol{x}_{\mathrm{t}}(t), \boldsymbol{x}_{\mathrm{r}}(t), \boldsymbol{u}_{\mathrm{t}}(t)\right) = \\ & \begin{pmatrix} -\frac{F_{\mathrm{z}}(t)}{m}\left(\vartheta(t)\cos(\psi(t)) + \phi(t)\sin(\psi(t))\right) - \frac{c_{\mathrm{x}}}{m}\dot{x}(t) \\ -\frac{F_{\mathrm{z}}(t)}{m}\left(\vartheta(t)\sin(\psi(t)) - \phi(t)\cos(\psi(t))\right) - \frac{c_{\mathrm{y}}}{m}\dot{y}(t) \\ g - \frac{F_{\mathrm{z}}(t)}{m} - \frac{c_{\mathrm{z}}}{m}\dot{z}(t) \\ \dot{x} \\ \dot{y} \\ \dot{z} \end{pmatrix} \\ \boldsymbol{y}_{\mathrm{t}}(t) =& \left(\boldsymbol{0}_{3\times3}\ \boldsymbol{I}_3\right)\boldsymbol{x}_{\mathrm{t}}(t), \qquad \boldsymbol{x}_{\mathrm{t}}(0) = \boldsymbol{x}_{\mathrm{t0}}.\end{aligned} \tag{10.7}$$

10.2.3 *Affine model*

The nonlinear model (10.2), (10.3) is linearised for the feedback controller design in Section 10.3.4. Due to the effect of the input variables on the system dynamics the state of the rotational subsystem is extended by the altitude and its derivative. The resulting state

$$\boldsymbol{x}_{\mathrm{rz}}(t) = \begin{pmatrix}\dot{\phi}(t) & \dot{\vartheta}(t) & \dot{\psi}(t) & \dot{z}(t) & \phi(t) & \vartheta(t) & \psi(t) & z(t)\end{pmatrix}^{\mathrm{T}}$$

describes the attitude and the altitude of the quadrotor. Hence, the affine model is given by

$$\begin{aligned}\dot{\boldsymbol{x}}_{\mathrm{rz}}(t) &= \begin{pmatrix}\begin{pmatrix}\boldsymbol{0}_{3\times3} & \boldsymbol{0}_{3\times1} \\ \boldsymbol{0}_{1\times3} & -\frac{c_{\mathrm{z}}}{m}\end{pmatrix} & \boldsymbol{0}_{4\times4} \\ \boldsymbol{I}_4 & \boldsymbol{0}_{4\times4}\end{pmatrix}\boldsymbol{x}_{\mathrm{rz}}(t) + \begin{pmatrix}\boldsymbol{I}_4 \\ \boldsymbol{0}_{4\times4}\end{pmatrix}\tilde{\boldsymbol{B}}_{\mathrm{r}}\boldsymbol{u}(t) + \begin{pmatrix}\begin{pmatrix}\boldsymbol{0}_{3\times1} \\ g\end{pmatrix} \\ \boldsymbol{0}_{4\times1}\end{pmatrix} \\ \boldsymbol{y}_{\mathrm{rz}}(t) &= \begin{pmatrix}\boldsymbol{0}_{4\times4} & \boldsymbol{I}_4\end{pmatrix}\boldsymbol{x}_{\mathrm{rz}}(t), \quad \boldsymbol{x}_{\mathrm{rz}}(0) = \begin{pmatrix}\boldsymbol{0}_{1\times6} & \bar{\psi} & \bar{z}\end{pmatrix}^{\mathrm{T}}\end{aligned} \tag{10.8}$$

with

$$\tilde{\boldsymbol{B}}_{\mathrm{r}} = \begin{pmatrix} -\alpha b_{\phi} & \alpha b_{\phi} & \alpha b_{\phi} & -\alpha b_{\phi} \\ \alpha b_{\vartheta} & \alpha b_{\vartheta} & -\alpha b_{\vartheta} & -\alpha b_{\vartheta} \\ b_{\psi} & -b_{\psi} & b_{\psi} & -b_{\psi} \\ -b_{\mathrm{z}} & -b_{\mathrm{z}} & -b_{\mathrm{z}} & -b_{\mathrm{z}} \end{pmatrix}, \quad b_{\mathrm{z}} = \frac{4\pi^2 c_{\mathrm{th}}}{m}. \tag{10.9}$$

A linear model for the translational subsystem without the altitude results from a linearisation around the equilibrium (10.5) with $\bar{\psi} = 0$. With the state

$$\boldsymbol{x}_{\mathrm{xy}}(t) = \begin{pmatrix}\dot{x}(t) & \dot{y}(t) & x(t) & y(t)\end{pmatrix}^{\mathrm{T}}$$

the model is given by

$$\dot{\boldsymbol{x}}_{\mathrm{xy}}(t) = \begin{pmatrix} \begin{pmatrix} -\frac{c_{\mathrm{x}}}{m} & 0 \\ 0 & -\frac{c_{\mathrm{y}}}{m} \end{pmatrix} & \mathbf{0}_{2\times 2} \\ \boldsymbol{I}_2 & \mathbf{0}_{2\times 2} \end{pmatrix} \boldsymbol{x}_{\mathrm{xy}}(t) + \begin{pmatrix} \boldsymbol{I}_2 \\ \mathbf{0}_{2\times 2} \end{pmatrix} \begin{pmatrix} 0 & -g \\ g & 0 \end{pmatrix} \begin{pmatrix} \phi(t) \\ \vartheta(t) \end{pmatrix}$$
$$\boldsymbol{y}_{\mathrm{xy}}(t) = \begin{pmatrix} \mathbf{0}_{2\times 2} & \boldsymbol{I}_2 \end{pmatrix} \boldsymbol{x}_{\mathrm{xy}}(t), \quad \boldsymbol{x}_{\mathrm{xy}}(0) = \begin{pmatrix} 0 & 0 & \bar{x} & \bar{y} \end{pmatrix}^{\mathrm{T}}. \tag{10.10}$$

In summary, the overall affine model of the quadrotor with the state (10.4) is given by the state-space model

$$\begin{aligned} \dot{\boldsymbol{x}}(t) &= \boldsymbol{A}\boldsymbol{x}(t) + \boldsymbol{B}\boldsymbol{u}(t) + \boldsymbol{g}, \quad \boldsymbol{x}(0) = \boldsymbol{x}_0, \\ \boldsymbol{y}(t) &= \boldsymbol{C}\boldsymbol{x}(t), \end{aligned} \tag{10.11}$$

with

$$\boldsymbol{A} = \begin{pmatrix} \mathbf{0}_{3\times 3} & \mathbf{0}_{3\times 3} & \mathbf{0}_{3\times 3} & \mathbf{0}_{3\times 3} \\ \boldsymbol{I}_3 & \mathbf{0}_{3\times 3} & \mathbf{0}_{3\times 3} & \mathbf{0}_{3\times 3} \\ \mathbf{0}_{3\times 3} & \begin{pmatrix} -g & 0 & 0 \\ 0 & g & 0 \\ 0 & 0 & 0 \end{pmatrix} & \begin{pmatrix} -\frac{c_{\mathrm{x}}}{m} & 0 & 0 \\ 0 & -\frac{c_{\mathrm{y}}}{m} & 0 \\ 0 & 0 & -\frac{c_{\mathrm{z}}}{m} \end{pmatrix} & \mathbf{0}_{3\times 3} \\ \mathbf{0}_{3\times 3} & \mathbf{0}_{3\times 3} & \boldsymbol{I}_3 & \mathbf{0}_{3\times 3} \end{pmatrix}, \quad \boldsymbol{g} = \begin{pmatrix} \mathbf{0}_{3\times 1} \\ \mathbf{0}_{3\times 1} \\ \begin{pmatrix} 0 \\ 0 \\ g \end{pmatrix} \\ \mathbf{0}_{3\times 1} \end{pmatrix}$$

$$\boldsymbol{B} = \begin{pmatrix} \begin{pmatrix} \boldsymbol{I}_3 & \mathbf{0}_{3\times 1} \end{pmatrix} & \mathbf{0}_{3\times 4} & \begin{pmatrix} 0 & 0 & 0 & 0 \\ 0 & 0 & 0 & 0 \\ 0 & 0 & 0 & 1 \end{pmatrix} & \mathbf{0}_{3\times 4} \end{pmatrix}^{\mathrm{T}} \cdot \tilde{\boldsymbol{B}}_{\mathrm{r}}, \; \boldsymbol{C} = \begin{pmatrix} \mathbf{0}_{3\times 3} & \boldsymbol{I}_3 \end{pmatrix}.$$

10.2.4 *Decomposition of the system behaviour by decoupling*

The MIMO system represented by the model (10.8) can be decomposed in a system that behaves like four SISO systems. For each of these systems a controller can be independently designed, which simplifies the design process. With the input

$$\boldsymbol{u}(t) = \tilde{\boldsymbol{B}}_{\mathrm{r}}^{-1}\tilde{\boldsymbol{u}}(t), \tag{10.12}$$

in which $\tilde{\boldsymbol{B}}_{\mathrm{r}}^{-1}$ is the inverse of (10.9), the model (10.8) results in the decoupled model

$$\dot{\boldsymbol{x}}_{\mathrm{rz}}(t) = \begin{pmatrix} \begin{pmatrix} \mathbf{0}_{3\times 3} & \mathbf{0}_{3\times 1} \\ \mathbf{0}_{1\times 3} & -\frac{c_{\mathrm{z}}}{m} \end{pmatrix} & \mathbf{0}_{4\times 4} \\ \boldsymbol{I}_4 & \mathbf{0}_{4\times 4} \end{pmatrix} \boldsymbol{x}_{\mathrm{rz}}(t) + \begin{pmatrix} \boldsymbol{I}_4 \\ \mathbf{0}_{4\times 4} \end{pmatrix} \tilde{\boldsymbol{u}}(t) + \begin{pmatrix} \begin{pmatrix} \mathbf{0}_{3\times 1} \\ g \end{pmatrix} \\ \mathbf{0}_{4\times 1} \end{pmatrix}$$
$$\boldsymbol{y}_{\mathrm{rz}}(t) = \begin{pmatrix} \mathbf{0}_{4\times 4} & \boldsymbol{I}_4 \end{pmatrix} \boldsymbol{x}_{\mathrm{rz}}(t), \quad \boldsymbol{x}_{\mathrm{rz}}(0) = \begin{pmatrix} \mathbf{0}_{1\times 6} & \bar{\psi} & \bar{z} \end{pmatrix}^{\mathrm{T}}. \tag{10.13}$$

Therefore, the condition $\tilde{\boldsymbol{B}}_\mathrm{r}\tilde{\boldsymbol{B}}_\mathrm{r}^{-1} = \boldsymbol{I}_4$ has to be fulfilled, which holds since $\operatorname{rank}(\tilde{\boldsymbol{B}}_\mathrm{r}) = 4$. With the input

$$\tilde{\boldsymbol{u}}(t) = \tilde{\boldsymbol{B}}_\mathrm{r}\boldsymbol{u}(t) = \begin{pmatrix}\tilde{u}_\phi(t) & \tilde{u}_\vartheta(t) & \tilde{u}_\psi(t) & \tilde{u}_\mathrm{z}(t)\end{pmatrix}^\mathrm{T} \tag{10.14}$$

and

$$\begin{aligned}
\tilde{u}_\phi(t) &= 4\alpha\left(n_2^2(t) + n_3^2(t) - n_1^2(t) - n_4^2(t)\right)\\
\tilde{u}_\vartheta(t) &= 4\alpha\left(n_1^2(t) + n_2^2(t) - n_3^2(t) - n_4^2(t)\right)\\
\tilde{u}_\psi(t) &= 4\left(n_1^2(t) + n_3^2(t) - n_2^2(t) - n_4^2(t)\right)\\
\tilde{u}_\mathrm{z}(t) &= 4\left(n_1^2(t) + n_2^2(t) + n_3^2(t) + n_4^2(t)\right)
\end{aligned} \tag{10.15}$$

the MIMO system (10.13) behaves like four, decoupled SISO systems. The control of the angles and the altitude is performed by all four motors, while each motor has the same effect. Hence, the resulting input on the system conforms to a single input with four times the gain. The model (10.10) can already be described by two decoupled SISO systems with $\tilde{u}_\mathrm{x}(t) = \phi(t)$ and $\tilde{u}_\mathrm{y}(t) = \vartheta(t)$. With (10.15) the decoupled SISO systems of the attitude are described by

$$\begin{aligned}
\dot{\boldsymbol{x}}_i(t) &= \boldsymbol{A}_i\boldsymbol{x}_i(t) + \boldsymbol{b}_i\tilde{u}_i(t), \quad i \in \{\phi, \vartheta, \psi, \mathrm{x}, \mathrm{y}, \mathrm{z}\}\\
y_i(t) &= \boldsymbol{c}_i^\mathrm{T}\boldsymbol{x}_i(t), \quad \boldsymbol{x}_i(0) = \boldsymbol{x}_{i0}
\end{aligned} \tag{10.16}$$

with

$$\begin{aligned}
\boldsymbol{x}_\phi(t) &= \begin{pmatrix}\dot{\phi}(t)\\ \phi(t)\end{pmatrix}, \quad \boldsymbol{x}_\vartheta(t) = \begin{pmatrix}\dot{\vartheta}(t)\\ \vartheta(t)\end{pmatrix}, \quad \boldsymbol{x}_\psi(t) = \begin{pmatrix}\dot{\psi}(t)\\ \psi(t)\end{pmatrix}\\
\boldsymbol{A}_i &= \begin{pmatrix}0 & 0\\ 1 & 0\end{pmatrix}, \quad \boldsymbol{b}_i = \begin{pmatrix}b_i\\ 0\end{pmatrix}, \quad \boldsymbol{c}_i^\mathrm{T} = \begin{pmatrix}0 & 1\end{pmatrix}, \quad i \in \{\phi, \vartheta, \psi\}
\end{aligned}$$

in which the index i describes the different SISO systems. The decoupled SISO systems for the position and the altitude are given by (10.16) with

$$\begin{aligned}
\boldsymbol{x}_\mathrm{x}(t) &= \begin{pmatrix}\dot{x}(t)\\ x(t)\end{pmatrix}, \quad \boldsymbol{x}_\mathrm{y}(t) = \begin{pmatrix}\dot{y}(t)\\ y(t)\end{pmatrix}, \quad \boldsymbol{x}_\mathrm{z}(t) = \begin{pmatrix}\dot{z}(t)\\ z(t)\end{pmatrix},\\
\boldsymbol{b}_\mathrm{x} &= \begin{pmatrix}g\\ 0\end{pmatrix}, \quad \boldsymbol{b}_\mathrm{y} = \begin{pmatrix}-g\\ 0\end{pmatrix}, \quad \boldsymbol{b}_\mathrm{z} = \begin{pmatrix}b_\mathrm{z}\\ 0\end{pmatrix},\\
\boldsymbol{A}_i &= \begin{pmatrix}-\frac{c_i}{m} & 0\\ 1 & 0\end{pmatrix}, \quad \boldsymbol{c}_i^\mathrm{T} = \begin{pmatrix}0 & 1\end{pmatrix}, \quad i \in \{\mathrm{x}, \mathrm{y}, \mathrm{z}\}.
\end{aligned}$$

10.2.5 *Motor dynamics*

For deriving an exact model of the quadrotor, the transient behaviour of the motors has to be considered, because the motor dynamics limit the gain of the feedback controller. The delay characteristic is modelled as a first-order lag element

$$\begin{aligned} \dot{x}_{\mathrm{M}}(t) &= -\frac{1}{T_{\mathrm{M}}} x_{\mathrm{M}}(t) + \frac{1}{T_{\mathrm{M}}} u(t), \quad x_{\mathrm{M}}(0) = x_{\mathrm{M}0}, \\ y_{\mathrm{M}}(t) &= x_{\mathrm{M}}(t) \end{aligned} \tag{10.17}$$

with the time constant given in Tab. 10.1 and the static gain $k_{\mathrm{s}} = 1$. The motor delay is connected upstream of the system model (10.2), (10.3).

The motor dynamics are taken into account in the feedback controller design but are not considered in the design of the feedforward controller and the trajectory planning method.

10.3 Local control of a quadrotor

10.3.1 *Control loop of the quadrotor*

A flatness-based two-degrees-of-freedom controller, which combines a feedforward controller with a feedback controller is used to control the quadrotor locally. The structure of the control loop is shown in Fig. 10.6. The control input to the quadrotor is given by $\boldsymbol{u}^*(t) = \boldsymbol{n}^{*2}(t)$, which is composed of the quadratic values of the actual rotor speeds $\boldsymbol{n}^*(t)$. The values $\boldsymbol{n}^*(t)$ follow the set-point values $\boldsymbol{w}_{\mathrm{n}}^*(t)$ delayed by the motor dynamics described by (10.17). Each entry in $\boldsymbol{w}_{\mathrm{n}}^*(t)$ is specified by a pulse width modulated signal (PWM) with the pulse width τ_{pw}, which is quantised and used in the form of τ_{pw}^* to control the motors using electronic speed controllers (ESC) that are incorporated in the microcontroller 'ARM Cortex-M4'. Hence, the PWM signal τ_{pw} represents the control input of the experimental quadrotor.

The feedback controller is realised as a cascaded control loop, in which the inner control loop consists of a state feedback and generates the control input $\boldsymbol{u}(t)$ in form of quadratic rotor speeds. The outer control loop is realised as a state feedback as well, where the position controller generates reference values $w_{\phi}(t)$ and $w_{\vartheta}(t)$ for the inner control loop. For the application of linear methods for the controller design, the linear model (10.8), (10.10) is used. Here, a linear correlation between the input $\boldsymbol{u}^*(t)$ and the output $\boldsymbol{y}_{\mathrm{xy}}(t)$ and $y_{\mathrm{z}}(t)$ is assumed.

Hence, in the following paragraphs some practical steps are described, which aim at ensuring the control signal $\boldsymbol{u}(t)$ to match the input of the quadrotor as $\boldsymbol{u}^*(t) = \boldsymbol{u}(t)$.

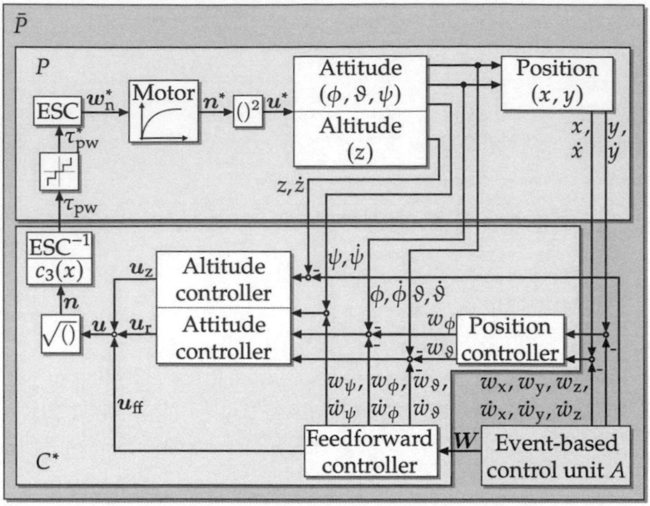

Fig. 10.6: Local control loop of the quadrotor.

Compensation of the effect of the quadratic rotor speeds. The nonlinear quadratic effect of the rotor speeds is linearised by a root element, which is connected upstream. However, this linearisation is only accurate in the stationary case, because the motor dynamics are located between the root element and the square function. Analyses have shown that this approximative linearisation models the system behaviour more closely than a linearisation of the square function. However, the linearisation would be required to consider the motor dynamics correctly in a linear state space model.

Quantisation effects. The microcontroller 'ARM Cortex-M4' of the quadrotor has a resolution of a PWM signal of 32 bit, which is sufficiently accurate so that the ESCs are able to generate accurate rotor speeds. Hence, in the remainder of this chapter the quantisation effects are neglected and $\tau_{pw}^* = \tau_{pw}$ is assumed.

Coding of the reference values of the rotor speeds. It is necessary to map the correlation between the input τ_{pw} of the ESCs and the set-points of the rotor speeds $w_n^*(t)$ as precisely as possible in order to provide the motors an accurate rotor speed. An exemplary PWM signal is shown in Fig. 10.7. The signal has a periodic time of $T = 2\,\text{ms}$, whereby the pulse width has to be at least half of the periodic time to be an acceptable signal for the ESCs. With the variable pulse width $0\,\text{ms} \leq \tau_{pw} \leq 0.9\,\text{ms}$ the set-points for the motors are specified. The maximum rotor speed is achieved with a pulse width of $\tau_{pw,max} = 0.9\,\text{ms}$.

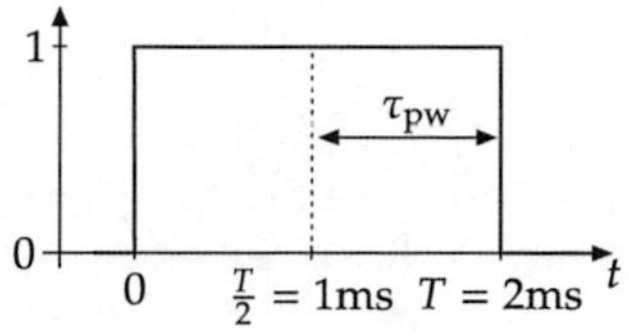

Fig. 10.7: Exemplary PWM signal.

The actuator characteristics of the individual motors are measured to convert a PWM signal into a corresponding rotor speed. The characteristic of a motor is shown in Fig. 10.8. The measurements have shown that the four motors have almost identical characteristics, which are approximated by a polynomial of third order as

$$c_3(x) = \sum_{i=0}^{3} a_i \, x^i .$$

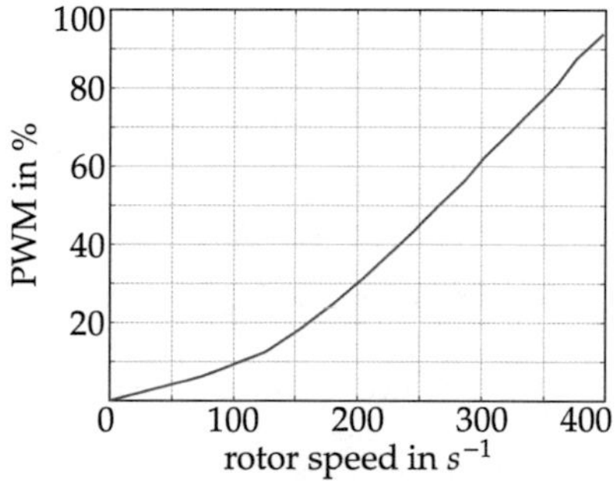

Fig. 10.8: Rotor speed characteristic of a motor.

It is assumed that the relation is approximated in a way that $\boldsymbol{w}_{\mathrm{n}}^{*}(t) = \boldsymbol{n}(t)$ applies. Thus, $\boldsymbol{n}^{*}(t) = \boldsymbol{n}(t)$ and $\boldsymbol{u}^{*}(t) = \boldsymbol{u}(t)$ holds. Hence, the nonlinear effects and the approximation of the actuator characteristics are neglected, which results in the linear control loop shown in Fig. 10.10 on p. 241 that models the behaviour of the quadrotor sufficiently. This control loop is used for the controller design.

Discretisation. For the implementation in the experiment the models are discretised with the sampling time $T = 8\,\mathrm{ms}$. As the sampling time is sufficiently small, a time-discrete description of the models is omitted.

10.3.2 *Flatness of the quadrotor model*

A flatness-based two-degrees-of-freedom controller is used to make the agents follow the planned trajectories. To this aim it is proven that the quadrotor model fulfils the flatness property. For the analysis the partially linearised model of the quadrotor (10.6), (10.7) is utilised. The design of the feedforward controller is based on the inverting of the complete model consisting of the series connection of the rotational subsystem and the translational subsystem. The flatness property is defined as follows.

Definition 10.1. (Flatness) A dynamical system

$$\dot{\boldsymbol{x}}(t) = \boldsymbol{f}(\boldsymbol{x}(t), \boldsymbol{u}(t)), \quad \boldsymbol{x}(0) = \boldsymbol{x}_0$$

with $\boldsymbol{x}(t) \in \mathbb{R}^n,\ \boldsymbol{u}(t) \in \mathbb{R}^m,\ m \leq n$ and

$$\operatorname{rank}\left(\frac{\partial \boldsymbol{f}(\boldsymbol{x}(t), \boldsymbol{u}(t))}{\partial \boldsymbol{u}(t)}\right) = m$$

is differentially flat, if there exists an output

$$\boldsymbol{y}_{\mathrm{f}}(t) = \boldsymbol{h}\left(\boldsymbol{x}(t), \boldsymbol{u}(t), \dot{\boldsymbol{u}}(t), \ldots, \boldsymbol{u}^{(\alpha)}(t)\right)$$

with $\alpha \in \mathbb{N}$ such that

1. the state $\boldsymbol{x}(t)$ is a function $\boldsymbol{\Psi}_1$ of $\boldsymbol{y}_{\mathrm{f}}(t)$ and its derivatives

$$\boldsymbol{x}(t) = \boldsymbol{\Psi}_1\left(\boldsymbol{y}_{\mathrm{f}}(t), \dot{\boldsymbol{y}}_{\mathrm{f}}(t), \ldots, \boldsymbol{y}_{\mathrm{f}}^{(\beta)}(t)\right), \text{ with } \beta \in \mathbb{N},$$

2. the input $\boldsymbol{u}(t)$ is a function $\boldsymbol{\Psi}_2$ of $\boldsymbol{y}_{\mathrm{f}}(t)$ and its derivatives

$$\boldsymbol{u}(t) = \boldsymbol{\Psi}_2\left(\boldsymbol{y}_{\mathrm{f}}(t), \dot{\boldsymbol{y}}_{\mathrm{f}}(t), \ldots, \boldsymbol{y}_{\mathrm{f}}^{(\beta+1)}(t)\right), \tag{10.18}$$

3. for the input and the output the relation

$$\dim(\boldsymbol{y}(t)) = \dim(\boldsymbol{u}(t))$$

holds. The output $\boldsymbol{y}_{\mathrm{f}}(t)$ is called a flat output of the system.

Proposition 10.1 (Flat output of a quadrotor)**.** *The nonlinear system* (10.6), (10.7) *is differentially flat with the flat output*

$$\boldsymbol{y}_{\mathrm{f}}(t) = \begin{pmatrix} x(t) & y(t) & z(t) & \psi(t) \end{pmatrix}^{\mathrm{T}}. \tag{10.19}$$

Proof. See Appendix B.15. ■

Relative degree of the quadrotor. As the motor dynamics are not considered for the design of the feedforward controller and the trajectory planning method it follows that with (B.25) – (B.28) from the proof of Proposition 10.1 the relative degree $\boldsymbol{r}$ of the quadrotor is given by

$$\boldsymbol{r} = \begin{pmatrix} r_{\mathrm{x}} \\ r_{\mathrm{y}} \\ r_{\mathrm{z}} \\ r_{\psi} \end{pmatrix} = \begin{pmatrix} 4 \\ 4 \\ 2 \\ 2 \end{pmatrix}.$$

If the motor dynamics are taken into account, the relative degree increases and is represented by

$$\boldsymbol{r} = \begin{pmatrix} r_{\mathrm{x}} \\ r_{\mathrm{y}} \\ r_{\mathrm{z}} \\ r_{\psi} \end{pmatrix} = \begin{pmatrix} 5 \\ 5 \\ 3 \\ 3 \end{pmatrix}.$$

10.3.3 *Feedforward controller*

The feedforward controller generates the control input $u_i(t) = n_i^2(t), (i = 1, \dots, 4)$ based on the planned trajectory $\boldsymbol{w}(t)$ so that the quadrotor follows the trajectory exactly if there are no disturbances or model uncertainties. Hence, with the controller the quadrotors are able to fulfil their local control aims (S2) and (G2) corresponding to control aim (A2) of the overall system. The reference values $\boldsymbol{W}(t)$ containing $\boldsymbol{w}(t)$ and its first four derivatives are provided by the event-based control units A_{S} and A_{G}, respectively.

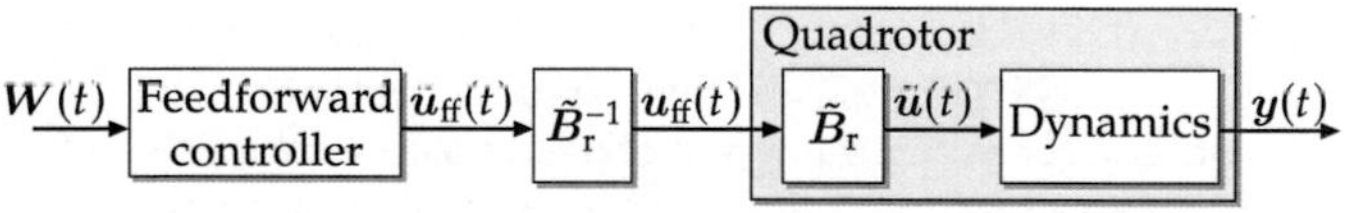

Fig. 10.9: Block diagram of the feedforward controller.

The computation of the quadratic rotor speeds is carried out in two steps, as shown in Fig. 10.9. First, the input $\tilde{\boldsymbol{u}}_{\mathrm{ff}}(t)$ as defined in (10.14) is determined based on the conditions on the trajectory of the flat output. With (10.18) it is parametrised by

$$\tilde{\boldsymbol{u}}_{\mathrm{ff}}(t) = \boldsymbol{\Psi}_2\left(\boldsymbol{y}_{\mathrm{f}}(t), \dot{\boldsymbol{y}}_{\mathrm{f}}(t), \dots, \boldsymbol{y}_{\mathrm{f}}^{(4)}(t)\right).$$

In order to achieve control aim (A2) $\boldsymbol{y}_{\mathrm{f}}(t) = \boldsymbol{w}(t)$, the flat output $\boldsymbol{y}_{\mathrm{f}}(t)$ is substituted by the reference values so that the feedforward controller is given by

$$\tilde{\boldsymbol{u}}_{\mathrm{ff}}(t) = \begin{pmatrix} \alpha_{\phi}(t) \\ \alpha_{\vartheta}(t) \\ \alpha_{\psi}(t) \\ a_{\mathrm{z}}(t) \end{pmatrix} = \begin{pmatrix} \ddot{w}_{\phi}(t) \\ \ddot{w}_{\vartheta}(t) \\ \ddot{w}_{\psi}(t) \\ g - \ddot{w}_{\mathrm{z}}(t) - \frac{c_{\mathrm{z}}}{m}\dot{w}_{\mathrm{z}}(t) \end{pmatrix}. \tag{10.20}$$

With (B.22), (B.23), (B.29) and (B.31) the reference values $\dot{w}_{\phi}(t)$, $\ddot{w}_{\phi}(t)$, $\dot{w}_{\vartheta}(t)$ and $\ddot{w}_{\vartheta}(t)$ are functions of the flat output.

Second, the input $\tilde{\boldsymbol{u}}_{\mathrm{ff}}(t)$ is transformed into the control input $\boldsymbol{u}_{\mathrm{ff}}(t)$ of the quadrotor model (10.10), (10.13) using the input transformation stated in (10.12).

As it is shown in the proof of the flatness property in Appendix B.15, the event-based control units have to generate reference trajectories for $\dot{\boldsymbol{w}}(t), \ldots, \boldsymbol{w}^{(4)}(t)$ in addition to the reference value $\boldsymbol{w}(t)$ so that the system can be controlled with (10.20) along the reference trajectory.

10.3.4 *Feedback controller*

The feedback controller has two tasks. First, it stabilises the system, since the quadrotor is an unstable system. Second, the controller becomes active in the presence of disturbances or model uncertainties for trajectory following. It just compensates deviations from the reference trajectory. Hence, with this controller the quadrotors are able to fulfil their local control aims (S3) and (G3), which correspond to control aim (A3) of the overall system. The feedback controller is composed of an altitude controller and an attitude and a position controller in a cascaded structure, as shown in Fig. 10.10. The outer control loop is closed by the position controller. It uses the set-points for the position $w_{\mathrm{x}}(t)$, $\dot{w}_{\mathrm{x}}(t)$ and $w_{\mathrm{y}}(t)$, $\dot{w}_{\mathrm{y}}(t)$ provided by the event-based control units A_{S} or A_{G} to generate the reference values $w_{\phi}(t)$ and $w_{\vartheta}(t)$ for the inner control loop. The further reference values $w_{\phi}(t)$, $\dot{w}_{\phi}(t)$, $w_{\vartheta}(t)$, $\dot{w}_{\vartheta}(t)$ and $w_{\psi}(t)$, $\dot{w}_{\psi}(t)$ of the three Euler angles are generated by the feedforward controller. The altitude controller is not part of the cascade, because the set-points $w_{\mathrm{z}}(t)$, $\dot{w}_{\mathrm{z}}(t)$ are provided by the control units A_{S} or A_{G} and are independent of the position controller. The controllers are realised as state feedbacks, which act as a PD controller and PID controllers, respectively. The reference values for the P term result from the reference trajectory $\boldsymbol{w}(t)$. In order to prevent the D term from controlling the speed to zero and thus counteracting the feedforward controller, the feedback controllers are also supplied with the reference values $\dot{\boldsymbol{w}}(t)$.

The event-based control units only generate reference values $\boldsymbol{w}(t)$ for the flat output and its derivatives. Due to the cascaded structure of the control loop (Fig. 10.10) it is necessary to supply reference values for the angles $\phi(t)$ and $\vartheta(t)$ and the angular speeds $\dot{\phi}(t)$ and

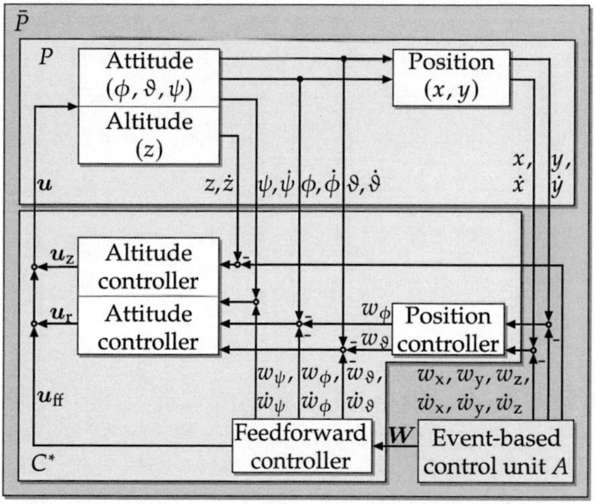

Fig. 10.10: Linear control loop of the quadrotor.

$\dot{\vartheta}(t)$. The reference values for $w_\phi(t)$, $\dot{w}_\phi(t)$ and $w_\vartheta(t)$, $\dot{w}_\vartheta(t)$ are determined from reference values of the flat output with (B.19), (B.20) and (B.22), (B.23).

The attitude and altitude controllers generate the control inputs $\boldsymbol{u}_{\mathrm{z}}(t) = \boldsymbol{n}_{\mathrm{z}}^2(t)$ and $\boldsymbol{u}_{\mathrm{r}}(t) = \boldsymbol{n}_{\mathrm{r}}^2(t)$, respectively. The overall control input $\boldsymbol{u}(t) = \boldsymbol{u}_{\mathrm{ff}}(t) + \boldsymbol{u}_{\mathrm{r}}(t) + \boldsymbol{u}_{\mathrm{z}}(t)$ is the result of the feedforward and the feedback controller. The root locus method is used for the controller design, because the system can be represented by multiple SISO systems, as stated in Section 10.2.4. The motor characteristics (10.17) are considered in the design process.

Attitude controller. The attitude controller acts as a PD controller given by

$$\boldsymbol{u}_{\mathrm{r}}(t) = -\boldsymbol{K}_{\mathrm{l}}\boldsymbol{e}_{\mathrm{r}}(t),$$
$$\boldsymbol{e}_{\mathrm{r}}(t) = \boldsymbol{x}_{\mathrm{r}}(t) - \begin{pmatrix} \dot{w}_\phi(t) & \dot{w}_\vartheta(t) & \dot{w}_\psi(t) & w_\phi(t) & w_\vartheta(t) & w_\psi(t) \end{pmatrix}^{\mathrm{T}}.$$

with

$$\boldsymbol{K}_{\mathrm{l}} = \begin{pmatrix} k_{\mathrm{D}\phi} & -k_{\mathrm{D}\vartheta} & -k_{\mathrm{D}\psi} & k_{\mathrm{P}\phi} & -k_{\mathrm{P}\vartheta} & -k_{\mathrm{P}\psi} \\ -k_{\mathrm{D}\phi} & -k_{\mathrm{D}\vartheta} & k_{\mathrm{D}\psi} & -k_{\mathrm{P}\phi} & -k_{\mathrm{P}\vartheta} & k_{\mathrm{P}\psi} \\ -k_{\mathrm{D}\phi} & k_{\mathrm{D}\vartheta} & -k_{\mathrm{D}\psi} & -k_{\mathrm{P}\phi} & k_{\mathrm{P}\vartheta} & -k_{\mathrm{P}\psi} \\ k_{\mathrm{D}\phi} & k_{\mathrm{D}\vartheta} & k_{\mathrm{D}\psi} & k_{\mathrm{P}\phi} & k_{\mathrm{P}\vartheta} & k_{\mathrm{P}\psi} \end{pmatrix}.$$

The controller gains for the attitude are determined separately from each other on the basis of the root locus and are summarised in the matrix $\boldsymbol{K}_{\mathrm{l}}$.

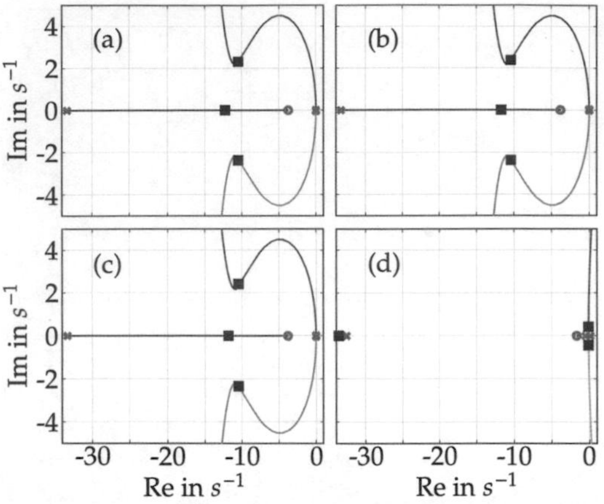

Fig. 10.11: Root locus of $\boldsymbol{x}_\phi(t)$ (a), $\boldsymbol{x}_\vartheta(t)$ (b), $\boldsymbol{x}_\psi(t)$ (c) and $\boldsymbol{x}_\mathrm{z}(t)$ (d).

In Fig. 10.11 (a), (b) and (c) the root locus of the SISO systems (10.16) controlled by a PD controller in addition with the motor characteristics (10.17) are shown. The system values of Tab. 10.1 are used. The poles of the closed-loop systems should be located in a way to ensure a fast transient response with a high damping. With the zero $T_\mathrm{D} = 0.2632\,\mathrm{s}$ the poles are placed to fulfil the desired properties, which are marked as black squares in the figure. The positions of the poles determine the proportional feedback gain k_P and the derivative feedback gain $k_\mathrm{D} = T_\mathrm{D}\, k_\mathrm{P}$. The determined values are summarised in Tab. 10.2.

Table 10.2: Values of the attitude controller.

$k_{\mathrm{P}\phi}$	$k_{\mathrm{D}\phi}$	$k_{\mathrm{P}\vartheta}$	$k_{\mathrm{D}\vartheta}$	$k_{\mathrm{P}\psi}$	$k_{\mathrm{D}\psi}$
12920	3400	15960	4200	60000	15789

Altitude controller. The altitude controller acts as a PID controller to ensure set-point following. It is represented by

$$
\begin{aligned}
\boldsymbol{u}_\mathrm{z}(t) &= -\begin{pmatrix}1 & 1 & 1 & 1\end{pmatrix}^\mathrm{T} \boldsymbol{k}_\mathrm{z}^\mathrm{T} \boldsymbol{e}_\mathrm{z}(t) \\
\boldsymbol{e}_\mathrm{z}(t) &= \begin{pmatrix}\boldsymbol{x}_\mathrm{z}(t) \\ x_{\mathrm{z,I}}(t)\end{pmatrix} - \begin{pmatrix}\dot{w}_\mathrm{z}(t) & w_\mathrm{z}(t) & 0\end{pmatrix}^\mathrm{T} \\
\dot{x}_{\mathrm{z,I}}(t) &= z(t) - w_\mathrm{z}(t),
\end{aligned}
$$

with $k_z^T = (k_{Dz} \quad k_{Pz} \quad k_{Iz})$. The root locus of the system (10.16) controlled by a PID controller in addition with the motor characteristics (10.17) is shown in Fig. 10.11 (d). With the zero $T_D = 0.6111$ s and the pole $T_I = 3.8158$ s of the controller the poles of the closed-loop system are placed to fulfil the desired properties, marked as black squares in the figure. The feedback gains are chosen as given in Tab. 10.3.

Table 10.3: Values of the altitude controller.

k_{Pz}	k_{Dz}	k_{Iz}
150800	92150	39520

Position controller. A PID controller given by

$$\begin{aligned} \boldsymbol{u}_{xy}(t) &= -\boldsymbol{K}_{xy}\boldsymbol{e}_{xy}(t) \\ \boldsymbol{e}_{xy}(t) &= \begin{pmatrix} \boldsymbol{x}_{xy}(t) \\ \boldsymbol{x}_{xy,I}(t) \end{pmatrix} - \begin{pmatrix} \dot{w}_x(t) & \dot{w}_y(t) & w_x(t) & w_y(t) & 0 & 0 \end{pmatrix}^T \\ \dot{\boldsymbol{x}}_{xy,I}(t) &= \begin{pmatrix} x_x(t) - w_x(t) \\ x_y(t) - w_y(t) \end{pmatrix} \end{aligned}$$

with

$$\boldsymbol{K}_{xy} = \begin{pmatrix} 0 & k_{Dy} & 0 & k_{Py} & 0 & k_{Iy} \\ k_{Dx} & 0 & k_{Px} & 0 & k_{Ix} & 0 \end{pmatrix}$$

is used to control the xy-position and to ensure set-point following. The controller generates the set-points for the attitude controller as $\boldsymbol{u}_{xy}(t) = (w_\phi(t) \quad w_\vartheta(t))^T$. The controller gains are determined separately and are summarised in the matrix $\boldsymbol{K}_{xy}$. The root locus of the SISO systems controlled by the PID controller are shown in Fig. 10.12 (a) and (b). With the zero $T_D = 0.8$ s and the pole $T_I = 14.6$ s the poles of the closed-loop systems are placed to fulfil the desired properties, which are marked as black squares in the figure. The feedback gains are summarised in Tab. 10.4.

Table 10.4: Values of the position controller.

k_{Px}	k_{Dx}	k_{Ix}	k_{Py}	k_{Dy}	k_{Iy}
0.735	0.6027	0.0503	-0.735	-0.6027	-0.0503

Since the design of the position controller is based on the linear model (10.10), (10.13), the set-points $w_\phi(t)$, $w_\vartheta(t)$ are only valid in the operating point $\bar{\psi}(t) = 0$. This means, the

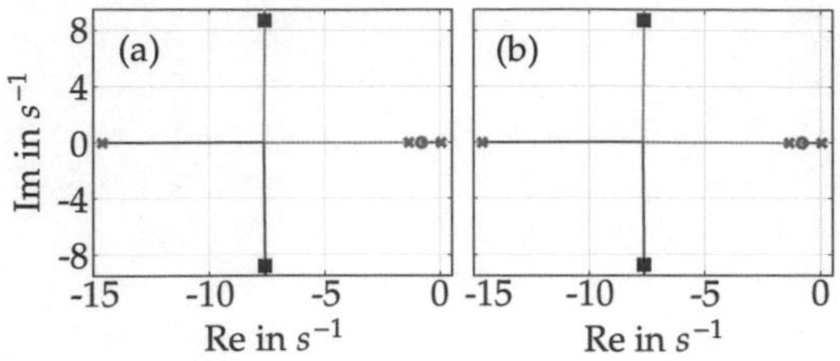

Fig. 10.12: Root locus of $x_x(t)$ (a) and $x_y(t)$ (b).

earth-fixed frame I and the body-fixed frame M have the same direction. In order to utilise the controller for $\bar{\psi}(t) \neq 0$ the set-points are transformed as

$$\begin{pmatrix} \tilde{w}_\phi(t) \\ \tilde{w}_\vartheta(t) \end{pmatrix} = \boldsymbol{R}(t) \begin{pmatrix} w_\phi(t) \\ w_\vartheta(t) \end{pmatrix}$$

with

$$\boldsymbol{R}(t) = \begin{pmatrix} \cos(\psi(t)) & \sin(\psi(t)) \\ -\sin(\psi(t)) & \cos(\psi(t)) \end{pmatrix}$$

from the earth-fixed frame to the body-fixed frame. The overall control input to the system is obtained as

$$\boldsymbol{u}(t) = \boldsymbol{u}_{\mathrm{ff}}(t) + \boldsymbol{u}_{\mathrm{r}}(t) + \boldsymbol{u}_{\mathrm{z}}(t)$$

with $\boldsymbol{u}_{\mathrm{ff}}(t) = \tilde{\boldsymbol{B}}_{\mathrm{r}}^{-1}\tilde{\boldsymbol{u}}_{\mathrm{ff}}(t)$. This ensures that the tracking errors $\boldsymbol{e}_{\mathrm{r}}(t)$, $\boldsymbol{e}_{\mathrm{z}}(t)$ and $\boldsymbol{e}_{\mathrm{xy}}(t)$ converge asymptotically towards zero for the choice of the control gains $\boldsymbol{K}_{\mathrm{l}}$, $\boldsymbol{k}_{\mathrm{z}}^{\mathrm{T}}$ and $\boldsymbol{K}_{\mathrm{xy}}$ in the preceding paragraphs.

10.3.5 *Maximum tracking error*

The tracking error due to bounded disturbances is bounded when the proposed controller is used. With the linear, undisturbed model (10.11) of the quadrotor and the linear, disturbed model of the quadrotor

$$\begin{aligned} \dot{\hat{\boldsymbol{x}}}(t) &= \boldsymbol{A}\hat{\boldsymbol{x}}(t) + \boldsymbol{B}\boldsymbol{u}(t) + \boldsymbol{g} + \boldsymbol{E}\boldsymbol{d}(t), \quad \hat{\boldsymbol{x}}(0) = \hat{\boldsymbol{x}}_0, \\ \hat{\boldsymbol{y}}(t) &= \boldsymbol{C}\hat{\boldsymbol{x}}(t) \end{aligned}$$

the tracking error model is given by

$$\begin{aligned} \dot{\boldsymbol{e}}(t) &= \dot{\hat{\boldsymbol{x}}}(t) - \dot{\boldsymbol{x}}(t) = \boldsymbol{A}\hat{\boldsymbol{x}}(t) + \boldsymbol{B}\boldsymbol{u}(t) + \boldsymbol{g} + \boldsymbol{E}\boldsymbol{d}(t) - \boldsymbol{A}\boldsymbol{x}(t) - \boldsymbol{B}\boldsymbol{u}(t) - \boldsymbol{g} \\ &= \boldsymbol{A}\boldsymbol{e}(t) + \boldsymbol{E}\boldsymbol{d}(t) \qquad (10.21) \\ \boldsymbol{f}(t) &= \boldsymbol{C}\boldsymbol{e}(t), \quad \boldsymbol{e}(0) = \boldsymbol{e}_0. \end{aligned}$$

The tracking error $\boldsymbol{f}(t)$ can be determined by the convolution of $\boldsymbol{d}(t)$ with the impulse response matrix $\boldsymbol{G}(t) = \boldsymbol{C}e^{\boldsymbol{A}t}\boldsymbol{E}$ of the model (10.21) as

$$\boldsymbol{f}(t) = \int_0^t \boldsymbol{G}(t-\tau) \cdot \boldsymbol{d}(\tau)\mathrm{d}\tau.$$

Hence, the following holds:

$$\begin{aligned} ||\boldsymbol{f}(t)|| &= \left\| \int_0^t \boldsymbol{G}(t-\tau) \cdot \boldsymbol{d}(\tau)\mathrm{d}\tau \right\| \\ &\leq \int_0^t ||\boldsymbol{G}(t-\tau)|| \cdot ||\boldsymbol{d}(\tau)||\mathrm{d}\tau. \end{aligned}$$

With a maximum disturbance $\bar{d}$ the tracking error is bounded by

$$||\boldsymbol{f}(t)|| \leq f_{\max}. \tag{10.22}$$

with

$$f_{\max} = \int_0^\infty ||\boldsymbol{G}(\tau)||\mathrm{d}\tau \cdot \bar{d}. \tag{10.23}$$

The result is stated in the following theorem:

Theorem 10.1 (Maximum tracking error). *For a bounded disturbance* $||\boldsymbol{d}(t)|| \leq \bar{d}$ *acting on the system* (10.11) *the maximum tracking error is bounded from above by* (10.22) *with* $f_{\max}$ *given by* (10.23).

10.4 Event-based control of two quadrotors

The cooperative control loop of the overall system is shown in Fig. 10.13. The system consists of the quadrotors P_S and P_G, which are controlled by the local flatness-based two-degrees-of-freedom controllers C_S^* and C_G^*. For simplicity in Fig. 10.13 only the linear feedback control loop depicted in Fig. 10.10 is shown. The attitude controller is denoted by 'AtC', the altitude controller by 'AlC', the position controller by 'PC'and 'FC' stands for the feedforward controller. The event-based control units A_S and A_G plan the local trajectories and exchange information over the communication network. In the figure, 'G_S' and 'G_G' are the prediction units of the stand-on agent and the give-way agent, 'E_S' and 'E_G' denote the event generators and 'T_S' and 'T_G' indicate the trajectory planning units. The delay estimators are denoted by 'D_S' and 'D_G', respectively.

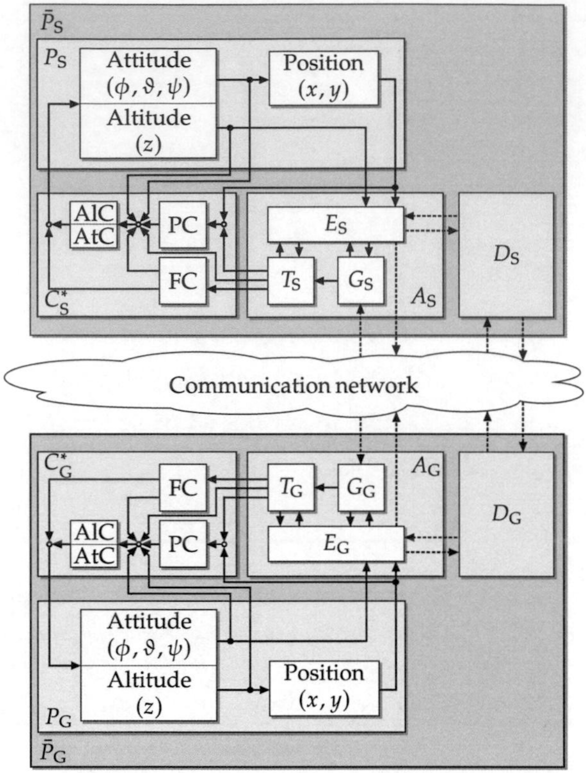

Fig. 10.13: Cooperative control loop of the overall system.

Simulation studies 11

The evaluation of the event-based control method in simulations is presented in this chapter. The aims of the simulation studies are introduced in Section 11.1. Three scenarios are considered: Estimation of the time delay in Section 11.2, ensuring collision avoidance in Section 11.3 and satisfaction of the maximum separation between the agents in Section 11.4. Section 11.5 states concluding remarks concerning the simulation results.

11.1 Aims

In this chapter the event-based control method is evaluated in simulations with three different scenarios. In all scenarios the method is applied to two quadrotors with the dynamics derived in Section 10.2 for a general movement. The simulation studies are focussed on an evaluation of the impact of an unreliable communication network on the control method. Hence, in the scenarios the distance between the agents varies significantly, because the properties of a communication channel vary only over large distances between two agents. The evaluation of the control method using an unreliable communication network is only carried out by simulations in this thesis, because the space at the experimental test bed 'MULAN', introduced in Section 12.2, is limited. Hence, the communication network used by the experimental quadrotors can be considered as almost ideal. Only very small time delays and only very few packet losses occur during the communication at the test bed and a reasonable experimental evaluation of the network effects is not possible.

The aim of the simulation study is to evaluate the impact of a delayed information transfer or lost data on the event-based control method. A comparison is stated of the ability to fulfil control aim (A1) when the quadrotors communicate over an ideal network or send information over an unreliable network. Furthermore, the results are compared for the cases when time delays and packet losses are taken into account in the control method and when they are not considered for the trajectory planning.

The event-based control method is examined in the following three scenarios that are described in detail in the following sections:

1. In the first scenario collision avoidance between two quadrotors that move in different directions is investigated. The focus is on the estimation of the time delay of the

delay estimator. It is compared whether a collision can be avoided in the case that the delay is not taken into account for the event-based control of the quadrotors and in the case that the delay is considered in the control method.

2. The second scenario deals with the collision avoidance of two quadrotors in the presence of packet losses induced by the network. It is investigated if collision avoidance is guaranteed when the agents communicate over an unreliable network. The results are compared to the case when the agents exchange data over an ideal network.

3. In the third scenario the influence of packet losses on the satisfaction of a maximum separation between the two quadrotors is evaluated. The results are compared for different movements of the stand-on agent in relation to the give-way agent.

11.2 Scenario 1: Estimation of the time delay depending on the movement of the agents

11.2.1 *Description of the scenario*

In the first scenario the focus is on the estimation of the time delay of the communication channel in dependence on the relative movement of the two agents. Its influence on the avoidance manoeuvre of the give-way agent to achieve collision avoidance (requirement (1.1) of control aim (G1)) is evaluated. The scenario is illustrated in Fig. 11.1. An application example for the scenario is the parcel delivery with drones. Here, the UAVs follow their trajectories autonomously to the delivery location and they have to avoid collisions. If the delivery location changes during the delivery, the trajectory has to be changed accordingly.

The stand-on agent starts moving on its initially planned trajectory $\boldsymbol{w}_{\mathrm{S}}(t)$ from a start point $S_{\mathrm{A,S}}$ to an end point $S_{\mathrm{B,S}}$ (dashed line), but changes its trajectory to arrive at the different destination $S'_{\mathrm{B,S}}$ (solid line). The give-way agent follows its initially planned trajectory $\boldsymbol{w}_{\mathrm{G}}(t)$ from the start point $S_{\mathrm{A,G}}$ to the end point $S_{\mathrm{B,G}}$. It reacts on the changed movement of the stand-on agent to ensure inequality (1.1) of control aim (G1) (solid curved line). The avoidance manoeuvre will be compared for the cases when the estimate of the delay estimator is taken into account in the control method and when it is not considered.

The safety distance $\underline{s}$ is defined by the geometry of the quadrotors. In order to simplify the determination of the distance between the agents, the quadrotors are assumed to be point masses. The safety distance is given by

$$\underline{s} = 2l + s_{\mathrm{r}} \qquad (11.1)$$

and results from the lengths l from the centre of gravity (COG) of the quadrotors to the outer edges of the rotors and a distance s_{r} so that the rotors do not touch. The bound $\bar{s}$ on the maximum separation is not considered, as the UAVs are not required to stay close to one another.

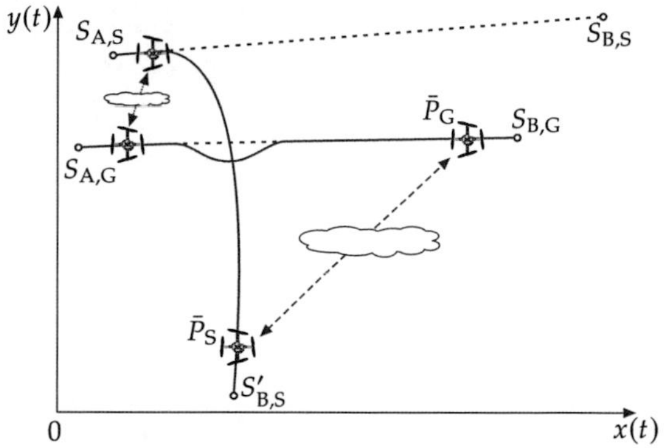

Fig. 11.1: Scenario of the network estimation.

11.2.2 *Results of the scenario*

In the simulation the two quadrotors communicate over a channel that is represented by the model stated in Section 5.3. For the signal propagation a dominant line-of-sight (LOS) path is assumed. The parameters of both the Rician fading model and the transmitters and the receivers of the UAVs are summarised in Tab. 11.1. The parameters of the Markov model are estimated by the delay estimator using Algorithm 15 on p. 189 and are given by

$$p_{\mathrm{gg}} = 0.995, \quad p_{\mathrm{bb}} = 0.5, \quad p_{\mathrm{gb}} = 0.005, \quad p_{\mathrm{bg}} = 0.5.$$

The initial state of the Markov model is selected to be in the state 'good'. According to Assumption 8.1 on p. 175, the statistics of the wireless channel are constant within the distance $\bar{d} = 10\,\mathrm{m}$, which conforms to the threshold used in condition (9.32). An analysis of the Worst Case Execution Time (WCET) of the components of the quadrotors states that the computation times are bounded by

$$\tau_{\mathrm{c,max}} = 33\,\mathrm{ms}. \tag{11.2}$$

In the scenario the quadrotors start on the positions $p_{\mathrm{S}} = (50 \;\; 400 \;\; 100)^{\mathrm{T}}$ and $p_{\mathrm{G}} = (50 \;\; 100 \;\; 100)^{\mathrm{T}}$, respectively and move on their trajectories in a constant height of $z = 100\,\mathrm{m}$, shown in Fig. 11.2. The stand-on agent $\bar{P}_{\mathrm{S}}$ moves constantly with its maximum speed of $v_{\mathrm{S}} = v_{\mathrm{S,max}} = 20\,\frac{\mathrm{m}}{\mathrm{s}}$, while the give-way agent $\bar{P}_{\mathrm{G}}$ moves with a speed of $v_{\mathrm{G}} = 14\,\frac{\mathrm{m}}{\mathrm{s}}$, which

Table 11.1: Parameters of the simulation analysis.

Description	Value
Transmission power level P_0	2 W
Reference distance d_0	1 m
Path loss exponent α	2
Packet size M	128 bytes
Bandwidth B	35 kbps
Rician factor $\mathcal{K}(d(\tilde{t}_i))$	10
Receiver sensitivity S_{rs}	−40 dBm = 100 nW
Bound on the mean packet loss probability $p_{e,max}$	0.03
Safety distance $\underline{s}$	10 m

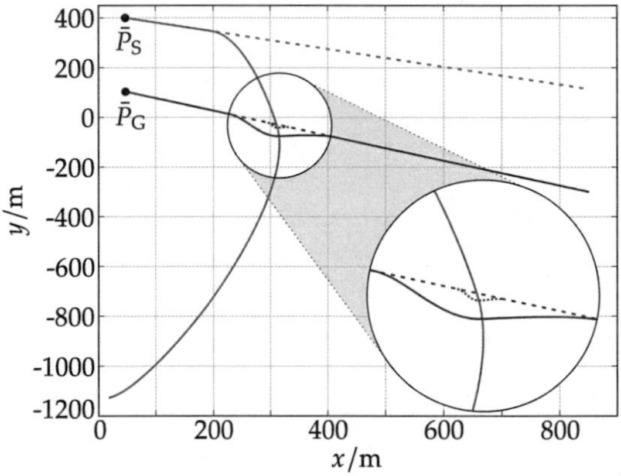

Fig. 11.2: Trajectories of the stand-on agent and the give-way agent.

increases during the avoidance manoeuvre to $v_{G,max} = 30\,\frac{m}{s}$. At time instant $t = 10$ s the stand-on agent changes its trajectory and continues moving on the blue, solid trajectory without notifying the give-way agent. Due to the event-based communication scheme, the give-way agent first detects the upcoming collision after the reception of the information of the stand-on agent at the next event time instant $t = 20$ s. The give-way agent has to change its trajectory in order to avoid the collision. The enlarged part in Fig. 11.2 illustrates the avoidance manoeuvre of the give-way agent when the communication delays are not taken into account (dotted line) and when the delays provided by the delay estimator are considered in the event generation (solid line). As it can be seen the avoidance manoeuvre starts later when the delay is not taken into account since the new information is received later. As a result there is not enough time to avoid the stand-on agent appropriately. In contrast, by considering the delay, the avoidance manoeuvre starts earlier and a suitable trajectory to avoid the stand-on agent is found. After that the give-way agent returns to its initially planned trajectory. In the first 50 s of the simulation no packet losses occurred. Hence, only the time delays have to be considered.

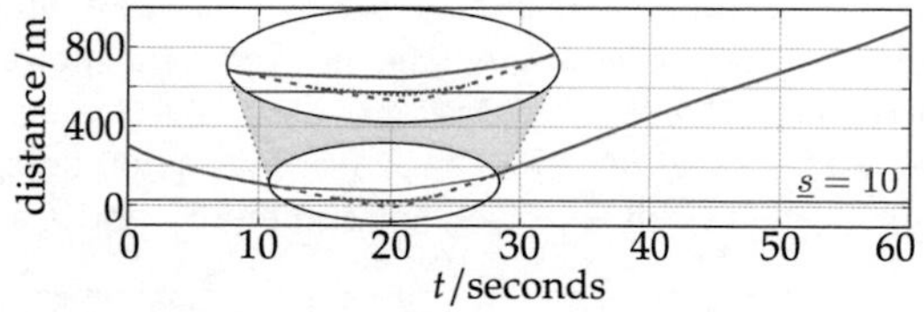

Fig. 11.3: Distance between the stand-on agent and the give-way agent.

Figure 11.3 shows the distance between the UAVs over time. The dashed line states the distance between the quadrotors if no avoidance manoeuvre is performed. It can be seen that the distance decreases rapidly after $t = 10$ s after the change of the trajectory of the stand-on agent. At $t = 20$ s the safety distance $\underline{s}$ (red line) is violated and the agents collide. The dotted line illustrates the distance between the agents when the avoidance manoeuvre is performed without considering the time delay. It can be seen that the avoidance manoeuvre is not sufficient and the agents collide. The solid line shows the distance between the quadrotors if the avoidance manoeuvre is executed under consideration of the estimate of the time delay of the delay estimator. The safety distance is never violated and the collision is avoided.

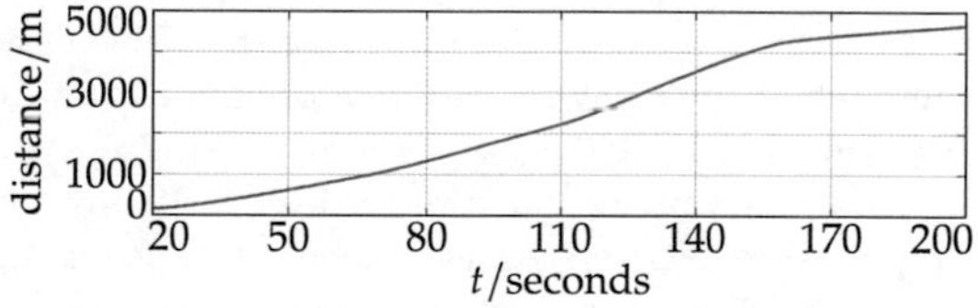

Fig. 11.4: Distance between the stand-on agent and the give-way agent when they follow their diverging trajectories.

In Fig. 11.4 the distance between the quadrotors is shown starting at time instant $t = 20$ s when they follow their diverging trajectories for a longer time. The distance between the agents increases rapidly. In Fig. 11.5 the variation of the received power level at the

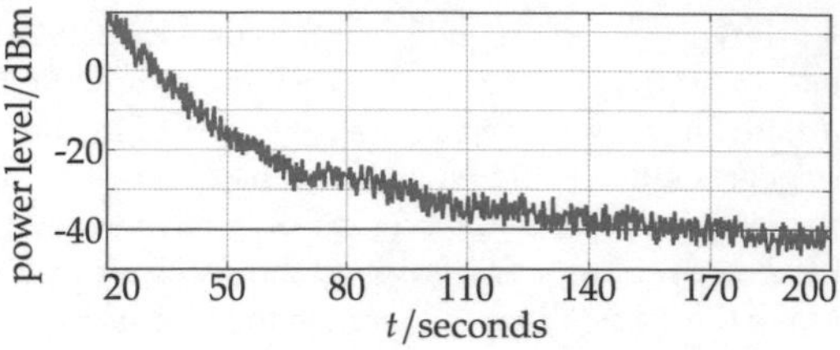

Fig. 11.5: Received power level of the give-way agent.

give-way agent is shown over time for the distance between the agents illustrated in Fig. 11.4. The red line indicates the receiver sensitivity. The received signal power decreases as the distance between the quadrotors increases. After $t = 150$ s the signal power is below the receiver sensitivity and the UAVs are so far away from one another that they are out of range of their communication modules. Hence, no communication is possible any more.

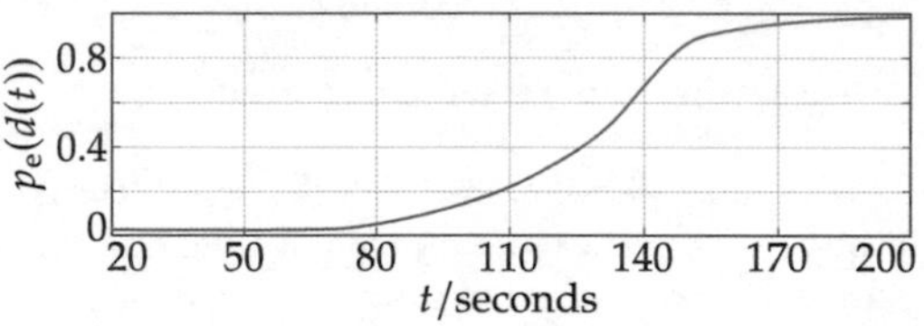

Fig. 11.6: Mean packet loss probability.

Figure 11.6 shows the packet loss probability (5.15), in Fig. 11.7 the time delay determined with (9.13) is illustrated. It can be seen in Fig. 11.6 that the packet loss probability increases as the distance between the UAVs increases. The packet loss probability tends to 1 when the received power level is close to or below the receiver sensitivity. At the same time the time delay increases in the same way (Fig. 11.7). The sent packets require more time to be received due to the reduced signal power for the transmission caused by the increased distance between the quadrotors. As the information is received with a power that equals the receiver sensitivity with (9.9), (9.10) and (9.13) the transmission delay tends to a constant value. Hence, information could be received within this time span despite the further increasing distance between the agents, but at the same time more data are lost due to the increasing packet loss probability shown in Fig. 11.6. As it is stated in Section 9.6 the time delay can be significantly reduced by increasing the bandwidth B of the channel or the transmission power P_0 of the communication modules.

The results of the simulation show that the proposed method provides a good estimate of the time delay, which depends on the distance between the agents and the transmitter and receiver properties. Collisions between two agents can only be avoided in an event-based fashion if the estimated time delay is taken into account in the event generation.

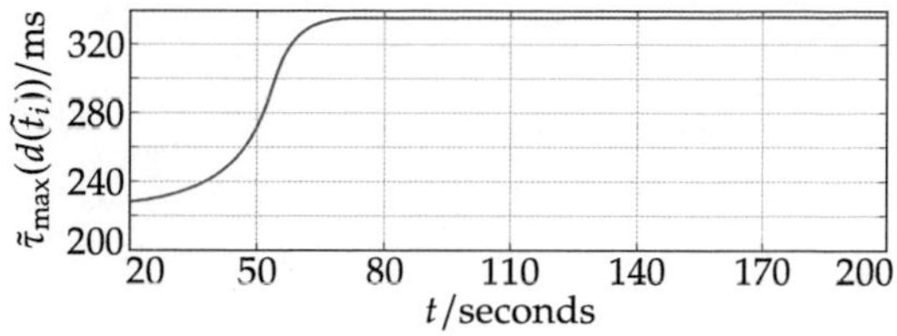

Fig. 11.7: Mean transmission delay.

11.3 Scenario 2: Ensuring collision avoidance in the presence of packet losses

11.3.1 *Description of the scenario*

The second scenario focusses on the event-based control method to guarantee collision avoidance despite the occurrence of packet losses. In the scenario, a movement of the stand-on agent is considered that leads to a collision with the give-way agent. A violation of requirement (1.2) of control aim (G1) does not threaten. The scenario is evaluated for three cases:

- The quadrotors are connected by an ideal network. Communication takes place without delays and packet losses.
- The quadrotors use an unreliable network in which transmission delays and packet losses occur, which are not taken into account in the event-based control method.
- The quadrotors are again connected by an unreliable network, but now the time delays and packet losses are considered in the control method.

An application example of the scenario is the communication relay over two UAVs acting as aerial base stations. They are required to avoid collisions and to keep a maximum separation to ensure requirements on the channel quality.

The scenario illustrated in Fig. 11.8 is as follows: The stand-on agent moves on its trajectory $w_S(t)$ (dashed line) from the start point $S_{A,S}$ to the end point $S_{B,S}$. It changes its trajectory at the time instant $t = t_c$ due to obstacles to the solid line, e.g. wind turbines (Fig. 11.8 (top)). The give-way agent starts from $S_{A,G}$ and should reach its end point $S_{B,G}$ by its initially planned trajectory $w_G(t)$ (dot-dashed line). As it is responsible to ensure the collision avoidance (1.1) it changes its trajectory to the proactive trajectory $w_{G,p}(t)$ (solid line) at time $t = t_c$ (Fig. 11.8 (bottom)). After the stand-on agent increases the distance to the give-way agent, the give-way agent returns to its initially planned trajectory at $t = t_r$ (right part in Fig. 11.8 (bottom)).

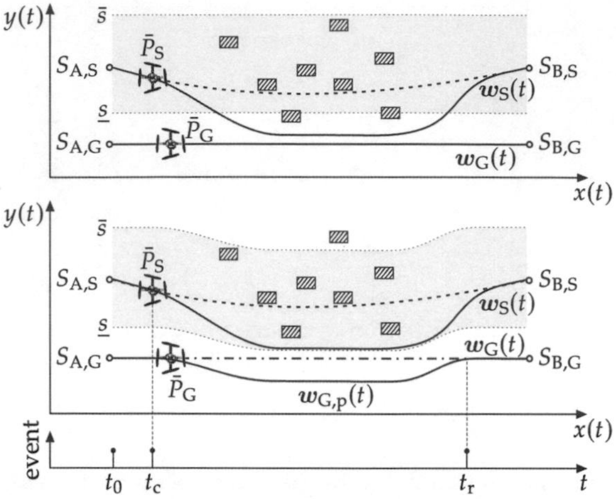

Fig. 11.8: Scenario of collision avoidance.

The safety distance $\underline{s}$ is determined with (11.1). The maximum separation $\bar{s}$ depends on the transmitters for the communication relay so that the required channel quality is maintained. The parameter is given by

$$\bar{s} = s_t \tag{11.3}$$

where s_t states the range of the transmitter to guarantee a desired channel quality.

11.3.2 *Results of the scenario*

The parameters of the control method are given in Tab. 11.2. The trajectories of the quadrotors need to be planned so as to fulfil the dynamic limitations (1.8) and (1.9). In all cases the quadrotors move in a constant height of 3 m. The stand-on agent has the start position $p_S(0) = (0 \;\; 5 \;\; 3)^T$ and moves on the red trajectory in Fig. 11.9, which is initially sent to the give-way agent. It reaches its end point $p_S(16) = (16 \;\; 6 \;\; 3)^T$ at time $t = 16$ s. The give-way agent starts from $p_G(0) = (0 \;\; 1 \;\; 3)^T$ and follows the blue trajectory to its destination $p_G(16) = (16 \;\; 1 \;\; 3)^T$.

Case 1. In this case the quadrotors are connected by an ideal communication network that does not induce transmission delays and packet losses. The computation times of the agents are neglected as well.

Table 11.2: Parameters of the event-based method.

Description	Value
Maximum speed v_{max}	$2\,\frac{m}{s}$
Event threshold $\bar{e}_G$	1.1 m
Event threshold $\bar{e}_{2d}(d(t_{k+1}))$	0.9 m
Event threshold $\bar{e}_d(d(t_{k+j}))$	0.45 m
Parameter t_{com}	0.43 s
Safety distance $\underline{s}$	0.4 m
Spatial distance $\bar{s}$	9 m
Rotor limitations $\underline{n}$; $\bar{n}$	$0\,s^{-1}$; $398\,s^{-1}$
Angle limitations $\underline{\phi}$, $\bar{\phi}$; $\underline{\vartheta}$, $\bar{\vartheta}$	± 60°

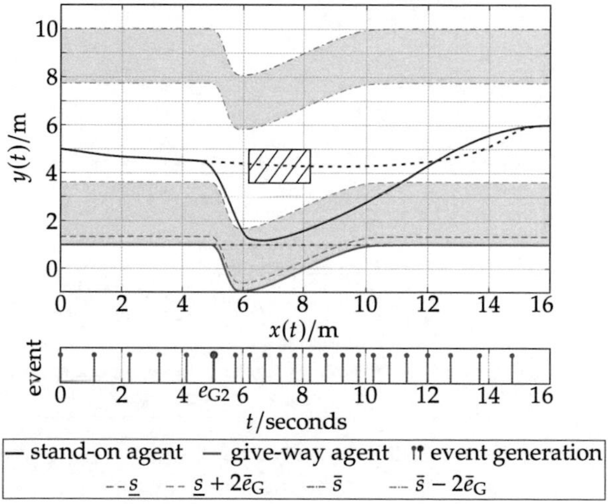

Fig. 11.9: Trajectories of the stand-on agent and the give-way agent for case 1.

Figure 11.9 shows the trajectories of the quadrotors in the xy-plane for the first case. The blue beams indicate an event time instant t_k. The light grey area corresponds to the distances for an event generation with condition (7.4) for an ideal network.

The stand-on agent changes its trajectory from the dashed line to the solid line directly after the event time instant at $t = 4.1$ s, while the give-way agent continues following its initial trajectory. At $t = 4.9$ s the next information is received by the give-way agent. The event e_{G2} is directly generated and the give-way agent evades the stand-on agent. It can be seen that the avoidance manoeuvre has been successful and the collision is safely avoided. In the time span between $t = 6$ s and $t = 11$ s the event time instants are separated from each other by the parameter $t_{com} = 0.43$ s. After $t = 11$ s the give-way agent returns to its initial trajectory.

Case 2. This case investigates the situation of an unreliable network for the communication that induces transmission delays and packet losses. The packet losses are not considered in the event-based control method and the event e_{G3} will not be generated. As the quadrotors are moving quite close to one another the transmission delays stay nearly constant and are given by $\tau_{n,max}(d(\tilde{t}_i)) = 216$ ms. The computation times of the agents are now considered as well and given by (11.2). Hence, the estimated time delay is with eqn. (9.13) determined to be $\tilde{\tau}_{max}(d(\tilde{t}_i)) = 260$ ms.

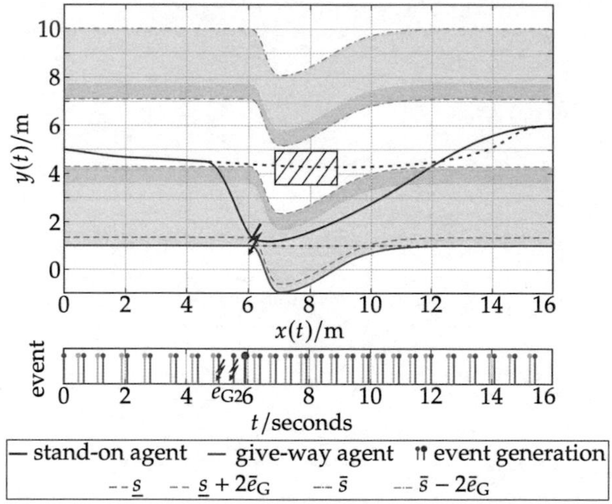

Fig. 11.10: Trajectories of the stand-on agent and the give-way agent for case 2.

Figure 11.10 shows the trajectories of the agents in the xy-plane. The grey beams state the communication time instants $t_{c,k}$, while the blue beams again indicate event time instants t_k. Again, the light grey area corresponds to the distances for an event generation for an ideal network. The dark grey area extends the distances by $\bar{e}_d(d(t_{k+j}))$ to consider the time delays. Both areas together give the condition in (9.24). It can be seen that a consideration of the delay reduces the allowed movement space of the stand-on agent. As a result, events are generated more often. In this case in the control method time delays are considered but not the occurrence of packet losses. Hence, the event e_{G3} will not be triggered if a packet gets lost.

The stand-on agent changes its trajectory from the dashed line to the solid line directly after the communication time instant at $t = 4.1$ s. At $t = 5$ s and at $t = 5.4$ s two consecutive packets get lost. As the event e_{G3} is not triggered the give-way agent continues following its initial trajectory while the stand-on agent gets close. First, at $t = 5.9$ s the next information is received by the give-way agent. The event e_{G2} is directly generated and the give-way agent evades the stand-on agent. This avoidance manoeuvre happens too late and the stand-on agent violates requirement (1.1) of control aim (G1) as illustrated by the flash in Fig. 11.10. Hence, the collision cannot be avoided if packet losses occur and they are not taken into account in the control method. Again in the time span between $t = 6$ s and $t = 11$ s the event time instants are separated from each other by the parameter $t_{com} = 0.43$ s.

Case 3. In this case the same network for the communication is used as in case 2. The difference is that now both, time delays and packet losses, are considered in the event-based control method. The time delay is still estimated to be $\tilde{\tau}_{max}(d(\tilde{t}_i)) = 260$ ms.

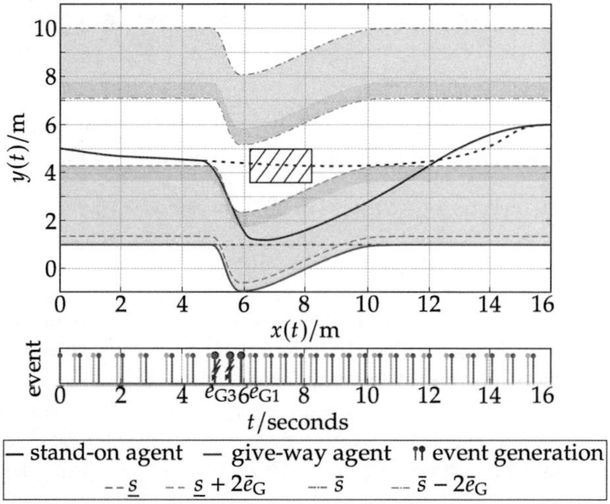

Fig. 11.11: Trajectories of the stand-on agent and the give-way agent for case 3.

The trajectories of the quadrotors are shown in Fig. 11.11 in the xy-plane. The stand-on agent changes its trajectory again at $t = 4.1$ s from the dashed line to the solid line. Two consecutive packets get lost at $t = 5$ s and at $t = 5.4$ s. At both time instants the event e_{G3} is generated with condition (9.30) and the give-way agent increases the distance to the stand-on agent by $2\,d_S$, because two packets got lost. It can be seen that the avoidance manoeuvre is successful and the collision is safely avoided. At $t = 5.9$ s the next information is received. Event e_{G1} is directly invoked and the trajectory of the give-way agent is planned so that it returns to its initial trajectory as the movement of the stand-on agent allows this.

11.4 Scenario 3: Fulfilment of the maximum separation in the presence of packet losses

11.4.1 *Description of the scenario*

The third scenario is similar to the second scenario. Again the event-based control method is applied to two quadrotors that should not exceed a given spatial separation $\bar{s}$ to one another, while satisfying the safety distance $\underline{s}$ to one another to ensure the collision avoidance. The difference is that now the evaluation is focussed on the fulfilment of the spatial separation $\bar{s}$ in the presence of packet losses. The scenario is executed for three cases:

- The quadrotors communicate over an unreliable network that induces transmission delays, but no packet losses occur and the stand-on agent changes its movement so that it diverges from the give-way agent.
- The quadrotors communicate over an unreliable network that induces transmission delays and packet losses and the stand-on agent changes its trajectory only slightly.
- The quadrotors communicate over an unreliable network that induces transmission delays and packet losses and the stand-on agent changes its movement so that it diverges from the give-way agent.

The scenario depicted in Fig. 11.12 is as follows: The two quadrotors move on their trajectories $\boldsymbol{w}_S(t)$ (dashed line) and $\boldsymbol{w}_G(t)$ (solid line) from their respective start points $S_{A,S}$ and $S_{A,G}$ to their respective end points $S_{B,S}$ and $S_{B,G}$ (Fig. 11.12 (top)). The agents should not exceed the distance $\bar{s}$ to one another, which means the stand-on agent has to stay inside the grey area in Fig. 11.12. At time instant $t = t_c$ the stand-on agent changes its trajectory due to obstacles (e.g. wind turbines). In the first and the third case of the scenario it changes its movement to the solid trajectory and in the second case its movement is adjusted to the dot-dashed line (Fig. 11.12 (top)). As the give-way agent is responsible to keep the spatial distance (1.2) fulfilled in the first and the third case it changes its trajectory to the solid line (Fig. 11.12 (bottom)). In the second case no change of the trajectory of the give-way agent is necessary (dot-dashed line in Fig. 11.12 (bottom)). After the avoidance manoeuvre the give-way agent returns to its initially planned trajectory (right part in Fig. 11.12 (bottom)).

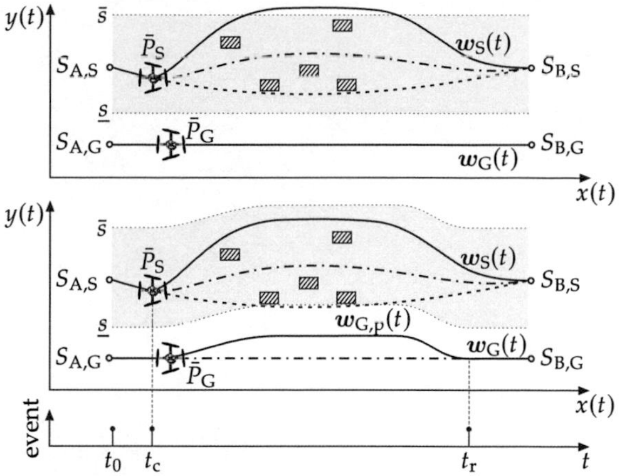

Fig. 11.12: Scenario of keeping a spatial separation.

The safety distance $\underline{s}$ is determined with (11.1), while the bound on the spatial separation $\bar{s}$ is obtained by (11.3).

11.4.2 *Results of the scenario*

The parameters of the control method are given in Tab. 11.2. The trajectories of the quadrotors need again to be planned so as to fulfil the dynamic limitations (1.8) and (1.9). In all cases the quadrotors move in a constant height of 3 m. The stand-on agent and the give-way agent start from the same positions and move to the same destinations as in the Scenario 2. In all cases the time delay is still estimated to be $\tilde{\tau}_{\max}(d(\tilde{t}_i)) = 260\,\text{ms}$ due to the close movement of the quadrotors.

Case 1. In this case the quadrotors are connected by a network that induces transmission delays but no packet losses occur.

The trajectories of the quadrotors in the xy-plane are shown in Fig. 11.13. Again, the grey beams state communication time instants $t_{c,k}$ and the black beams represent event time instants t_k. The stand-on agent changes its trajectory from the dashed line to the solid line after $t = 2\,\text{s}$. The give-way agent detects a violation of requirement (1.2) of control aim (G1) at $t = 5.8\,\text{s}$ and generates event e_{G1} at $t = 5\,\text{s}$ with condition (7.11). It leaves its initial trajectory (dashed line) and continues following the proactive trajectory (solid line). At $t = 15\,\text{s}$ the give-way agent returns to its initially planned trajectory. It can be seen that the spatial separation between the agents is satisfied all the time due to the avoidance manoeuvre of the give-way agent.

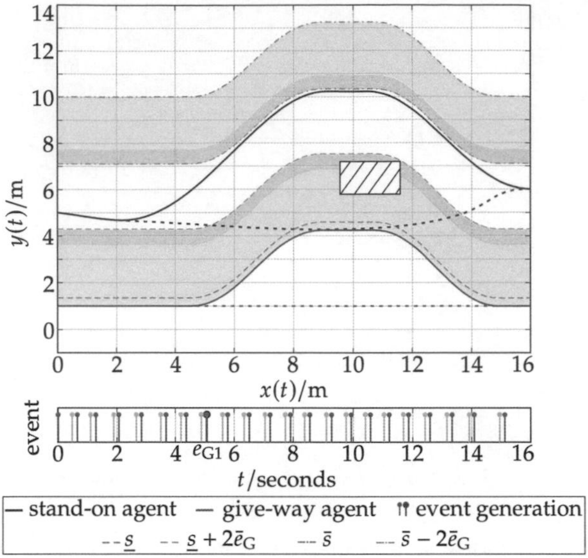

Fig. 11.13: Trajectories of the stand-on agent and the give-way agent for case 1.

Case 2. This case investigates the situation in which an unreliable network is used for the communication that induces transmission delays and packet losses. Both effects are now taken into account in the event-based control method.

The trajectories of the quadrotors in the xy-plane are shown in Fig. 11.14. The stand-on agent changes its trajectory again from the dashed line to the solid line after $t = 2\,\mathrm{s}$. At $t = 5\,\mathrm{s}$ and at $t = 5.4\,\mathrm{s}$ two consecutive packets get lost. Hence, the event e_{G3} is generated twice with condition (9.30), which leads the give-way agent to increase the distance to the stand-on agent by $2\,d_{\mathrm{S}}$ due to the lack of information. It can be seen in the figure that the maximum separation (1.2) between the agents is fulfilled at any time. This is possible, because the stand-on agent moves on a trajectory that keeps it close to the give-way agent. At time $t = 6.1\,\mathrm{s}$ condition (7.11) is violated. The give-way agent notices this fact at $t = 5.9\,\mathrm{s}$ when it receives the current information of the stand-on agent. The event e_{G1} is generated but the trajectory remains unchanged, because the give-way agent already moves with its maximum speed and no violation of requirement (1.2) threatens.

Case 3. In this case the same situation is considered as in case 2. The trajectories of the quadrotors in the xy-plane are shown in Fig. 11.15.

The stand-on agent changes its trajectory at the same time instant under the same circumstances from the dashed line to the solid line as in case 2. Due to the packet losses,

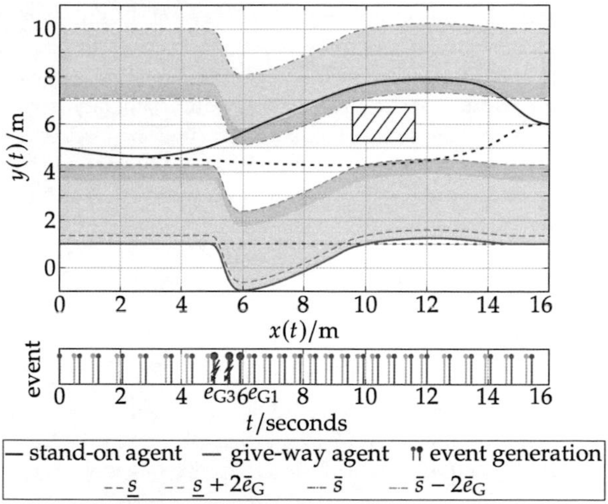

Fig. 11.14: Trajectories of the stand-on agent and the give-way agent for case 2.

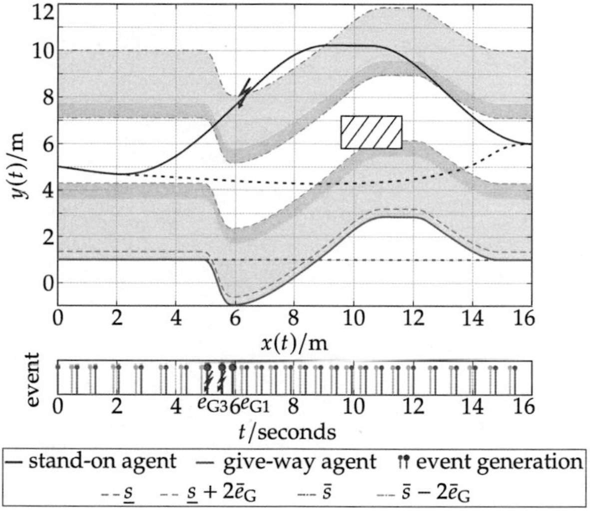

Fig. 11.15: Trajectories of the stand-on agent and the give-way agent for case 3.

the event e_{G3} is generated twice with condition (9.30), which leads the give-way agent to increase the distance to the stand-on agent by $2\,d_S$. It can be seen that the maximum separation cannot be satisfied any more due to the change of the trajectory of the give-way agent. The trajectory replanning has to be performed due to the lack of information from the stand-on agent to ensure requirement (1.1). It is necessary, because the collision avoidance has a higher priority than keeping the maximum separation between the agents. Hence, the maximum separation is violated for a short time span. After receiving the next data from the stand-on agent at time $t = 5.9\,\mathrm{s}$, event e_{G1} is generated and the give-way agent changes its trajectory so as to fulfil both requirements (1.1) and (1.2) again and to reach its given destination.

11.5 Concluding remarks

The simulations have shown that the event-based control method is able to guarantee the collision avoidance between the agents even in the presence of time delays and packet losses. The maximum separation between the agents can be guaranteed in the presence of time delays, but the control aim cannot be ensured if packet losses occur. In this case the fulfilment of (1.2) depends on the movement of the stand-on agent. Due to the lack of information caused by the packet losses, the give-way agent generates the event e_{G3} to guarantee requirement (1.1) even though this causes a violation of requirement (1.2).

In the control method the time delays and the packet losses need to be considered, because otherwise no satisfaction of control aim (G1) can be achieved. However, this reduces the movement space of the stand-on agent and leads to a slightly increased number of event generations and thus of communication between the agents.

The event generation depends strongly on the quality of the estimate of the delay estimator. Since a packet is considered to be lost if the current delay τ_k is greater than the estimated delay $\tilde{\tau}_{\max}(d(\tilde{t}_i))$, the event e_{G3} can be generated unnecessarily. This also happens when packets are received after the time span $\tilde{\tau}_{\max}(d(\tilde{t}_i))$ and not have been lost. The avoidance trajectory is also planned unnecessarily in this cases, which could lead to a violation of requirement (1.2).

Experiments 12

This chapter presents the experimental evaluation of the event-based control method. The aims of the experiments are stated in Section 12.1. Section 12.2 introduces the test bed MULAN, which has been used for the cooperative flights of the quadrotors. The scenario to be investigated and the experimental results are presented in Section 12.3. In Section 12.4 computational aspects of the experiments are summarised.

12.1 Aims

In this chapter the event-based control method is first evaluated in experiments with two quadrotors. The quadrotors have the dynamics derived in Section 10.2 and move on circular trajectories. The aim of the experiments is to investigate whether the control aim (A1) for a circular movement can be satisfied in the presence of external disturbances that lead to an inaccurate trajectory tracking of the quadrotors. Furthermore, it is examined whether the complex algorithms are suitable for real-time applications.

Due to the limited space at the test bed the distance between the quadrotors is quite small and does not vary significantly. Hence, the properties of the utilised communication channel do not change and can be considered to be ideal without transmission delays and packet losses. For this reason, in the scenarios the communication frequency is evaluated depending on different values for the tolerance range γ of requirement (1.3) of control aim (A1) for a circular movement.

The event-based control method is examined in the following scenario that correspond to control aim (A1) for a circular movement and consists of two parts:

1. In the first part of the scenario the quadrotors maintain a phase difference between them, while moving on different heights. Height changes are not investigated. The focus is on the evaluation of the communication frequency for different values of the tolerance range γ.

2. In the second part of the scenario the quadrotors maintain still a phase difference between them, but now they do change their heights. The influence of height changes on the communication frequency is investigated.

Furthermore, the event-based control method is applied to two robots that have the dynamics given in Section 2.5.2 and move also on circular trajectories. The aim of the experiment is to evaluate whether the control method is suitable to control a different type of agent so as to fulfil control aim (A1) for a circular movement. To this aim the first part of the scenario is considered with the difference that the robots move in one plane.

12.2 Experimental test bed

Setup of the test bed. The event-based control method is tested with two quadrotors moving in the test bed 'MULAN' in the laboratory of the Ruhr University Bochum, shown in Fig. 12.1.

Fig. 12.1: Experimental test bed 'MULAN'.

The overall structure of the test bed is illustrated in Fig. 12.2. The airspace is a $4\,\mathrm{m} \times 4\,\mathrm{m} \times 7\,\mathrm{m}$ cuboid with net walls along with padded flooring, which provides a safe and enclosed setting. A 8 – camera infrared motion capture system with cameras of type 'Flex 13' from the company 'OptiTrack' delivers precise quadrotor position, altitude and speed measurements at 120 Hz to a ground station.

Implementation of the control loop. The quadrotors are controlled from the ground station consisting of a desktop computer with an Intel Core i7-3770 processor by a self-developed control scheme. The structure of the implemented control loop is shown in Fig. 12.3. The attitude controllers generating the input $u_{\mathrm{r}}(t)$ are implemented locally on the quadrotors. In contrast, the position and the altitude controllers as well as the

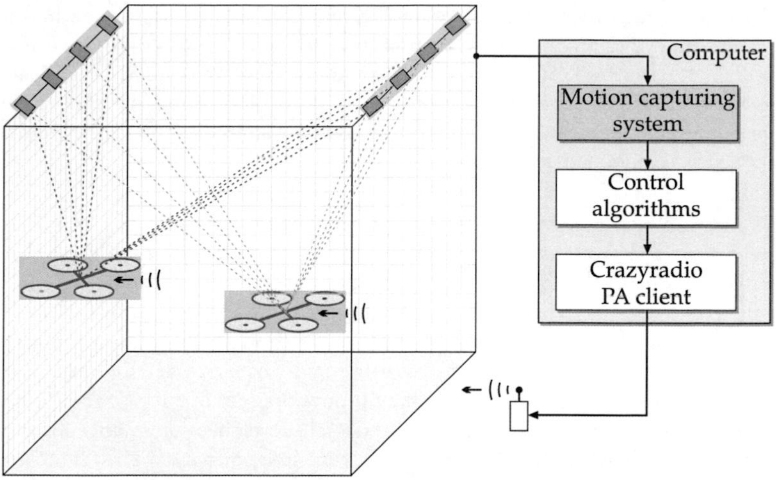

Fig. 12.2: Structure of the experimental test bed.

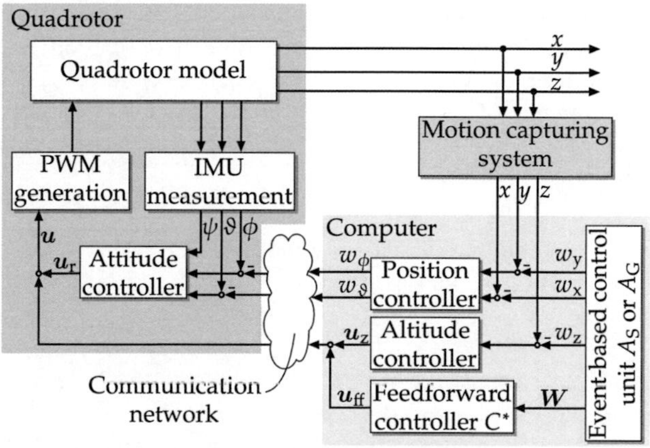

Fig. 12.3: Implementation of the control loop.

feedforward controllers C^* are carried out on the computer. Furthermore, the control units A_S of the stand-on agent and A_G of the give-way agent for the trajectory planning run on the computer. The setpoints $w_\phi(t)$ and $w_\vartheta(t)$ given by the position controllers, the rotor speeds $\boldsymbol{u}_z(t)$ generated by the altitude controllers and the input $\boldsymbol{u}_{ff}(t)$ of the feedforward controllers are sent to the corresponding UAVs. The overall rotor speeds $\boldsymbol{u}(t)$ are generated with the local attitude controllers using the local measurements and the communicated information from the PC to achieve the trajectory following. Even though parts of the method are implemented on the PC, the information structure is still decentralised and corresponds to the structure shown in Fig. 10.13 on p. 246. Thus, the chosen type of implementation has no influence on the results and only simplifies the implementation of the event-based components.

Communication between ground station and quadrotors. Wireless communication is required for two tasks: on the one hand to control the quadrotors from the ground station and on the other hand to exchange data of the control method between the agents.

For the communication between the ground station and the quadrotors, both the computer and the UAVs are equipped with 'Crazyradio PA' communication modules. These modules are 2.4 GHz antennas, which consist of a power amplifier (PA) and a transmission power of 20 dBm with a maximum data rate of $20\frac{\text{Mbit}}{\text{s}}$. The 'Enhanced Shockburst' (ESB) as a physical radio channel and a self-developed communication protocol are used by the Crazyradio. The ESB handles acknowledgements (ACK) and retries. The Crazyradio sends a packet on a given channel and waits for an ACK. If a quadrotor of type 'Crazyflie' receives the packet without an error it sends an ACK packet. If an ACK is received, the Crazyradio sends the next packet. If no ACK is received, the Crazyradio will automatically retry by sending the same packet again.

The proposed communication protocol is split into an initialisation phase and an execution phase. In the initialisation phase the message, given in Tab. 12.1, is sent to the quadrotors, which can be distinguished by their individual identification numbers (ID). After a successful initialisation the quadrotors send the same message back to the PC to complete the initialisation phase.

Table 12.1: Structure of the initial message.

Byte	1	2	3	4	5
Sign	#	ID	,	1	;

In the execution phase the quadrotors expect a new message every 10 ms. If an agent did not receive a message after 50 ms, an emergency landing of this quadrotor is executed for safety reasons. The data to be sent is structured as follows: The character 'd' indicates the beginning of the information. The information consists only of digits separated by the letter 'x'. The first four digits indicate the quadratic rotor speeds $n_i(t)$, $(i = 1, \ldots, 4)$ generated by the altitude controller on the ground station. Subsequently, the set-points for the three Euler angles $w_\phi(t)$, $w_\vartheta(t)$, $w_\psi(t)$ are transmitted in the message. The structure of the message is stated in Tab. 12.2.

The quadratic rotor speeds are transmitted as a number of three digits, the angles are sent as a number of two digits, since the angles are limited to $\pm 60°$.

Table 12.2: Structure of the regular message.

Byte	1	2	3	4	5	6	7	8	9	10	11	12	13	14	15
Sign	#	ID	,	d	u_1			x	u_2			x	u_3		
Byte	16	17	18	19	20	21	22	23	24	25	26	27	28	29	
Sign	x	u_4			x	w_ϕ		x	w_ϑ		x	w_ψ		;	

Communication between the quadrotors. The communication between the two quadrotors is not established as a physical network. As the controller on the Crazyflie is only able to receive data but not to send data on its own the communication between the agents is imitated with the network architecture on the computer using the model derived in Section 5.3 and the information between the agents are exchanged internally in the ground station. The communication protocol used in this virtual network is described in Section 7.5.1.

12.3 Experimental results

12.3.1 *Description of the scenario*

In the scenario the event-based control method is applied to the two quadrotors to fulfil control aim (A1) for a circular movement under experimental conditions. An application example of the scenario is the communication relay over two UAVs acting as aerial base stations. In order to reduce their energy consumption the relay stations move on circular trajectories while forwarding the data packets.

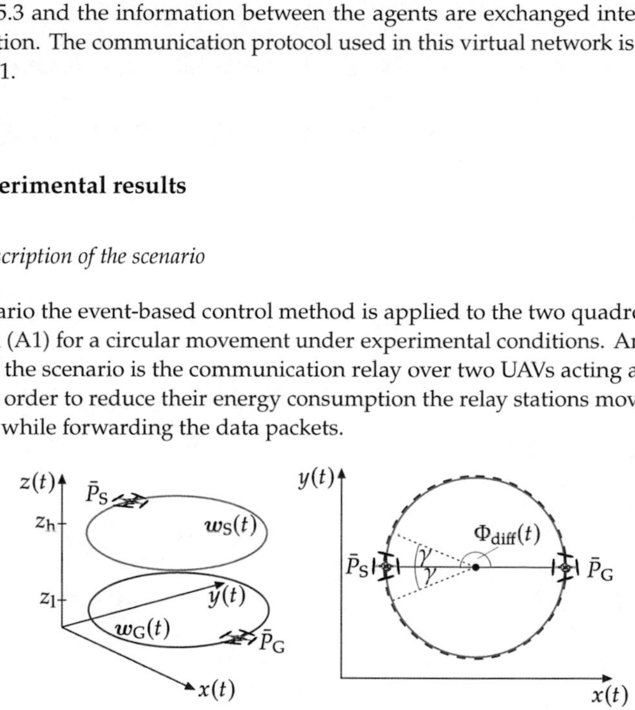

Fig. 12.4: Problem of formation preservation to be evaluated.

The scenario illustrated in Fig. 12.4 is as follows: Two quadrotors move on circular trajectories in two different heights z_l and z_h. The stand-on agent is able to change its trajectory $w_S(t)$ at any time by varying the speed $v_S(t)$ on the circular path or by changing the height $z_S(t)$ of its circular movement. The give-way agent is responsible to ensure the requirements on the phase difference (1.3) and on the height difference (1.6) of control aim (A1) for a circular movement. In the scenario, this means that the quadrotors should always be located opposite to one another and should move on different heights. The give-way agent has to react on the movement of the stand-on agent in order to keep control aim (A1) for a circular movement satisfied. The phase difference has always to be fulfilled within the tolerance range, while the height difference has to be satisfied at any time except during height changes of the UAVs.

The scenario is evaluated in two parts. In the first part only the satisfaction of the phase difference between the quadrotors is investigated for different values of the tolerance range γ. The UAVs change only the speed along their circular path and do not change their heights during the scenario. In the second part of the scenario the quadrotors change both the speed on their circular path and the height of their movement.

For an evaluation of the event-based control method that is applied to two robots a similar scenario is considered. The robots should maintain a phase difference between them while moving on circular trajectories. The only difference to the scenario depicted in Fig. 12.4 is that the agents move in one plane and do not execute any height changes.

12.3.2 *Part 1: Maintenance of a phase difference between two quadrotors*

In the first scenario only a change of speed of the stand-on agent is investigated. Hence, the event threshold $\bar{e}_Z$ is not reached. The experiment is evaluated for three values of the tolerance range: $\gamma = 45°$, $\gamma = 90°$ and $\gamma = 135°$. The parameters of the event-based control method are stated in Tab. 12.3.

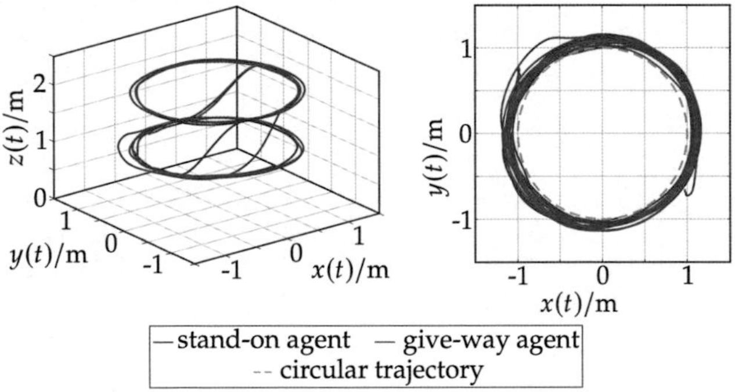

Fig. 12.5: Movement of the two quadrotros for the second part of the experiment.

The trajectories of the quadrotors need to be planned so as to fulfil the dynamic limitations (1.8) and (1.9). The stand-on agent moves on the red circular trajectory, while the give-way

Table 12.3: Parameters of the event-based control method.

Description	Value
Radius of the circle r	$1\,\mathrm{m}$
Lower flight altitude z_l	$1\,\mathrm{m}$
Upper flight altitude z_h	$2\,\mathrm{m}$
Minimum speed $v_{\mathrm{G,min}}$	$0\,\frac{\mathrm{m}}{\mathrm{s}}$
Minimum speed $v_{\mathrm{S,min}}$	$0.2\,\frac{\mathrm{m}}{\mathrm{s}}$
Maximum speed $v_{\mathrm{G,max}}$	$1.2\,\frac{\mathrm{m}}{\mathrm{s}}$
Maximum speed $v_{\mathrm{S,max}}$	$1\,\frac{\mathrm{m}}{\mathrm{s}}$
Event threshold $\bar{e}_\Phi$	4.58°
Event threshold $\bar{e}_z$	$0.5\,\mathrm{m}$
Parameter t_{com}	$1.41\,\mathrm{s}$
Angle limitations $\underline{\phi},\ \bar{\phi};\ \underline{\vartheta},\ \bar{\vartheta}$	$\pm\,60^\circ$
Rotor limitations $\underline{n};\ \bar{n}$	$0\,\mathrm{s}^{-1};\ 398\,\mathrm{s}^{-1}$

agent follows the blue circular trajectory shown in Fig. 12.5. Both trajectories are planned with a radius of $r = 1\,\mathrm{m}$.

Results for $\gamma = 45^\circ$. In the first experiment the quadrotors should maintain the phase difference with a tolerance range of $\gamma = 45^\circ$. The stand-on agent starts moving on its circular trajectory at a height $z_\mathrm{S}(t) = 2\,\mathrm{m}$ with the speed $v_\mathrm{S}(0) = 0.4\frac{\mathrm{m}}{\mathrm{s}}$. The give-way agent starts opposite to the stand-on agent on its circular trajectory at a height $z_\mathrm{G}(t) = 1\,\mathrm{m}$ with the speed $v_\mathrm{G}(0) = 0.4\frac{\mathrm{m}}{\mathrm{s}}$. The stand-on agent changes during the experiment its speed three times as summarised in Tab 12.4.

In Fig. 12.6 the phases $\Phi_\mathrm{S}(t)$, $\Phi_\mathrm{G}(t)$ of the agents, the phase difference $\Phi_{\mathrm{diff}}(t)$, the height difference $z_{\mathrm{diff}}(t)$ and the generation of the events e_{G0} and e_{G1} are indicated. The change of the speed of the agents can be seen in the upper part of the figure by a flattening or by a rise of the slope of the phase of the agents. Each time the stand-on agent changes its speed, the give-way agent reacts on this change by an adaptation of its speed after the next event time instant. The tolerance range is chosen so small that whenever the stand-on agent communicates a change of its trajectory, the give-way agent directly generates the event

Table 12.4: Time instants for a change of the speed.

Time instants $t_{c,si}$	Speeds $v_{S,i}$
$t_{c,s0} = 0\,\mathrm{s}$	$v_{S,0} = 0.4\frac{\mathrm{m}}{\mathrm{s}}$
$t_{c,s1} = 7.85\,\mathrm{s}$	$v_{S,1} = 0.5\frac{\mathrm{m}}{\mathrm{s}}$
$t_{c,s2} = 20.73\,\mathrm{s}$	$v_{S,2} = 0.3\frac{\mathrm{m}}{\mathrm{s}}$
$t_{c,s3} = 45.41\,\mathrm{s}$	$v_{S,3} = 0.4\frac{\mathrm{m}}{\mathrm{s}}$

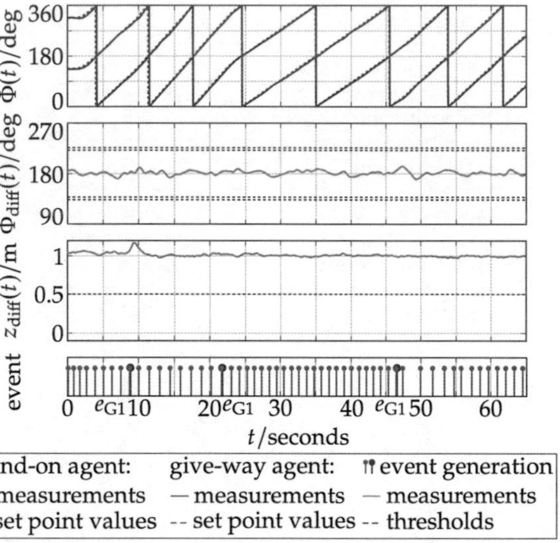

Fig. 12.6: Results of the first experiment of the first part of the scenario.

e_{G1}. It can be seen that the give-way agent changes its speed accordingly by a change of the rise of the slope of its phase.

In the middle part of Fig. 12.6 it can be seen that the event generation of the give-way agent is appropriate to keep the phase difference $\Phi_{\mathrm{diff}}(t)$ inside the tolerance range. Furthermore, the phase difference equals nearly at any time the desired set point. In the lower part of the figure it is shown that the height difference equals $\bar{z} = 1\,\mathrm{m}$ all the time since no height changes are performed.

The last part of the figure indicates that the average time between two consecutive event time instants is $\Delta t_k = 1.16\,\mathrm{s}$. Hence, the communication effort is reduced considerably compared to a continuous communication scheme. In the experiment the average loop time for the execution of the algorithms is $\Delta t_{\mathrm{loop}} = 20.99\,\mathrm{ms}$.

Results for $\gamma = 90°$. In the second experiment the phase difference between the two quadrotors should be maintained with a tolerance range of $\gamma = 90°$. The UAVs start on the same trajectories with the same speeds as in the first experiment. In addition the stand-on agent changes its speed again at the time instants given in Tab. 12.4.

In Fig. 12.7 again the phases $\Phi_{\mathrm{S}}(t)$, $\Phi_{\mathrm{G}}(t)$ of the agents, the phase difference $\Phi_{\mathrm{diff}}(t)$, the height difference $z_{\mathrm{diff}}(t)$ and the generation of the events e_{G0} and e_{G1} are indicated. As the stand-on agent changes its speed at $t_{\mathrm{c,s1}}$ and $t_{\mathrm{c,s3}}$ only slightly, the event e_{G1} is generated about 10 s after the changes were communicated at $t = 19.45\,\mathrm{s}$ and $t = 51.45\,\mathrm{s}$, because the violation of control aim (A1) for a circular movement is far in the future. After the generation of the events the give-way agent accelerates as it can be seen by the rise of the slope of the phase of the agent. Afterwards it continues moving with the speed of the stand-on agent. In contrast, at $t_{\mathrm{c,s2}}$ the event e_{G1} is generated only about 3 s after the change of speed was communicated at $t = 23.89\,\mathrm{s}$.

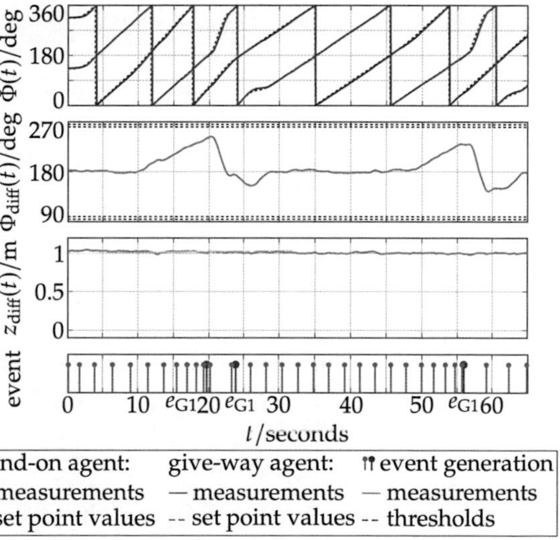

Fig. 12.7: Results of the second experiment of the first part of the scenario.

In the middle part of Fig. 12.7 it can be seen that the event generation of the give-way agent is again appropriate to keep the phase difference $\Phi_{\mathrm{diff}}(t)$ inside the tolerance range and control aim (A1) is satisfied at any time. The maximum deviation from the desired phase difference amounts to 50° and is much larger compared to the first experiment. In the lower part of the figure it is shown that the height difference equals $\bar{z} = 1\,\mathrm{m}$ all the time since again no height changes are performed.

The last part of the figure indicates that the average time between two consecutive event time instants is $\Delta t_k = 2.03\,\mathrm{s}$. Hence, the communication effort is smaller compared to the first experiment. In the experiment the average loop time for the execution of the algorithms is again $\Delta t_{\mathrm{loop}} = 20.99\,\mathrm{ms}$.

Results for $\gamma = 135°$. In the third experiment it is demanded that the phase difference between the two UAVs remains inside a tolerance range of $\gamma = 135°$. Again the quadrotors start on the same trajectories and with the same speeds as in the first experiment. Furthermore, the stand-on agent changes its speed again at the time instants given in Tab. 12.4.

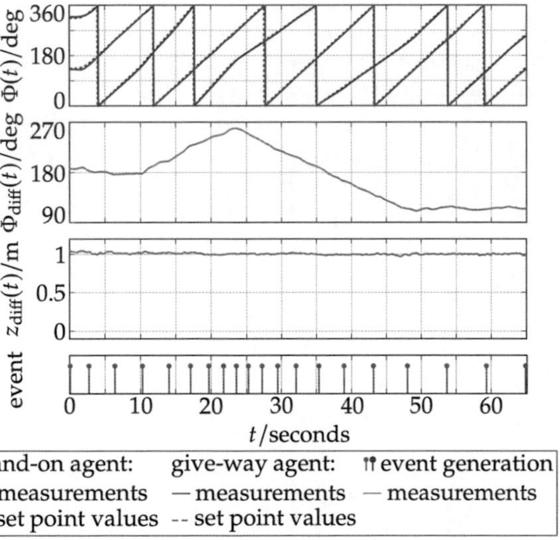

Fig. 12.8: Results of the third experiment of the first part of the scenario.

In Fig. 12.8 again the phases $\Phi_S(t)$, $\Phi_G(t)$ of the agents, the phase difference $\Phi_{\mathrm{diff}}(t)$, the height difference $z_{\mathrm{diff}}(t)$ and the generation of the events e_{G0} and e_{G1} are indicated. The event thresholds are not shown as they are outside of the illustrated range of the plot. The tolerance range is chosen to be so large that no event e_{G1} is generated during the experiment. Due to the large tolerance range the changes of the speed of the stand-on agent lead to a variation of the phase difference but not to a violation of requirement (1.3). The first change of speed of the stand-on agent leads to an increasing phase difference, the next change of speed leads to a decreasing phase difference. From the time instant

$t_{c,s3}$ on both quadrotors move with the same speed so that a constant phase difference of $\Phi_{diff}(t) = 117.9° \neq 180°$ results.

In the middle part of Fig. 12.8 it can be seen that the deviation from the desired phase difference is larger than in the first two experiments, since not event e_{G1} is generated. The maximum deviation amounts to 78°. In the lower part of the figure it is shown that the height difference equals $\bar{z} = 1\,\text{m}$ all the time since again no height changes are performed.

The last part of the figure states that the average time between two consecutive event time instants is $\Delta t_k = 3.25\,\text{s}$. Hence, the communication effort is smaller compared to the first two experiments. In the experiment the average loop time for the execution of the algorithms is again $\Delta t_{loop} = 20.99\,\text{ms}$.

12.3.3 *Part 2: Maintenance of a phase difference and a height difference between two quadrotors*

In the second scenario the change of both the speed and the height of the stand-on agent are investigated for a tolerance range of $\gamma = 90°$.

Table 12.5: Time instants for a change of the height and the speed.

Time instants $t_{c,hi}$, $t_{c,si}$	Heights $z_{S,i}$, speeds $v_{S,i}$
$t_{c,s0} = 0\,\text{s}$	$z_{S,0} = z_h$
$t_{c,h1} = 16.57\,\text{s}$	$z_{S,1} = z_l$
$t_{c,h2} = 33.00\,\text{s}$	$z_{S,2} = z_h$
$t_{c,h3} = 47.38\,\text{s}$	$z_{S,3} = z_l$
$t_{c,s0} = 0\,\text{s}$	$v_{S,0} = 0.5\frac{\text{m}}{\text{s}}$
$t_{c,s1} = 6.28\,\text{s}$	$v_{S,1} = 0.6\frac{\text{m}}{\text{s}}$
$t_{c,s2} = 13.82\,\text{s}$	$v_{S,2} = 0.53\frac{\text{m}}{\text{s}}$
$t_{c,s3} = 19.55\,\text{s}$	$v_{S,3} = 0.57\frac{\text{m}}{\text{s}}$
$t_{c,s4} = 27.90\,\text{s}$	$v_{S,3} = 0.63\frac{\text{m}}{\text{s}}$
$t_{c,s5} = 35.50\,\text{s}$	$v_{S,3} = 0.69\frac{\text{m}}{\text{s}}$
$t_{c,s6} = 44.68\,\text{s}$	$v_{S,3} = 0.48\frac{\text{m}}{\text{s}}$
$t_{c,s7} = 47.38\,\text{s}$	$v_{S,3} = 0.63\frac{\text{m}}{\text{s}}$
$t_{c,s8} = 55.17\,\text{s}$	$v_{S,3} = 0.51\frac{\text{m}}{\text{s}}$
$t_{c,s9} = 57.84\,\text{s}$	$v_{S,3} = 0.72\frac{\text{m}}{\text{s}}$

The agents begin their movements with the same speed as in the first part of the scenario. The stand-on agent changes its speed at time instants $t_{c,si}$ and its height at time instants $t_{c,hi}$ several times as summarised in Tab. 12.5. The results of the experiment are shown in Fig. 12.9. As in the first three experiments the change of speed of the agents can be seen as a change of the slope of the phases of the agents shown in the upper part of the figure. In the middle part of the figure it is shown that the control aim (A1) for a circular movement is satisfied at any time except during the height changes of the agents. Hence, the generation of the events is appropriate to fulfil the control aim.

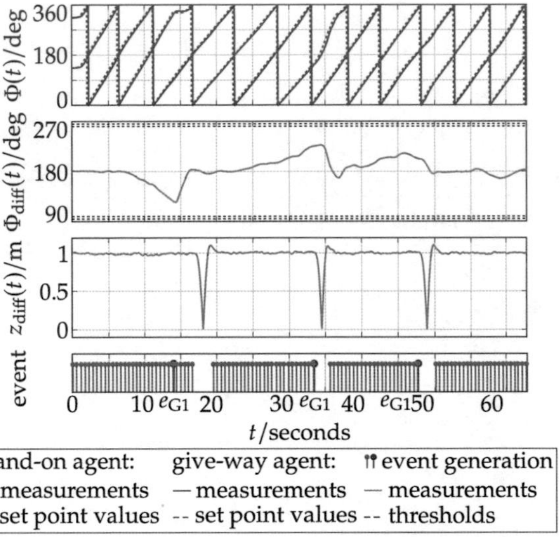

Fig. 12.9: Results of the second experiment.

In the lower part of the figure it can be seen that more communication is invoked as in the first three experiments. Now the average time between two consecutive event time instants is $\Delta t_k = 0.53$ s. This shorter time interval is caused by the height changes in this experiment, which cause the main number of communication invocations. Hence, the uncertainty about the phase and the height of the stand-on agent remains small compared to the first scenario. Furthermore, due to the height changes at time instants $t_{c,hi}$ the event time instants are separated by the time span t_{com} to avoid Zeno behaviour. The average loop time for the execution of the algorithms is $\Delta t_{loop} = 21.45$ ms. Hence, despite taking height changes into account the loop time remains nearly unchanged.

12.3.4 *Maintenance of a phase difference between two robots*

In this section the event-based control method is applied to two robots. The scenario to be considered is similar to the scenario in Section 12.3.2, where the two agents should maintain a phase difference of $\Phi_{\text{diff}}(t) = 180°$ to one another. However, a change of the height is not considered, because the robots move in one plane. The experiment is carried out at the test bed 'SAMS' in the laboratory at the Ruhr University Bochum, which is described in detail in [167]. It is again evaluated for three values of the tolerance range ($\gamma = 45°$, $\gamma = 90°$ and $\gamma = 135°$), to compare the results to the experimental results given in Section 12.3.2. The aim of the experiments is to show that the proposed method is also suitable for the control of different types of agents by only changing the parametrisation of the method with respect to the different dynamics of the agents.

Fig. 12.10: Robots of type 'M3PI'.

Two robots of type 'M3PI' from the company 'Pololu' are used, which are shown in Fig. 12.10 and represented by the state space model (2.22) on p. 41. The robots move by using a differential drive consisting of two independently powered wheels. The marker balls on top of the robots are required to track them with a camera system.

Table 12.6: Parameters of the event-based control method for the robots.

Description	Value
Event threshold $\bar{e}_\Phi$	3.44°
Parameter t_{com}	1.84 s

As the robot model (2.22) is differentially flat [119], a flatness-based two degrees of freedom controller C^* is derived, where the feedforward controller is obtained in [119] and the feedback controller is determined in [167]. The event-based control units of the robots use the components stated in Chapter 7, where the trajectories are planned using the method given in Chapter 3. The parameters of the control method are newly determined using the dynamics of the robots and are summarised in Tab. 12.6. The radius of the circle as well as the minimum speed and the maximum speed of the robots are identical to the quadrotors and can be seen in Tab. 12.3.

Results for $\gamma = 45°$. In the first experiment the two robots should maintain the phase difference with a tolerance range of $\gamma = 45°$. The stand-on agent starts its circular movement with the speed $v_S(t) = 0.18\frac{m}{s}$ and changes its speed three times as stated in Tab. 12.7. The give-way agent starts opposite to the stand-on agent with the speed $v_G(t) = 0.18\frac{m}{s}$.

Table 12.7: Time instants for a change of the speed.

Time instants $t_{c,si}$	Speeds $v_{S,i}$
$t_{c,s0} = 0\,s$	$v_{S,0} = 0.18\frac{m}{s}$
$t_{c,s1} = 13.44\,s$	$v_{S,1} = 0.76\frac{m}{s}$
$t_{c,s2} = 22.64\,s$	$v_{S,2} = 0.62\frac{m}{s}$
$t_{c,s3} = 25.29\,s$	$v_{S,3} = 0.35\frac{m}{s}$

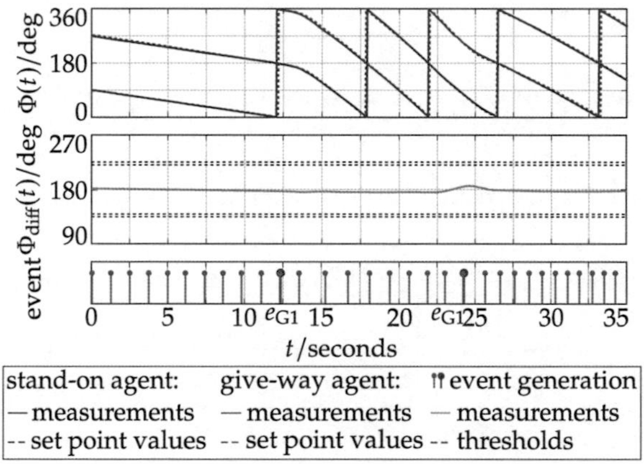

Fig. 12.11: Results of the first experiment using two robots.

In Fig. 12.11 the phases $\Phi_S(t)$, $\Phi_G(t)$ of the agents, the phase difference $\Phi_{diff}(t)$ and the generation of the events e_{G0} and e_{G1} are shown. It can be seen that the give-way agent reacts with a change of its speed whenever the stand-on agent changes its speed. In the experiment, the tolerance range is chosen so small that the give-way agent changes its speed directly when new data of the stand-on agent are communicated by a generation of the event e_{G1}.

In the middle part of Fig. 12.11 it can be seen that the phase difference is inside the tolerance range at any time and it stays almost constant despite the changes of the speed of the stand-on agent. Compared to the first experiment with the quadrotors, shown in Fig. 12.6, the results are nearly identical. Furthermore, in the lower part of the figure it can be seen that the communication effort is reduced considerably and it is even lower compared to the experiment with the quadrotors (Fig. 12.6). The reason for this effect is the slower dynamics of the robots compared to the quadrotors, which leads to a slower expansion of the set of predicted positions of the stand-on agent.

Results for $\gamma = 90°$. In the second experiment the phase difference between the robots should be maintained with a tolerance range of $\gamma = 90°$. The agents move on the same circular paths as in the first experiment and the stand-on agent changes its speed at the time instants given in Tab. 12.7.

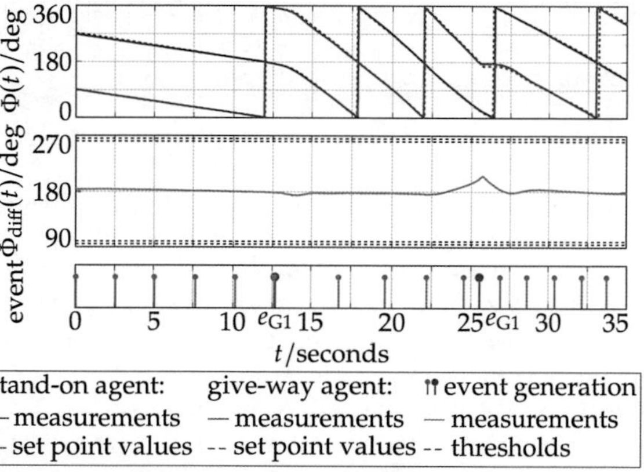

Fig. 12.12: Results of the second experiment using two robots.

In Fig. 12.12 the phases $\Phi_S(t)$, $\Phi_G(t)$ of the agents, the phase difference $\Phi_{\text{diff}}(t)$ and the generation of the events e_{G0} and e_{G1} are given. It can be seen at $t = 25.1$ s that the generation of the event e_{G1} is independent of triggering the communication event e_{G0}. The event generation is again appropriate to keep the phase difference inside the tolerance range, while the deviation from the desired phase difference is slightly increased compared to the first experiment. Compared to the results of the experiment with the quadrotors, shown in Fig. 12.7, the deviation of the phase difference from the desired value is smaller, which is caused by the slower dynamics of the agents. In addition, the slower dynamics of the agents cause the reduction of the communication effort compared to the first experiment with the robots and to the experiment with the quadrotors.

Results for $\gamma = 135°$. In the third experiment it is demanded that the phase difference between the robots remains inside the tolerance range of $\gamma = 135°$. The agents move on the same circular paths as in the first two experiments. The stand-on agent changes its speed again at the time instants given in Tab. 12.7.

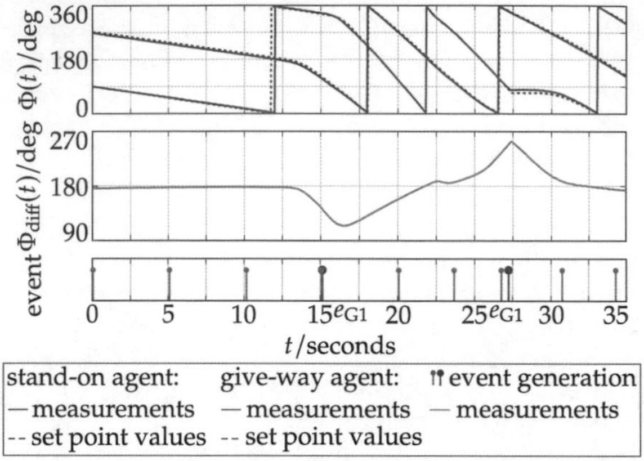

Fig. 12.13: Results of the third experiment using two robots.

Figure 12.13 shows the phases $\Phi_S(t)$, $\Phi_G(t)$ of the agents, the phase difference $\Phi_{\text{diff}}(t)$ and the generation of the events e_{G0} and e_{G1}. The event thresholds are not shown, because they are outside of the illustrated range of the plot. Similar to the third experiment with the quadrotors, depicted in Fig. 12.8, the phase difference is maintained inside the tolerance range, while the deviation to the desired value is even more increased compared to the second experiment with the robots. However, the deviation to the desired value is smaller compared to the third experiment with the quadrotors. This effect results from the generation of the event e_{G1}, which leads to a change of the speed of the give-way agent. Hence, in contrast to the quadrotor experiment the phase difference is equal to the desired value for $t > 32$ s. The event generation results from a greater change of the speed of the stand-on agent at the last time instant $t_{c,s3}$ compared to the change of the speed of the quadrotors. In the lower part of the figure it can be seen that the communication frequency is reduced compared to the first two experiments and the experiment with the quadrotors.

12.4 Computational aspects and concluding remarks

The experiments have shown that the proposed event-based control method is suitable to satisfy control aim (A1) for a circular movement when the trajectory tracking is fulfilled with a reasonable accuracy according to Fig. 12.5. The communication effort compared to a continuous communication scheme is reduced significantly. As the loop times of the control algorithms last in average $t_{\text{loop}} = 21\,\text{ms}$ in the first scenario, communication is reduced by a factor of 6 – 16 depending on the chosen tolerance range γ. In the second scenario, communication is still reduced by a factor of 3. Here, communication needs to be invoked more often, because the height changes of the stand-on agent are taken into account.

In particular it can be seen that the method ensures the control aims even in the presence of small time delays resulting from the computation times of the algorithms. If the control algorithms were implemented completely locally on the quadrotors, this would have caused longer calculation times due to the lower computing capacity onboard the agents. In addition, as the prediction method requires knowledge about the model of the quadrotors, model uncertainties influence the prediction result. As a consequence of both effects, more events must be generated to keep the uncertainty about the communicated data small. If the time delays become larger, e.g. due to a greater distance between the agents or larger computation times of the agents, the delay estimator must be utilised to cope with the delays.

In the experiments for a change of the trajectory of the give-way agent only the event e_{G1} was triggered. No situation occurred where the generation of the event e_{G2} was necessary. Hence, to fulfil the control aims there was no need for the give-way agent to exploit its dynamical limitations. Thus, the deviation of the phase difference between the agents from the set point mostly remains in the middle of the tolerance range. During height changes only for short time spans the parameter t_{com} has to be utilised for the communication to avoid Zeno behaviour.

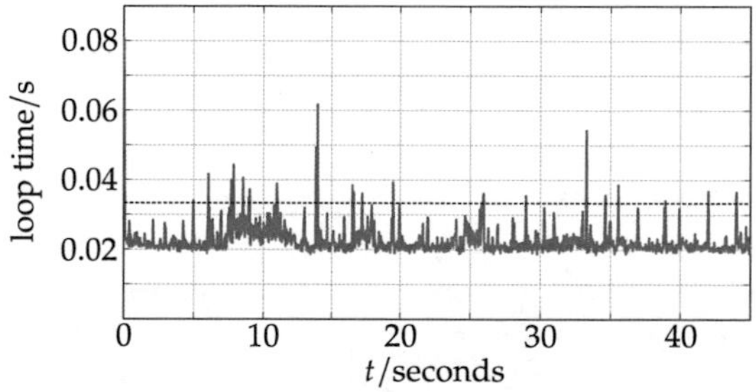

Fig. 12.14: Loop times of the algorithms for the second experiment.

In Fig. 12.14 the loop times for an execution of the algorithms for the second part of the scenario are shown. It can be seen that most of the loop times are small with a few higher peaks. As only 1.16% of the loop times are larger than the desired loop time, which is indicated by the dashed line in Fig. 12.14, the stability of the control loops of the quadrotors is not threatened.

The experiments with two robots have shown that the event-based control method is also able to control other types of agents so as to fulfil their control aims. To this aim only the parameters of the control method need to be changed, while the event-based control units can be applied unchanged. The changed parameters result from the different dynamics of the agents, which results in a less frequent communication between the agents when the dynamics of the agents are slower.

Conclusion 13

13.1 Summary

This thesis has presented an event-based control method for mobile agents that are able to fulfil individual tasks and cooperative tasks in a group by ensuring collision avoidance. The agents use only locally measured data and communicated information. Communication over an unreliable network that induces transmission delays and packet losses is only invoked if the local data becomes too uncertain. The control problem occurs for two agents, which have to maintain a safety distance $\underline{s}$ for collision avoidance and a maximum separation $\bar{s}$ to fulfil different functions as the communication relay over aerial base stations. Depending on the current situation the agents are subdivided into a stand-on agent and a give-way agent. The stand-on agent can change its trajectory at any time and moves without regard to the give-way agent. The give-way agent has to adapt its movement online so as to keep the distance $s(t)$ between the agents in the interval $s(t) \in [\underline{s}, \bar{s}]$.

For the control method the control aims of the networked system are subdivided into local control aims of the agents. The stand-on agent has just to fulfil its individual task, while the give-way agent is responsible for keeping the distance between them in the desired interval. To this aim the agent is provided with an event-based control unit consisting of three parts and a delay estimator:

1. The delay estimator uses a network model from communication technology to derive an estimate of the current transmission delay induced by the network. As the delay varies with the distance between the agents, a new estimate is generated in an event-based fashion if a threshold is violated, which indicates that the current estimate became too inaccurate. The event based delay estimation is given in Algorithm 15 on p. 189.

2. The prediction unit generates a set that includes every possible future position of the stand-on agent for a certain time span. The set is only based on the communicated position and speed of the stand-on agent and is determined with respect to the dynamics of the agents. The prediction is executed with Algorithm 11 on p. 88 and Algorithm 12 on p. 92 and stated in Theorem 9.1 on p. 191 and Theorem 9.2 on p. 193.

3. The event generator acts as supervisor and monitors the distance between the agents. It determines by an evaluation of event thresholds when it is necessary to invoke

communication and when to change the trajectory of the agent in order to keep the control aim satisfied. Appropriate thresholds with which events are triggered so as to fulfil the control aims are given in Theorem 7.1 on p. 140 and Theorem 7.2 on p. 147. In contrast to the classical event-based control, the event generator triggers five different types of events for communication and a trajectory replanning depending on the current situation. Furthermore, by evaluating the communicated trajectory, future events can already be determined.

4. The trajectory planning unit determines the trajectories of the agents based on Bézier curves. The unit is subdivided into the algorithm section, which plans the trajectories and the planning task section that specifies the boundary conditions on the trajectory depending on the current situation in order to fulfil the control aim. Hence, different trajectories are planned by only changing the planning task. With Algorithm 13 on p. 157 the planning task is derived, with Algorithm 14 on p. 168 appropriate trajectories are planned that respect the dynamic limitations of the agent. The result of the event-based control method is stated in Theorem 9.5 on p. 219.

It has been proven that with the proposed event-based control method collisions are avoided even in the presence of time delays and packet losses. The maximum separation between the agents can be guaranteed in the presence of time delays. If packet losses occur, the separation can only be satisfied with a probability given in Theorem 9.5. The communication effort is significantly reduced compared to a continuous communication scheme. It has been proven that two consecutive communication events are separated by a minimum time span (Theorem 9.3 on p. 217 and Theorem 9.4 on p. 218) so that no Zeno behaviour can occur.

The proposed control method has been applied to two quadrotors in two different scenarios and to two robots. It has been shown by simulations that the control method is suitable for the cooperative control of the two agents for a general movement, where the properties of the communication network vary. In a second scenario the control method has been successfully applied to control two quadrotors for a circular movement in experiments. The results have indicated that the approach is still applicable when the agents do not exactly follow their trajectories due to external disturbances. The control aims have been still satisfied. Furthermore, the successful application of the control method to two robots has shown that the method is not limited to the control of quadrotors but also other types of agents can be controlled in order to satisfy their control aims.

13.2 Outlook

The proposed event-based control method can be extended in two directions. First, for the trajectory planning the specific individual tasks of the agents can be more closely considered. Currently the agents are able to fulfil individual tasks that are based on the assumption that they just have to keep a minimum and a maximum distance between them. However, this limits the scope of the tasks. Using more flexible event conditions compared to the currently applied static conditions and improving the trajectory planning method enables a broader application range of the agents.

In [1] and [12] approaches were presented in order to improve the trajectory planning for UAVs that act as aerial communication relay stations. The trajectories of the UAVs are planned so as to provide several ground objects with communication links, which have different quality-of-service requirements. The ground objects follow locally generated trajectories, which are communicated to the aerial base stations and can be changed at any time. The UAVs possess network estimators that are an extension of the delay estimator introduced in this thesis. They determine the channel properties between the UAVs and each ground object. Whenever the estimated QoS properties violate the requirements, the trajectory of an UAV is changed online. The aim is to keep all individual requirements of the ground objects fulfilled during their movement. When it is not possible for a group of UAVs to provide all ground objects with appropriate communication links due to diverging movements, another UAV is requested to ensure the coverage of all objects. This approach extends the delay estimator, the event generator and the trajectory planning unit compared to this thesis in order to satisfy the control aim.

Second, the method can be applied to more than two agents by extending the hierarchical structure. Then, the networked system consists of a stand-on agent and several give-way agents, which have to avoid the stand-on agent and give-way agents with a higher priority. The prediction method needs to be extended to predict the movement of more agents and the event thresholds have to be modified accordingly. By using more agents, more complex scenarios could be handled, e.g. several UAVs acting as aerial communication relay stations, which provide communication links to several moving ground agents over long distances. However, this extension requires the development of time-efficient algorithms due to the increased computational complexity of the prediction method and the event generation.

Bibliography

Contributions of the Author

[1] Schwung, M. and Lunze, J. (2021). Control of an UAV acting as a communication base station to satisfy data requirements. *Proc. of the 19th European Control Conference*, pp. 183–188.

[2] Schwung, M., Hagedorn, F., and Lunze, J. (2019). Networked event-based collision avoidance of mobile objects. *Proc. of the 17th European Control Conference*, pp. 63–70.

[3] Schwung, M. and Lunze, J. (2018). Networked event-based trajectory planning for mobile objects. *Proc. of the 7th IFAC Workshop on Distributed Estimation and Control in Networked Systems*, pp. 170–175.

[4] Schwung, M. and Lunze, J. (2019). Networked event-based control of moving objects with improved position estimation. *Proc. of the 8th IFAC Workshop on Distributed Estimation and Control in Networked Systems*, pp. 133–138.

[5] Schwung, M. and Lunze, J. (2019). Vernetzte ereignisbasierte Kollisionsvermeidung mobiler Objekte. *GMA-FA 1.50 Grundlagen vernetzter Systeme*,

[6] Schwung, M. and Lunze, J. (2020). Event-based trajectory planning for 3D collision avoidance in a leader-follower formation. *Proc. of the 18th European Control Conference*, pp. 1929–1936.

[7] Schwung, M. and Lunze, J. (2020). Experimental evaluation of an event-based collision avoidance method with application to quadrotors. *Proc. of the 4th IEEE Conference on Control Technology and Applications*, pp. 633–639.

[8] Schwung, M. and Lunze, J. (2021). Cooperative control of UAVs over an unreliable communication network. *IEEE Transactions on Aerospace and Electronic Systems (submitted)*,

[9] Schwung, M. and Lunze, J. (2021). Event-based control of mobile objects over an unreliable network. *Proc. of the 60th Conference on Decision and Control*, pp. 1986–1993.

[10] Schwung, M. and Lunze, J. (2021). Kooperative ereignisbasierte Steuerung von mobilen Objekten über ein unzuverlässiges Kommunikationsnetzwerk. *at-Automatisierungstechnik*, 70, pp. 105–118.

[11] Schwung, M. and Lunze, J. (2021). Networked event-based collision avoidance of mobile objects with trajectory planning based on Bézier curves. *European Journal of Control*, 58, pp. 327–339.

[12] Schwung, M. and Lunze, J. (2021). Online trajectory planning with application to an UAV relay station. *Proc. of the 9th International Conference on Unmanned Aircraft Systems*, pp. 708–713.

[13] Schwung, M., Vey, D., and Lunze, J. (2019). Quadrotor tracking control: design and experiments. *Proc. of the 3rd IEEE Conference on Control Technology and Applications*, pp. 768–775.

[14] Schwung, M. et al. (2020). Event-based collision avoidance utilising a channel estimation method. *Proc. of the 21st IFAC World Congress*, pp. 2793–2800.

[15] Schwung, M. et al. (2020). Event-based quality-of-service parameter estimation of a wireless channel. *Proc. of the 6th International Conference on Event-Based Control, Communication and Signal Processing*, pp. 1–7.

Supervised Theses

[16] Gürpinar, S. (2018). *Entwicklung einer OnBoard-Lageregelung für einen fernsteuerbaren Hexakopter*. Bachelorarbeit. Ruhr-Universität Bochum, Lehrstuhl für Automatisierungstechnik und Prozessinformatik.

[17] Gürpinar, S. and Seckin, L. (2017). *Aufbau und Inbetriebnahme eines fernsteuerbaren Hexacopters*. Praxisprojekt. Ruhr-Universität Bochum, Lehrstuhl für Automatisierungstechnik und Prozessinformatik.

[18] Habersang, T. (2020). *Entwicklung einer graphischen Oberfläche zur Planung von Trajektorien für Quadrokopter mit Bézierkurven*. Bachelorarbeit. Ruhr-Universität Bochum, Lehrstuhl für Automatisierungstechnik und Prozessinformatik.

[19] Hagedorn, F. (2018). *Ereignisbasierte Trajektorienplanung zur Kollisionsvermeidung mobiler Objekte mit Anwendung an Quadrokoptern*. Masterarbeit. Ruhr-Universität Bochum, Lehrstuhl für Automatisierungstechnik und Prozessinformatik.

[20] Hinsen, P. (2018). *Anpassung eines Simulationsmodells und Entwurf einer Regelung für einen Quadrokopter*. Bachelorarbeit. Ruhr-Universität Bochum, Lehrstuhl für Automatisierungstechnik und Prozessinformatik.

[21] Kart, O. (2019). *Modellbildung, Parameteridentifikation und Regelung des Temperaturverhaltens eines Kochmixers*. Masterarbeit. Ruhr-Universität Bochum, Lehrstuhl für Automatisierungstechnik und Prozessinformatik.

[22] Littek, B. (2018). *Entwicklung einer OnBoard-Lageregelung für einen Quadrokopter*. Bachelorarbeit. Ruhr-Universität Bochum, Lehrstuhl für Automatisierungstechnik und Prozessinformatik.

[23] Littek, B. (2021). *Modifikation und experimentelle Erprobung einer Methode zur ereignisbasierten Steuerung mobiler Objekte*. Masterarbeit. Ruhr-Universität Bochum, Lehrstuhl für Automatisierungstechnik und Prozessinformatik.

[24] Littek, B. and Hinsen, P. (2018). *Experimentelle Auslegung eines Lagereglers für einen Quadrokopter*. Praxisprojekt. Ruhr-Universität Bochum, Lehrstuhl für Automatisierungstechnik und Prozessinformatik.

[25] Nagel, L. (2021). *Ereignisbasierte Kollisionsvermeidung von Robotern an der Versuchsanlage SAMS*. Bachelorarbeit. Ruhr-Universität Bochum, Lehrstuhl für Automatisierungstechnik und Prozessinformatik.

[26] Nagel, L., Kallweit, J., and Szkaradek, O. (2021). *Trajektorienplanung für Roboter an der Anlage SAMS*. Praxisprojekt. Ruhr-Universität Bochum, Lehrstuhl für Automatisierungstechnik und Prozessinformatik.

[27] Serif-Oglu, D. (2019). *Ansteuerung und Positionsregelung eines Quadrokopters an der Anlage MULAN*. Praxisprojekt. Ruhr-Universität Bochum, Lehrstuhl für Automatisierungstechnik und Prozessinformatik.

[28] Serif-Oglu, D. (2020). *Modellierung und Vorsteuerungsentwurf für einen Quadrokopter*. Bachelorarbeit. Ruhr-Universität Bochum, Lehrstuhl für Automatisierungstechnik und Prozessinformatik.

Further Literature

[29] Abdi, A. et al. (2001). On the estimation of the K Parameter for the Rice fading distribution. *IEEE Communications Letters*, 5, (3), pp. 92–94.

[30] Abichandani, P., Benson, H. Y., and Kam, M. (2011). Decentralized multi-vehicle path coordination under communication constraints. *Proc. of the International Conference on Intelligent Robots and Systems*, pp. 2306–2313.

[31] Abualhaol, I. Y. and Matalgah, M. M. (2010). Performance analysis of multicarrier relay-based UAV network over fading channels. *Proc. of the IEEE Global Communications Conference*, pp. 1811–1815.

[32] Ahmed, N., Kanhere, S. S., and Jha, S. (2016). On the importance of link characterization for aerial wireless sensor networks. *IEEE Communication Magazine*, 54(5), pp. 52–57.

[33] Akhtar, A., Waslander, S. L., and Nielsen, C. (2013). Fault tolerant path following for a quadrotor. *Proc. of the 52nd Conference on Decision and Control*, pp. 847–852.

[34] Amoozadeh, M. et al. (2015). Platoon management with cooperative adaptive cruise control enabled by VANET. *Vehicular Communications*, 2, (2), pp. 110–123.

[35] Arreola, L. et al. (2018). Improvement in the UAV position estimation with low-cost GPS, INS and vision-based system: Application to a quadrotor UAV. *Proc. of the International Conference on Unmanned Aircraft Sytems*, pp. 1248–1254.

[36] Årzén, K.-E. (1999). A simple event-based PID controller. *Proc. of the 14th IFAC World Congress*, pp. 423–428.

[37] Astrom, K. and Bernhardsson, B. (1999). Comparison of periodic and event based sampling for first-order stochastic systems. *Proc. of the 14th IFAC World Congress*, pp. 301–306.
[38] Beard, K. and Palancioglu, H. M. (2000). Estimating positions and paths of moving objects. *Proc. of the 7th International Workshop on Temporal Representation and Reasoning*, pp. 1–8.
[39] Beard, R. W. and McLain, T. W. (2003). Multiple UAV cooperative search under collision avoidance and limited range communication constraints. *Proc. of the 42nd Conference on Decision and Control*, pp. 25–30.
[40] Beard, R. W. et al. (2006). Decentralized cooperative aerial surveillance using fixed-wing miniature UAVs. *Proc. of the IEEE*, 94, (7), pp. 1306–1324.
[41] Behrends, E. (2002). *Introduction to Markov Chains*. Vieweg.
[42] Bergen, G. van den (1999). A fast and robust GJK implementation for collision detection of convex objects. *Journal of Graphics Tools*, 4, (2), pp. 7–25.
[43] Boban, M., Barros, J., and Tonguz, O. K. (2014). Geometry-Based Vehicle-to-Vehicle Channel Modeling for Large-Scale Simulation. *IEEE Transactions on Vehicular Technology*, 63, (9), pp. 4146–4164.
[44] Borgers, D. P. and Heemels, W. P. M. H. (2014). Event-separation properties of event triggered control systems. *IEEE Transactions on Automatic Control*, 59, (10), pp. 2644–2656.
[45] Boyd, S. and Vandenberghe, L. (2004). *Convex Optimization*. Cambridge University Press.
[46] Bupe, P., Haddad, R., and Rios-Gutierrez, F. (2015). Relief and emergency communication network based on an autonomous decentralized UAV clustering network. *Proc. of the Southeast Regional Conference*, pp. 1–8.
[47] Byrnes, C. and Isidori, A. (1991). Asymptotic stabilization of minimum phase nonlinear-systems. *IEEE Transactions on Automatic Control*, 36, (10), pp. 1122–1137.
[48] Caballero, F. et al. (2007). Homography Based Kalman Filter for Mosaic Building. Applications to UAV position estimation. *Proc. of the IEEE Conference on Robotics and Automation*, pp. 2004–2009.
[49] Cai, X. et al. (2017). Low altitude UAV propagation channel modelling. *Proc. of the 11th European Conference on Antennas and Propagation*, pp. 1443–1447.
[50] Camacho, E. and Bordons, C. (2004). *Model predictive control*. Advanced Textbooks in Control and Signal Processing. Springer.
[51] Chang, J.-W. et al. (2011). Computation of the minimum distance between two Bézier curves/surfaces. *Computers & Graphics*, 35, pp. 677–684.
[52] Chen, Y., Feng, W., and Zheng, G. (2018). Optimum placement of UAV as relays. *IEEE Communication Letters*, 22, (2), pp. 248–251.
[53] Choi, D. H., Kim, S. H., and Sung, D. K. (2014). Energy-efficient maneuvering and communication of a single UAV-based relay. *IEEE Transactions on Aerospace and Electronic Systems*, 50, (3), pp. 2320–2327.

[54] Choi, H. H. et al. (2016). Collision avoidance scheme for micro UAVs delivering information. *Proc. of the International Conference on Information Networking*, pp. 45–50.

[55] Chung, H.-M. et al. (2020). Edge intelligence empowered UAVs for automated wind farm monitoring in smart grids. *Proc. of the IEEE Global Communication Conference*, pp. 1–6.

[56] Cichella, V. et al. (2013). A 3D path-following approach for a multirotor UAV on SO(3). *IFAC Proceedings Volumes*, 46, pp. 13–18.

[57] Cormen, T. H. et al. (2009). *Introduction to algorithms*. The MIT Press.

[58] Cuenca, A. et al. (2019). Periodic event-triggered sampling and dual-rate control for a wireless networked control system with applications to UAVs. *IEEE Transactions on Industrial Electronics*, 66, (4), pp. 3157–3166.

[59] Danial, J., Feldmann, D., and Hutterer, A. (2019). Position estimation of moving objects: Practical provable approximation. *IEEE Robotics and Automation Letters*, 4(2), pp. 1985–1992.

[60] Derler, P. et al. (2013). Cyber-physical system design contracts. *Proc. of the ACM/IEEE International Conference on Cyber-Physical Systems*, pp. 109–118.

[61] Dimagoronas, D. V., Frazoli, E., and Johansson, K. H. (2012). Distributed event triggered control for multi-agent systems. *IEEE Transactions on Automatic Control*, 57, (5), pp. 1291–1297.

[62] Ding, L., Han, Q. L., and Guo, G. (2013). Network-based leader-following consensus for distributed multi-agent systems. *Automatica*, 49, (7), pp. 2281–2286.

[63] Dolk, V., Borgers, D., and Heemels, W. P. M. H. (2017). Output-based and decentralized dynamic event-triggered control with guaranteed $\mathcal{L}_\mathrm{p}$-gain performance and Zeno-freeness. *IEEE Transactions on Automatic Control*, 62, (1), pp. 34–49.

[64] Donkers, M. and Heemels, W. P. M. H. (2012). Output-based event-triggered control with guaranteed $\mathcal{L}_\infty$ gain and improved and decentralized event triggering. *IEEE Transactions on Automatic Control*, 57, (6), pp. 128–142.

[65] Elliott, E. O. (1963). Estimates of error rate for codes of burst-noise channels. *Bell System Technical Journal*, 42, pp. 1977–1997.

[66] Farouki, R. T. (2008). *Pythagorean-Hodograph curves*. Springer.

[67] Feng, Q. et al. (2006). Path loss models for air-to-ground radio channels in urban environments. *Proc. of the IEEE Vehicular Technology Conference*, pp. 2901–2905.

[68] Garcia, C. E., Prett, D. M., and Morari, M. (1989). Model predictive control: Theory and practice-a survey. *Automatica*, 25, (3), pp. 335–348.

[69] Ge, X., Han, Q. L., and Yang, F. (2017). Event-based set-membership leader-following consensus of networked multi-agent systems subject to limited communication resources and unknown-but-bounded noise. *IEEE Transactions of Industrial Electronics*, 64, (6), pp. 5045–5054.

[70] Ge, X. and Han, Q.-L. (2017). Distributed formation control of networked multi-agent systems using a dynamic event-triggered communication mechanism. *IEEE Transactions on Industrial Electronics*, 64, (10), pp. 8118–8127.

[71] Gezici, S. (2008). A survey on wireless position estimation. *Wireless Personal Communications*, 44, pp. 263–282.
[72] Gilbert, E. N. (1960). Capacity of a burst-noise channel. *Bell System Technical Journal*, 39, pp. 1253–1265.
[73] Gilbert, E. G., Johnson, D. W., and Keerth, S. S. (1988). A fast procedure for computing the distance between complex objects in three-dimensional space. *IEEE Journal of Robotics and Automation*, 4(2), pp. 193–203.
[74] Girard, A. (2015). Dynamic triggering mechanisms for event-triggered control. *IEEE Transactions on Automatic Control*, 60, (7), pp. 1992–1997.
[75] Grami, A. (2015). *Introduction to Digital Communications*. Academic Press.
[76] Gu, J. et al. (2018). Multiple moving targets surveillance based on a cooperative network for multi-UAV. *IEEE Communications Magazine*, 56, (4), pp. 82–89.
[77] Gu, W. et al. (2020). Autonomous wind turbine inspection using a quadrotor. *Proc. of the International Conference on Unmanned Aircraft Systems*, pp. 709–715.
[78] Gu, Z., Yue, D., and Tian, E. (2018). On designing of an adaptive event-triggered communication scheme for nonlinear networked interconnected control systems. *Information Sciences*, 422, pp. 257–270.
[79] Guinaldo, M. (2016). *Contributions to networked and event-triggered control of linear systems*. Springer.
[80] Guinaldo, M. et al. (2012). Distributed event-triggered control with network delays and packet losses. *Proc. of the 51st Conference on Decision and Control*, pp. 1–6.
[81] Gupta, L., Jain, R., and Vaszkun, G. (2016). Survey of important issues in UAV communication networks. *IEEE Communication Surveys & Tutorials*, 18(2), pp. 1123–1152.
[82] Han, J. and Chen, Y. (2014). Multiple UAV formations for cooperative source seeking and contour mapping of a radiative signal field. *Journal of Inelligent & Robotic Systems*, (74), pp. 323–332.
[83] Han, Z. et al. (2019). Integrated relative localization and leader–follower formation control. *IEEE Transactions on Automatic Control*, 64(1), pp. 20–34.
[84] Haßlinger, G. and Hohlfeld, O. (2008). The Gilbert-Elliott model for packet loss in real time services on the internet. *Proc. of the 14th GI/ITG Conference on Measurement, Modeling and Evaluation of Computer and Communication Systems*, pp. 269–286.
[85] He, S. et al. (2019). Leader–follower formation control of UAVs with prescribed performance and collision avoidance. *IEEE Transactions on Industrial Informatics*, 15(1), pp. 572–281.
[86] Heemels, W. P. M. H., Donkers, M. C. F., and Teel, A. R. (2013). Periodic event-triggered control for linear systems. *IEEE Transactions on Automatic Control*, 58, (4), pp. 847–861.
[87] Heemels, W. P. M. H., Johansson, K. H., and Tabuada, P. (2012). An introduction to event-triggered and self-triggered control. *Proc. of the 51st Conference on Decision and Control*, pp. 3270–3285.

[88] Heemels, W. P. M. H. et al. (1999). Asynchronous measurement and control: a case study on motor synchronization. *Control Engineering Practice*, 7(12), pp. 1467–1482.

[89] Heemels, W. P. M. H. et al. (2010). Networked control systems with communication constraints: Tradeoffs between transmission intervals, delays and performance. *IEEE Transactions on Automatic Control*, 55, (8), pp. 1781–1796.

[90] Henkel, P. and Sperl, A. (2016). Real-time kinematic positioning for unmanned air vehicles. *Proc. of the IEEE Airspace Conference*, pp. 1–7.

[91] Hochstenbach, M. et al. (2015). Design and control of an unmanned aerial vehicle for autonomous parcel delivery with transition from vertical take-off to forward flight – vertiKUL, a quadcopter tailsitter. *International Journal of Micro Air Vehicles*, 7, (4), pp. 395–405.

[92] Horn, R. A. and Johnson, C. R. (1985). *Matrix analysis*. Cambridge University Press.

[93] Al-Hourani, A., Kandeepan, S., and Lardner, S. (2014). Optimal LAP altitude for maximum coverage. *IEEE Wireless Communications Letters*, 3(6), pp. 569–572.

[94] Huo, M., Duan, H., and Fan, Y. (2021). Pigeon-inspired circular formation control for multi-UAV system with limited target information. *Guidance, Navigation and Control*, 1(1), pp. 1–23.

[95] International Maritime Organization (2003). *COLREG: Convention on the international regulations for preventing collisions at sea, 1972*. London: Int. Maritime Organization.

[96] Jackson, B. E. et al. (2020). Scalable cooperative transport of cable-suspended loads with UAVs using distributed trajectory optimization. *IEEE Robotics and Automation Letters*, 5, (2), pp. 3368–3374.

[97] Kaadan, A., Refai, H. H., and LoPresti, P. G. (2014). Multielement FSO transceivers alignment for inter-UAV communications. *Journal of Lightwave Technology*, 32(24), pp. 4183–4193.

[98] Kaplansky, I. (2001). *Set Theory and Metric Spaces*. American Mathematical Society Chelsea Publishing.

[99] Karacora, Y. and Sezgin, A. (2020). An energy-efficient event-based MIMO communication scheme for UAV formation control. *Proc. of the Asilomar Conference on Signals, Systems and Computers*, pp. 1–6.

[100] Karaman, S. and Frazzoli, E. (2011). Sampling-based algorithms for optimal motion planning. *International Journal of Robotics Research*, 30, (7), pp. 846–894.

[101] Kato, S. et al. (2002). Vehicle control algorithms for cooperative driving with automated vehicles and intervehicle communications. *IEEE Transactions on Intelligent Transportation Systems*, 3, (3), pp. 155–161.

[102] Khawaja, W., Guvenc, I., and Matolak, D. (2016). UWB channel sounding and modeling for UAV air-to-ground propagation channels. *Proc. of the IEEE Global Communications Conference*, pp. 1–7.

[103] Khuwaja, A. A. et al. (2018). A survey of channel modeling for UAV communications. *IEEE Communications Surveys & Tutorials*, 20(4), pp. 2804–2821.

[104] Kim, J., Koo, Y., and Kim, S. (2018). MOD: Multi-camera based local position estimation for moving objects detection. *Proc. of the IEEE International Conference on Big Data and Smart Computing*, pp. 642–643.

[105] Kim, S., Lee, J., and Lee, J. (2005). Trajectory estimation of a moving object using Kohonen networks. *Artificial Life and Robotics*, 9(1), pp. 36–40.

[106] Kingston, D. B. and Beard, R. W. (2004). Real-Time Attitude and Position Estimation for Small UAVs Using Low-Cost Sensors. *Proc. of the AIAA 3rd Unmanned unlimited Technical Conference, Workshop and Exhibit*, pp. 1–9.

[107] LaValle, S. M. (1998). Rapidly-exploring random trees: a new tool for path planning. *The Annual Mathematics Research Report*, pp. 1–4.

[108] LaValle, S. M. (2006). *Planning algorithms*. Cambridge University Press.

[109] LaValle, S. M. and Kuffner, J. J. (2001). Rapidly-exploring random trees: Progress and prospects. *Workshop on the algorithmic foundations of robotics; Algorithmic and computational robotics, new directions*, pp. 293–308.

[110] Lee, G. and Chwa, D. (2018). Decentralized behavior-based formation control of multiple robots considering obstacle avoidance. *Intelligent Service Robotics*, 11(1), pp. 127–138.

[111] Lehmann, D. and Lunze, J. (2012). Event-based control with communication delays and packet losses. *International Journal of Control*, 85(5), pp. 563–577.

[112] Leonard, N. E. and Fiorelli, E. (2001). Virtual leaders, artificial potentials and coordinated control of groups. *Proc. of the 40th Conference on Decision and Control*, pp. 2968–2973.

[113] Likhachev, M., Gordon, G. J., and Thrun, S. (2004). *ARA**: anytime *A** with provable bounds on sub-optimality. *Advances in Neural Information Processing Systems*, pp. 767–774.

[114] Lin, Y. and Saripalli, S. (2016). Sampling based collision avoidance for UAVs. *Proc. of the American Control Conference*, pp. 1353–1358.

[115] Linsenmayer, S. et al. (2019). Integration of communication networks and control systems using a slotted transmission classification model. *Proc. of the 16th Annual Consumer Communciations & Networking Conference*, pp. 1–6.

[116] Lunze, J. (2017). *Ereignisdiskrete Systeme*. De Gruyter Oldenbourg.

[117] Lunze, J. and Lehmann, D. (2009). A state-feedback approach to event-based control. *Automatica*, 46, (1), pp. 211–215.

[118] Luo, C. et al. (2013). UAV Position Estimation and Collision Avoidance Using the Extended Kalman Filter. *IEEE Transactions on Vehicular Technology*, 62(6), pp. 2749–2762.

[119] Luviano-Juarez, A., Cortes-Romero, J., and Sira-Ramirez, H. (2015). Trajectory tracking control of a mobile robot through a flatness-based exact feedforward linearization scheme. *Journal of Dynamcic Systems, Measurement and Control*, 137, (5), pp. 1–8.

[120] Ma, L. and Hovakimyan, N. (2013). Cooperative target tracking in balanced circular formation: multiple UAVs tracking a ground vehicle. *Proc. of the American Control Conference*, pp. 5386–5391.

[121] Malone, N. et al. (2017). Hybrid dynamic moving obstacle avoidance using a stochastic reachable set-based potential field. *IEEE Transactions on Robotics*, 33(5), pp. 1124–1138.

[122] Marinho, T. et al. (2018). Guaranteed collision avoidance based on line-of-sight angle and time-to-collision. *Proc. of the American Control Conference*, pp. 4305–4310.

[123] Matolak, D. W. and Sun, R. (2017). Air-ground channel characterization for unmanned aircraft systems—Part I: Methods, measurements, and models for over-water settings. *IEEE Transactions on Vehicular Technology*, 66(1), pp. 26–44.

[124] Mazo, M., Anta, A., and Tabuada, P. (2010). An ISS self-triggered implementation of linear controllers. *Automatica*, 46, (8), pp. 1310–1314.

[125] Mehdi, S. B., Choe, R., and Hovakimyan, N. (2015). Avoiding multiple collisions through trajectory replanning using piecewise Bézier curves. *Proc. of the 54th Conference on Decision and Control*, pp. 2755–2760.

[126] Mehdi, S. B. et al. (2017). Collision avoidance in multi-vehicle cooperative missions using speed adjustment. *Proc. of the 56th Conference on Decision and Control*, pp. 2152–2157.

[127] Mellinger, D. and Kumar, V. (2011). Minimum snap trajectory generation and control for quadrotors. *Proc. of the IEEE Conference on Robotics and Automation*, pp. 2520–2525.

[128] Merino, L. et al. (2006). Vision-based multi-UAV position estimation. *IEEE Robotics & Automation Magazine*, 13, (3), pp. 53–62.

[129] Miller, S. L. and Childers, D. G. (2004). *Probability and random processes: With applications to signal processing and communications*. Elsevier Academic Press.

[130] Nelson, D. R. et al. (2007). Vector field path following for miniature air vehicles. *IEEE Transactions on Robotics*, 23, (3), pp. 519–529.

[131] Newhall, W. G. and Reed, J. H. (2002). A geometric air-to-ground radio channel model. *Proc. of the IEEE Military Communications Conference*, pp. 632–636.

[132] Paris, D. T. and Hurd, F. K. (1969). *Basic Electromagnetic Theory (Physical & Quantum Electronics)*. Mc Graw-Hill Inc.

[133] Peng, C. and Han, Q. L. (2013). A novel event-triggered transmission scheme and $\mathcal{L}_2$ control co-design for sampled-data control systems. *IEEE Transactions on Automatic Control*, 58, (10), pp. 2620–2626.

[134] Punnoose, R. J., Nikitin, P., and Stancil, D. D. (2000). Efficient simulation of Ricean fading within a packet simulator. *52nd IEEE Vehicular Technology Conference*, 2, pp. 764–767.

[135] Rappaport, T. S. (1996). *Wireless Communications: Principles and Practice*. Prentice-Hall Inc.

[136] Ren, W. and Beard, R. W. (2004). Decentralized scheme for spacecraft formation flying via the virtual structure approach. *Journal of Guidance, Control and Dynamics*, 27(1), pp. 73–82.

[137] Rice, S. O. (1944). Mathematical analysis of random noise. *The Bell System Technical Journal*, 23, pp. 282–332.
[138] Richter, C., Bry, A., and Roy, N. (2013). Polynomial trajectory planning for aggressive quadrotor flight in dense indoor environments. *Proc. of the International Symposium of Robotics Research*, pp. 649–666.
[139] Ross, S. M. (2010). *Introduction to Probability Models*. Academic Press.
[140] Roth, S., Kariminezhad, A., and Sezgin, A. (2019). Base-stations up in the air: Multi-UAV trajectory control for min-rate maximization in uplink C-RAN. *Proc. of the IEEE International Conference on Communications*, pp. 1–6.
[141] Rudol, P. and Doherty, P. (2008). Human body detection and geolocalization for UAV search and rescue missions using color and thermal imagery. *Proc. of the IEEE Aerospace Conference*, pp. 1–8.
[142] Ruetten, L. et al. (2020). Area-optimized UAV swarm network for search and rescue operations. *Proc. of the 10th Annual Computing and Communication Workshop and Conference*, pp. 613–618.
[143] Sabol, C., Burns, R., and McLaughlin, C. A. (2001). Satellite formation flying design and evolution. *Journal of Spacecraft and Rockets*, 38(2), pp. 270–278.
[144] Scherer, J. et al. (2015). An autonomous multi-UAV system for search and rescue. *Proc. of the 1st Workshop on Micro Aerial Vehicle Networks, Systems and Applications for Civilian Use*, pp. 33–38.
[145] Schmidt, V. (2010). Markov Chains and Monte-Carlo Simulation. *Lecture Notes Ulm University, Institute of Stochastics*,
[146] Schulman, J. et al. (2014). Motion planning with sequential convex optimization and convex collision checking. *International Journal of Robotics Research*, 33, (9), pp. 1251–1270.
[147] Seneta, E. (1981). *Non-negative Matrices and Markov Chains*. Springer.
[148] Seyboth, G. S., Dimarogonas, D. V., and Johansson, K. H. (2011). Control of multi-agent systems via event-based communication. *Proc. of the 18th IFAC World Congress*, pp. 10086–10091.
[149] Simeon, T., Laumond, J.-P., and Nissoux, C. (2000). Visibility-based probabilistic roadmaps for motion planning. *Advanced Robotics*, 14, (6), pp. 477–493.
[150] Simunek, M., Fontan, F. P., and Pechac, P. (2013). The UAV low elevation propagation channel in urban areas: Statistical analysis and timeseries generator. *IEEE Transactions on Antennas and Propagation*, 61(7), pp. 3850–3858.
[151] Simunek, M., Pechac, P., and Fontan, F. P. (2011). Excess loss model for low elevation links in urban areas for UAVs. *Radioengineering*, 20(3), pp. 561–568.
[152] Sira-Ramirez, H. and Agrawal, S. K. (2004). *Differentially flat systems*. Taylor & Francis.
[153] Sivaprakasam, S. and Shanmugan, K. S. (1995). An equivalent Markov model for burst errors in digital channels. *IEEE Transactions on Communications*, 43(2/3/4), pp. 1347–1355.

[154] Skrjanc, I. and Klancar, G. (2010). Optimal cooperative collision avoidance between multiple robots based on Bernstein-Bézier curves. *Robotics and Autonomous Systems*, 58, pp. 1–9.

[155] Song, B. D., Park, K., and Kim, J. (2018). Persistent UAV delivery logistics: MILP formulation and efficient heuristic. *Computers & Industrial Engineering*, 120, pp. 418–428.

[156] Stöcker, C., Vey, D., and Lunze, J. (2013). Decentralized event-based control: Stability analysis and experimental evaluation. *Nonlinear Analysis: Hybrid Systems*, 10, pp. 141–155.

[157] Sun, R., Matolak, D. W., and Liu, P. (2013). Parking Garage Channel Characteristics at 5 GHz for V2V Applications. *IEEE 78th Vehicular Technology Conference*, pp. 1–5.

[158] Tabuada, P. (2007). Event-triggered real-time scheduling of stabilizing control tasks. *IEEE Transactions on Automatic Control*, 52, (9), pp. 1680–1685.

[159] Takahashi, S. et al. (2017). Real-time monitoring for structure deformations using hand-held rtk-gnss receivers on the wall. *Proc. of the International Conference on Indoor Positioning and Indoor Navigation*, pp. 1–7.

[160] Talak, R., Karaman, S., and Modiano, E. (2016). Speed limits in autonomous vehicular networks due to communication constraints. *Proc. of the IEEE 55th Conference on Decision and Control*, pp. 4998–5003.

[161] Turin, W. (2004). *Performance analysis and modeling of digital transmission systems*. Springer.

[162] Villa, D. K. D., Brandao, A. S., and Sarcinelli-Filho, M. (2020). A survey on load transportation using multirotor UAVs. *Journal of Intelligent & Robotic Systems*, 98, pp. 267–296.

[163] Viriyasitavat, W. et al. (2015). Vehicular Communications: Survey and Challenges of Channel and Propagation Models. *IEEE Vehicular Technology Magazine*, 10, (2), pp. 55–66.

[164] Walle, D. van der et al. (2008). Non-hierarchical UAV formation control for surveillance tasks. *Proc. of the American Control Conference*, pp. 777–782.

[165] Wang, H. S. and Moayeri, N. (1995). Finite-state Markov channel: A useful model for radio communciation channels. *IEEE Transactions on Vehicular Technology*, 44, pp. 163–171.

[166] Wang, Y. and Ishii, H. (2020). Resilient Consensus through event-based communication. *IEEE Transactions on Control of Network Systems*, 7, (1), pp. 471–482.

[167] Welz, P., Fischer, M., and Lunze, J. (2020). Experimentelle Erprobung einer kollisionsfreien Abstandsregelung für mobile Roboter. *at - Automatisierungstechnik*, 68, (1), pp. 32–43.

[168] Wu, D. and Nagi, R. (2003). Effective capacity: A wireless link model for support of quality of service. *IEEE Transactions on Wireless Communications*, 2(4), pp. 630–643.

[169] Wu, D. and Nagi, R. (2004). Effective capacity-based quality of service measures for wireless networks. *Proc. of the 1st International Conference on Broadband Networks*, pp. 527–536.

[170] Xiao, Z., Xia, P., and Xia, X.-G. (2016). Enabling UAV cellular with millimeterwave communication: Potentials and approaches. *IEEE Communication Magazine*, 54(5), pp. 66–73.

[171] Xie, N., Lin, X., and Yu, Y. (2016). Position estimation and control for quadrotor using optical flow and GPS sensors. *Proc. of the 31st Youth Academic Annual Conference of Chines Association of Automation*, pp. 181–186.

[172] Yang, S., Uthoff, E., and Wortman, K. (2015). Collision detection of two fast moving objects. *Proc. of the IEEE Aerospace Conference*, pp. 1–7.

[173] Yang, Y., Pan, J., and Wan, W. (2019). Survey of optimal motion planning. *IET Cyber-Systems and Robotics*, 1, (1), pp. 13–19.

[174] Yanmaz, E., Kuschnig, R., and Bettstetter, C. (2011). Channel measurements over 802.11a-based UAV-to-ground links. *Proc. of the IEEE Global Communications Conference*, pp. 1280–1284.

[175] Yel, E., Lin, T. X., and Bezzo, N. (2017). Reachability-based self-triggered scheduling and replanning of UAV operations. *Proc. of the NASA/ESA Conference on Adaptive Hardware and Systems*, pp. 221–228.

[176] Yoo, J. and Johansson, K. H. (2017). Learning communication delay patterns for remotely controlled UAV networks. *Proc. of the 20th IFAC World Congress*, pp. 13758–13763.

[177] Zeng, Y., Zhang, R., and Lin, T. J. (2016). Throughput maximization for UAV-enabled mobile relaying systems. *IEEE Transactions on Communications*, 64(12), pp. 4983–4996.

[178] Zeng, Y., Wu, Q., and Zhang, R. (2019). Accessing from the sky: A tutorial on UAV communications for 5G and beyond. *Proc. of the IEEE*, 107, (12), pp. 2327–2375.

[179] Zhang, M. and Liu, H. H. T. (2016). Cooperative tracking a moving target using multiple fixed-wing UAVs. *Journal of Intelligent & Robotic Systems*, (81), pp. 505–529.

[180] Zhou, D., Wang, Z., and Schwager, M. (2018). Agile coordination and assistive collision avoidance for quadrotor swarms using virtual structures. *IEEE Transactions on Robotics*, 34(4), pp. 916–923.

[181] Zhou, Y., Raghavan, A., and Baras, J. S. (2016). Time varying control set design for UAV collision avoidance using reachable tubes. *Proc. of the 55th Conference on Decision and Control*, pp. 6857–6862.

[182] Zhou, Y. et al. (2012). A Markov-based packet dropout model for UAV wireless communications. *Journal of Communications*, 7(6), pp. 418–426.

[183] Zucker, M. et al. (2013). CHOMP: covariant Hamiltonian optimization for motion planning. *International Journal of Robotics Research*, 32, pp. 1164–1193.

Appendix

Introduction to Markov Models

A

A.1 Mathematical basics

As the delay estimator uses a Markov chain to model the random packet losses induced by the communication network, in this section the mathematical background for the Markov models is reviewed.

Eigenvalue and eigenvector. Let $\boldsymbol{A}$ be a $N \times N$ matrix with elements in $\mathbb{R}$ or $\mathbb{C}$, let $\boldsymbol{u}, \boldsymbol{v} \neq \boldsymbol{0}$ be two vectors in $\mathbb{R}^N$ or $\mathbb{C}^N$ such that for each of them at least one of their components is different from 0 and λ be an arbitrary real or complex integer. If

$$\boldsymbol{A}\boldsymbol{v} = \lambda \boldsymbol{v} \quad \text{or} \quad \boldsymbol{u}^{\mathrm{T}}\boldsymbol{A} = \lambda \boldsymbol{u}^{\mathrm{T}} \tag{A.1}$$

holds, then λ is an eigenvalue of $\boldsymbol{A}$ and $\boldsymbol{v}$ and $\boldsymbol{u}^{\mathrm{T}}$ are right and left eigenvectors of $\boldsymbol{A}$ for λ. As

$$(\boldsymbol{u}^{\mathrm{T}}\boldsymbol{A})^{\mathrm{T}} = (\lambda \boldsymbol{u}^{\mathrm{T}})^{\mathrm{T}} \Leftrightarrow \boldsymbol{A}^{\mathrm{T}}\boldsymbol{u} = \lambda \boldsymbol{u} \tag{A.2}$$

holds, the left eigenvectors of $\boldsymbol{A}$ are equal to the right eigenvectors of $\boldsymbol{A}^{\mathrm{T}}$. As (A.1) is equivalent to

$$(\boldsymbol{A} - \lambda \boldsymbol{I})\boldsymbol{u} = \boldsymbol{0} \quad \text{and} \quad \boldsymbol{v}^{\mathrm{T}}(\boldsymbol{A} - \lambda \boldsymbol{I}) = \boldsymbol{0}^{\mathrm{T}}.$$

λ is an eigenvalue of $\boldsymbol{A}$ if and only if λ is a solution of the characteristic equation

$$\det(\boldsymbol{A} - \lambda \boldsymbol{I}) = 0.$$

Spectral representation of a matrix. Let

$$\boldsymbol{V} = \begin{pmatrix} \boldsymbol{v}_1 & \cdots & \boldsymbol{v}_N \end{pmatrix} \quad \text{and} \quad \boldsymbol{U} = \begin{pmatrix} \boldsymbol{u}_1^{\mathrm{T}} \\ \vdots \\ \boldsymbol{u}_N^{\mathrm{T}} \end{pmatrix}$$

be the $N \times N$ matrices consisting of the right eigenvectors $\boldsymbol{v}_1, \ldots, \boldsymbol{v}_N$ and left eigenvectors $\boldsymbol{u}_1^{\mathrm{T}}, \ldots, \boldsymbol{u}_N^{\mathrm{T}}$. By definition of the eigenvectors the equations

$$\boldsymbol{A}\boldsymbol{V} = \boldsymbol{V}\operatorname{diag}(\boldsymbol{\lambda}) \quad \text{and} \quad \boldsymbol{U}\boldsymbol{A} = \operatorname{diag}(\boldsymbol{\lambda})\boldsymbol{U} \tag{A.3}$$

hold, where $\boldsymbol{\lambda} = \begin{pmatrix} \lambda_1 & \cdots & \lambda_N \end{pmatrix}^{\mathrm{T}}$ and diag($\boldsymbol{\lambda}$) denotes the diagonal matrix with diagonal elements $\lambda_1, \ldots, \lambda_N$. If the right eigenvectors $\boldsymbol{v}_1, \ldots, \boldsymbol{v}_N$ are linearly independent, the inverse $\boldsymbol{V}^{-1}$ of the matrix $\boldsymbol{V}$ exists and $\boldsymbol{U} = \boldsymbol{V}^{-1}$ can be set. Furthermore, in this case (A.3) implies

$$\boldsymbol{A} = \boldsymbol{V}\mathrm{diag}(\boldsymbol{\lambda})\boldsymbol{V}^{-1} = \boldsymbol{V}\mathrm{diag}(\boldsymbol{\lambda})\boldsymbol{U}$$

and hence

$$\boldsymbol{A}^k = \boldsymbol{V}(\mathrm{diag}(\boldsymbol{\lambda}))^k\boldsymbol{V}^{-1} = \boldsymbol{V}(\mathrm{diag}(\boldsymbol{\lambda}))^k\boldsymbol{U}.$$

This yields the spectral representation of $\boldsymbol{A}$ as

$$\boldsymbol{A}^k = \sum_{i=1}^{N} \lambda_i^k \, \boldsymbol{v}_i \, \boldsymbol{u}_i^{\mathrm{T}}. \tag{A.4}$$

Lemma A.1 (Linear independence of eigenvalues [92]). *If all eigenvalues $\lambda_1, \ldots, \lambda_N$ of $\boldsymbol{A}$ are distinct, the right eigenvectors $\boldsymbol{v}_1, \ldots, \boldsymbol{v}_N$ of $\boldsymbol{A}$ are linearly independent.*

Remark. Since in this report the transition matrix $\boldsymbol{G}_{\mathrm{M}}$ of the Markov chain is a 2×2 column stochastic matrix, its two eigenvalues are distinct, which results in linearly independent right eigenvectors.

A.2 Discrete probability theory

As Markov chains are discrete random processes, some results of the discrete probability theory are introduced, which are taken from [41].

Probability space. The probability theory is based on a *probability space* as a triple $(\Omega, \Sigma, \mathrm{Prob})$ consisting of:

- **Sample space Ω:** It is an arbitrary non-empty set

$$\Omega = \{\omega_1, \omega_2, \ldots, \omega_q\}$$

  of elementary events ω_i, $(i = 1, \ldots, q)$ of a stochastic experiment. The result of any experiment is one element $\omega \in \Omega$. The relative frequency of the occurrence of the elementary events is given by the probability distribution $\mathrm{Prob}(\omega)$, $\omega \in \Omega$.

- **Set of events Σ:** It concerns a σ-algebra over Ω (σ-field), which is a set $\Sigma = \{\mathcal{A}_1, \mathcal{A}_2, \ldots, \mathcal{A}_m\}$ of random events $\mathcal{A}_i$, $(i = 1, \ldots, m)$, that are subsets $\mathcal{A}_i \subseteq \Omega$ of the sample space. The following rules have to be satisfied:

  - Σ contains the sample space and the empty set:

$$\Omega \in \Sigma, \quad \emptyset \in \Sigma.$$

- Considering two events $\mathcal{A}_1, \mathcal{A}_2 \in \Sigma$, the union $\mathcal{A}_1 \cup \mathcal{A}_2$, the intersection $\mathcal{A}_1 \cap \mathcal{A}_2$ and the difference $\mathcal{A}_1 \setminus \mathcal{A}_2$ of $\mathcal{A}_1$ and $\mathcal{A}_2$ are events as well:

$$\mathcal{A}_1 \cup \mathcal{A}_2 \in \Sigma, \quad \mathcal{A}_1 \cap \mathcal{A}_2 \in \Sigma, \quad \mathcal{A}_1 \setminus \mathcal{A}_2 \in \Sigma.$$

- Considering any sequence $\mathcal{A}_i$, $(i = 1, \ldots, m)$ of events of length m, the union of the events is an event:

$$\bigcup_{i=1}^{m} \mathcal{A}_i \in \Sigma.$$

- **Probability** Prob**:** It is a set function, which assigns a number in the interval $[0, 1]$ to each event $\mathcal{A}_i \in \Sigma$, (Prob : $\Omega \to [0, 1]$) so that the following is satisfied:

$$\begin{aligned} &0 \leq \text{Prob}(\mathcal{A}) \leq 1, \quad \forall \mathcal{A} \in \Sigma \\ &\text{Prob}(\Omega) = 1 \\ &\text{Prob}(\mathcal{A}_1 \cap \mathcal{A}_2) = 0 \quad \text{for } \mathcal{A}_1 \cap \mathcal{A}_2 = \emptyset. \end{aligned} \tag{A.5}$$

The relations (A.5) are named Kolmogorov axioms of probability theory and result in the following relations

$$\begin{aligned} &\sum_{\omega \in \Omega} \text{Prob}(\omega) = 1 \\ &\text{Prob}(\bar{\mathcal{A}}) = 1 - \text{Prob}(\mathcal{A}) \end{aligned}$$

with $\bar{\mathcal{A}} = \Omega \setminus \mathcal{A}$ as the complementary event of $\mathcal{A}$.

Probability distribution. A probability distribution is a mathematical function that gives the probabilities of the occurrence of different possible elementary events for an experiment. The probability Prob : $\Omega \to [0, 1]$ needs to be defined for all elements of the sample space Ω. To clarify that the rules for Prob(ω) are not only valid for a single argument ω but for all $\omega \in \Omega$, the term of the *probability distribution* is used that links each elementary event of an experiment to the probability

$$\text{Prob}(\omega), \quad \omega \in \Omega.$$

Discrete random variable. A *discrete random variable* X is a measurable function

$$X : \Omega \to \mathcal{X}$$

from the sample space Ω to a measurable space $\mathcal{X}$. The probability that X takes on a value $x \in \mathcal{X} \subseteq \mathbb{R}$ is given by

$$\text{Prob}(X = x) = \text{Prob}(\{\omega \in \Omega | X(\omega) = x\}).$$

As $\omega_i \in \Omega$ represent independent events, it applies:

$$\text{Prob}(\{\omega_1, \ldots, \omega_q\}) = \sum_{i=1}^{q} \text{Prob}(\omega_i).$$

Expected value. The *expected value* $E(X)$ of a discrete random variable X defined by

$$E(X) = \sum_{x \in \mathcal{X}} x \cdot \text{Prob}(X = x)$$

states the arithmetic mean of a large number of independent realisations of X.

Variance. The *variance* $V(X)$ of a discrete random variable X defined by

$$V(X) = E(X^2) - (E(X))^2 = \sum_{x \in \mathcal{X}} (x - E(X))^2 \cdot \text{Prob}(X = x)$$

states the expectation of the squared deviation of a discrete random variable from its mean.

Conditional probability. The conditional probability $\text{Prob}(\mathcal{A}_1|\mathcal{A}_2)$ denotes the probability of the occurrence of an event $\mathcal{A}_1$ under the condition that a different event $\mathcal{A}_2$ simultaneously occurred. For two arbitrary events $\mathcal{A}_1$ and $\mathcal{A}_2$ with $\text{Prob}(\mathcal{A}_1) > 0$ the conditional probability is defined with

$$\text{Prob}(\mathcal{A}_1|\mathcal{A}_2) = \frac{\text{Prob}(\mathcal{A}_1 \cap \mathcal{A}_2)}{\text{Prob}(\mathcal{A}_2)}, \quad \text{for Prob}(\mathcal{A}_2) > 0.$$

Stochastic process. A sequence of random variables

$$\{X(k,\omega), k \geq 0\} = (X(0,\omega), X(1,\omega), \dots, X(N,\omega))$$

with values in a set $\mathcal{X}$ is called a *stochastic process* with state space $\mathcal{X}$. If the elementary events ω are specified, $\{X(k,\omega), k \geq 0\}$ is a deterministic sequence

$$\{X(k,\omega), k \geq 0\} = (x(0), x(1), \dots, x(N))$$

and the process is named a realisation of the random sequence $\{X(k,\omega)\}$ with

$$X(k,\omega) = x(k), \quad k = 0, 1, \dots, N \text{ for a specified } \omega.$$

A.3 Markov chains

Markov property. A stochastic process $\{X(k,\omega)\}, (k = 0, \dots, N)$ with state space $\mathcal{X}$ is called a Markov chain if for every $k \in \mathbb{N}$ and all states $x_k \in \mathcal{X}$ the equation

$$\begin{aligned} \text{Prob}(X(k+1,\omega) = x_{k+1}|X(k,\omega) = x_k, \dots, X(0,\omega) = x_0) \\ = \text{Prob}(X(k+1,\omega) = x_{k+1}|X(k,\omega) = x_k) \\ \text{for Prob}(X(k,\omega) = x_k) \end{aligned} \tag{A.6}$$

holds. The property (A.6) is named *Markov property* or *memoryless property*, because the conditional probability of $X(k+1,\omega)$ under the conditions $X(0,\omega),\ldots,X(k,\omega)$ depends only on $X(k,\omega)$. A Markov chain is named a *homogenous* Markov chain if the right side of (A.6) is independent of k.

Representation forms of Markov chains. A Markov chain can be represented by the triple

$$\mathcal{S} = (\mathcal{X}, G, \boldsymbol{p}(0,x))$$

with

- $\mathcal{X} = \{1,2,\ldots,N\}$ as the set of states.
- G as the state transition probability given by

$$G(x(k+1)|x_k) = \text{Prob}(X(k+1,\omega) = x_{k+1}|X(k,\omega) = x_k), \qquad x_k, x_{k+1} \in \mathcal{X},\ k = 0,1,\ldots,N.$$

- $\boldsymbol{p}_0$ as the probability distribution of the initial state.

A discrete Markov chain can be specified by the Chapman-Kolmogorov equation as

$$\text{Prob}(X(k+1,\omega) = x_{k+1}) = \sum_{x\in\mathcal{X}} G(x_{k+1}|x_k)\cdot \text{Prob}(X(k,\omega) = x_k).$$

In matrix notation the Markov chain is represented by

$$\boldsymbol{p}(k+1) = \boldsymbol{G}\boldsymbol{p}(k), \quad \boldsymbol{p}(0) = \boldsymbol{p}_0$$

with $\boldsymbol{G}$ as the transition matrix described in the next paragraph.

Transition matrix. The $N\times N$ matrix $\boldsymbol{G} = (g_{ij}), (i,j \in \mathcal{X})$ is defined as

$$g_{ij} := \text{Prob}(X(k+1,\omega) = i|X(k,\omega) = j) \text{ with}$$
$$g_{ij} \geq 0 \text{ and } \sum_{i=1}^{N} g_{ij} = 1, \quad \forall i,j \in \mathcal{X}. \tag{A.7}$$

It indicates the probability of moving from state j to state i in the next step. As the transition matrix fulfils the property (A.7), $\boldsymbol{G}$ is a column stochastic matrix.

Distribution of Markov chains. The initial probability distribution of a Markov chain $\{X(0,\omega)\}$ is a vector $\boldsymbol{p}(0) \in \mathbb{R}^N$ given by

$$\boldsymbol{p}(0) = \begin{pmatrix} p_1(0) & \cdots & p_N(0) \end{pmatrix}^{\mathrm{T}} = \begin{pmatrix} \mathrm{Prob}(X(0,\omega) = 1) & \cdots & \mathrm{Prob}(X(0,\omega) = N) \end{pmatrix}^{\mathrm{T}}.$$

The random variable $X(0,\omega)$ is called the initial state. As $\boldsymbol{p}(0)$ is a probability distribution

$$\sum_{i=1}^{N} p_i(0) = 1$$

holds. Using the equation of motion of discrete Markov chains

$$\boldsymbol{p}(k) = \boldsymbol{G}^k \, \boldsymbol{p}(0)$$

the probability distribution of the Markov chain

$$\boldsymbol{p}(k) = \begin{pmatrix} p_1(k) & \cdots & p_N(k) \end{pmatrix}^{\mathrm{T}} = \begin{pmatrix} \mathrm{Prob}(X(k,\omega) = 1) & \cdots & \mathrm{Prob}(X(k,\omega) = N) \end{pmatrix}^{\mathrm{T}}$$

at the time step k can be determined.

A.4 Properties of Markov chains

Irreducibility. A Markov chain $\{X(k,\omega)\}$, $(k = 0, \ldots, N)$ with state space $\mathcal{X}$ is said to be *irreducible* if there exists for every pair $i, j \in \mathcal{X}$ with $i \neq j$ a number k, for which the element $\tilde{g}_{ij}$ of the matrix $\boldsymbol{G}^k = (\tilde{g}_{ij})$ is different from zero.

Aperiodicity. The period of a state $i \in \mathcal{X}$ is the greatest common divisor (gcd) of the set of points in time at which the Markov chain returns to i when starting in i. The period is denoted with $p_{\mathrm{M},i} = \gcd\{k \geq 1 \in \mathbb{N} | \tilde{g}_{ii}^k > 0\}$, where $\tilde{g}_{ii}$ is an element of $\boldsymbol{G}^k$. The state $i \in \mathcal{X}$ is called aperiodic if $p_{\mathrm{M},i} = 1$ holds.

If all states $i \in \mathcal{X}$ have the same period p_{M}, then p_{M} is called the period of the Markov chain. A Markov chain is named *aperiodic* if all states of $\{X(k,\omega)\}$ are aperiodic.

Ergodicity. An irreducible and aperiodic Markov chain $\{X(k,\omega)\}$, $(k = 0, \ldots, N)$ is called *ergodic*. A Markov chain with transition matrix $\boldsymbol{G}$ is ergodic if and only if there is an integer $k \in \mathbb{N}$ so that $\boldsymbol{G}^k$ has strictly positive components.

Ergodicity states that every state is reached over time, so that the state space is completely filled over time. This means that the expected value does not depend on the initial state.

Reversibility. A Markov chain $\{X(k,\omega)\}$, $(k = 0, \ldots, N)$ with state space $\mathcal{X}$ is called *reversible* if the Detailed-Balance-Equation

$$g_{ij}\,\pi_i = g_{ji}\,\pi_j, \quad \forall i, j \in \mathcal{X} \tag{A.8}$$

is fulfilled, where g_{ij} and g_{ji} are elements of $\boldsymbol{G}$ and $\pi_i, \pi_j \in \mathbb{R}^N$ denote the stationary probability distribution of the Markov chain, introduced in the next section.

Reversibility states the property that it cannot be distinguished whether the Markov chain runs forward or backward in time. Hence, it is invariant under time reversal.

Sojourn time. The mean time duration that the Markov chain remains in one state is named *sojourn time*. The mean sojourn times of the states are geometrically distributed [139] and can be computed with the transition probabilities g_{ii} as

$$t_i = \frac{1}{1 - g_{ii}}. \tag{A.9}$$

A.5 Stationary probability distribution

Definition of a stationary distribution. The vector

$$\boldsymbol{\pi} = \begin{pmatrix} \pi_1 & \cdots & \pi_N \end{pmatrix}^{\mathrm{T}} \in \mathbb{R}^N \tag{A.10}$$

states the *stationary distribution* of a Markov chain $\{X(k,\omega)\}$, $(k = 0, \ldots, N)$ with state space $\mathcal{X}$ and transition matrix $\boldsymbol{G}$ if the following holds:

- $$\pi_i \geq 0, \quad \forall i \in \mathcal{X}. \tag{A.11}$$
- $$\sum_{i=1}^{N} \pi_i = 1. \tag{A.12}$$
- $$\boldsymbol{G}\boldsymbol{\pi} = \boldsymbol{\pi}, \text{ i.e. } \sum_{j=1}^{N} g_{ij} p_{ij} = \pi_i, \; \forall j \in \mathcal{X}. \tag{A.13}$$

Equations (A.11) and (A.12) imply $\boldsymbol{\pi}$ to be a probability distribution on $\mathcal{X}$. Equation (A.13) is denoted in literature as Global-Balance-Equation.

Existence of a stationary distribution. As it is proven in [41] for an ergodic Markov chain $\{X(k,\omega)\}$, $(k = 0,\ldots,N)$ with state space $\mathcal{X}$ and transition matrix $\boldsymbol{G}$ the following holds:

- $\{X(k,\omega)\}$ has a uniquely defined stationary distribution $\pi \in \mathbb{R}^N$.
- $\boldsymbol{G}^k$ converges componentwise for $k \longrightarrow \infty$ towards the $N \times N$ matrix

$$\boldsymbol{D} = \begin{pmatrix} \pi & \cdots & \pi \end{pmatrix}. \tag{A.14}$$

- $$\pi_i = \frac{1}{\mu_{ii}}, \quad \forall i \in \mathcal{X},$$

where $\mu_{ii} = \sum_{k=1}^{\infty} k \cdot f_{ii}^k$ denotes the mean number of steps to get from state i to state i. $f_{ii}^k = \text{Prob}(X(k,\omega) \neq i, \ldots, X(k-1,\omega) = i | X(0,\omega) = i)$ is the probability that the Markov chain starting in i reaches the state i for the first time in the k-th step.

Determination of a stationary distribution. The stationary distribution of a Markov chain can be determined iteratively or directly if the state space $\mathcal{X}$ of $\{X(k,\omega)\}$ is not too large. With the direct calculation method the stationary distribution $\pi \in \mathbb{R}^N$ can be derived by [116]

$$\pi = (\boldsymbol{I} - \boldsymbol{G} + \mathbb{1}\mathbb{1}^{\mathrm{T}})^{-1}\mathbb{1} \tag{A.15}$$

with

- $\boldsymbol{I} - \boldsymbol{G} + \mathbb{1}\mathbb{1}^{\mathrm{T}}$ is invertible.
- $\mathbb{1} = \begin{pmatrix} 1 & 1 & \cdots & 1 \end{pmatrix}^{\mathrm{T}}$ is a vector with N ones.
- $\mathbb{1}\mathbb{1}^{\mathrm{T}}$ is a $N \times N$ matrix with all elements equal to one.
- $\boldsymbol{I}$ as the identity matrix.

The stationary distribution can be determined iteratively since the matrix $\boldsymbol{G}^k$ converges for $k \longrightarrow \infty$ towards the matrix $\boldsymbol{D}$, which columns consist of π as defined in (A.14).

Convergence rate towards a stationary distribution. The convergence rate of a Markov chain towards its stationary distribution can be estimated using the eigenvalues of the transition matrix $\boldsymbol{G}$. This requires the Perron-Frobenius theorem:

Theorem A.1 (Perron-Frobenius [147]). *Let $\boldsymbol{G}$ be an irreducible non-negative $N \times N$ matrix with eigenvalues $\lambda_1, \ldots, \lambda_N$ such that $|\lambda_1| \geq \ldots \geq |\lambda_N|$. Then the following holds:*

- *The eigenvalue λ_1 is real and positive.*
- $\lambda_1 > |\lambda_i|, \forall i = 2, \ldots, N$.

- *The N – dimensional left eigenvector* $\boldsymbol{u}_1^{\mathrm{T}} = \begin{pmatrix} u_{1,1} & \cdots & u_{1,N} \end{pmatrix}$, $\boldsymbol{u}_1 > \mathbf{0}$ *and the N – dimensional right eigenvector* $\boldsymbol{v}_1 = \begin{pmatrix} v_{1,1} & \cdots & v_{1,N} \end{pmatrix}^{\mathrm{T}}$, $\boldsymbol{v}_1 > \mathbf{0}$ *can be chosen such that all components are positive and* $\boldsymbol{u}_1^{\mathrm{T}}\boldsymbol{v}_1 = 1$ *holds.*

For irreducible non-negative matrices there exist a $k \in \mathbb{N}$ such that $\boldsymbol{G}^k > 0$ holds.

In Theorem A.2 an estimate of the convergence rate of an ergodic Markov chain towards its stationary distribution is derived using Corollary A.1 and the second largest eigenvalue $|\lambda_2|$ of $\boldsymbol{G}$ according to amount.

Corollary A.1 (Largest eigenvalue of the transition matrix of an ergodic Markov chain [145]). *Let* $\{X(k,\omega)\}$, $(k = 0,\ldots,N)$ *be an ergodic Markov chain with state space* $\mathcal{X}$, *transition matrix* $\boldsymbol{G}$ *and stationary distribution* $\boldsymbol{\pi} \in \mathbb{R}^N$. $\boldsymbol{G}$ *has the eigenvalues* $\lambda_1,\ldots,\lambda_N$ *with* $|\lambda_1| \geq \ldots \geq |\lambda_N|$, $\boldsymbol{v}_i$ *is the right eigenvector of* λ_i *and* $\boldsymbol{u}_i$ *is the left eigenvector of* λ_i. *Then*

$$\lambda_1 = 1,\ \boldsymbol{u}_1^{\mathrm{T}} = \begin{pmatrix} 1 & \ldots & 1 \end{pmatrix} \text{ and } \boldsymbol{v}_1 = \boldsymbol{\pi}$$

holds and $|\lambda_i| < 1$, $(\forall i = 2,\ldots,N)$.

Proof. As $\boldsymbol{G}$ is a stochastic matrix, $\boldsymbol{G}\boldsymbol{v}_1 = \boldsymbol{v}_1$ holds and (A.13) implies $\boldsymbol{G}\boldsymbol{\pi} = \boldsymbol{\pi}$. Thus, 1 is an eigenvalue of $\boldsymbol{G}$ and $\boldsymbol{u}_1^{\mathrm{T}}$ and $\boldsymbol{\pi}$ are left and right eigenvectors of this eigenvalue.

Let λ_i be an arbitrary eigenvalue of $\boldsymbol{G}$ and let $\boldsymbol{v}_i = \begin{pmatrix} v_{i,1} & \cdots & v_{i,N} \end{pmatrix}^{\mathrm{T}}$ be an eigenvector corresponding to λ_i and let $|v_m| = \max_{j\in\mathcal{X}} |v_{i,j}|$. By Definition (A.1) of λ_i and $\boldsymbol{v}_i$ it results that $\lambda_i\, v_m = \sum_{j=1}^{N} g_{mj}\, v_{i,j}$ and

$$\begin{aligned} |\lambda_i||v_m| = |\lambda_i\, v_m| &= \left|\sum_{j=1}^{N} g_{mj} v_{i,j}\right| \\ &\leq \sum_{j=1}^{N} g_{mj}|v_{i,j}| \leq \sum_{j=1}^{N} g_{mj} v_m = |v_m| \sum_{j=1}^{N} g_{mj} = |v_m| \end{aligned}$$

holds. Consequently $|\lambda_i| \leq 1$ and therefore $\lambda_1 = 1$ is the largest eigenvalue of $\boldsymbol{G}$. Theorem A.1 implies $|\lambda_i| < 1$ for $i = 2,\ldots,l$, which proves the corollary. ■

In Theorem A.2 the Landau symbol O is used. Generally $f(n) = O(g(n))$ means $|f(n)| \leq c\,|g(n)|$, where f and g are real-valued functions and c is a constant.

Theorem A.2 (Convergence rate of an ergodic Markov chain [145]). *Let* $\{X(k,\omega)\}$, $(k = 0,\ldots,N)$ *be an ergodic Markov chain with state space* $\mathcal{X}$, *transition matrix* $\boldsymbol{G}$ *and stationary distribution* $\boldsymbol{\pi} = \begin{pmatrix} \pi_1 & \cdots & \pi_N \end{pmatrix}^{\mathrm{T}} \in \mathbb{R}^N$. *All eigenvalues* $\lambda_1,\ldots,\lambda_N$ *of* $\boldsymbol{G}$ *are distinct and* $|\lambda_1| \geq \ldots \geq |\lambda_N|$ *holds. Then*

$$|g_{ij}^k - \pi_j| = O(|\lambda_2|^k), \quad \forall i,j \in \mathcal{X},\ k \in \mathbb{N}$$

applies.

For the proof of the theorem it is defined that for two arbitrary $N \times N$ matrices $\boldsymbol{A}$ and $\boldsymbol{B}$, $\boldsymbol{A} \leq \boldsymbol{B}$ holds if $a_{ij} \leq b_{ij}$, $\forall i, j = 1, \ldots, N$.

Proof. [145] Since all eigenvalues of $\boldsymbol{G}$ are distinct, $\boldsymbol{G}^k$ can be specified in spectral representation (A.4) as

$$\boldsymbol{G}^k = \sum_{i=1}^{N} \lambda_i^k \, v_i \, u_i^{\mathrm{T}}.$$

Corollary A.1 implies $\lambda_1 = 1$, $\boldsymbol{u}_1^{\mathrm{T}} = \begin{pmatrix} 1 & \cdots & 1 \end{pmatrix}$ and $\boldsymbol{v}_1 = \boldsymbol{\pi}$, being the left and right eigenvectors of λ_1, respectively. As $|\lambda_2|$ is the second largest eigenvalue of $\boldsymbol{G}$ according to amount it applies

$$\begin{aligned} \boldsymbol{G}^k - \begin{pmatrix} \boldsymbol{\pi} & \cdots & \boldsymbol{\pi} \end{pmatrix} &= \boldsymbol{G}^k - \lambda_1^k \boldsymbol{v}_1 \boldsymbol{u}_1^{\mathrm{T}} = \sum_{i=1}^{N} \lambda_i^k \boldsymbol{v}_i \boldsymbol{u}_i^{\mathrm{T}} - \lambda_1^k \boldsymbol{v}_1 \boldsymbol{u}_1^{\mathrm{T}} = \sum_{i=2}^{N} \lambda_i^k \boldsymbol{v}_i \boldsymbol{u}_i^{\mathrm{T}} \\ &\leq \sum_{i=2}^{N} |\lambda_2|^k \boldsymbol{v}_i \boldsymbol{u}_i^{\mathrm{T}} \leq |\lambda_2|^k \left| \sum_{i=2}^{N} \boldsymbol{v}_i \boldsymbol{u}_i^{\mathrm{T}} \right|. \end{aligned}$$

Due to Corollary A.1 $|\lambda_2| < 1$ holds and hence $|\lambda_2|^k$ converges to zero for $k \to \infty$. Finally

$$\begin{aligned} |g_{ij}^k - \pi_j| &\leq |\lambda_2|^k \left| \sum_{i=2}^{N} v_{li} \, u_{lj} \right| \\ \Rightarrow |g_{ij}^k - \pi_j| &= O(|\lambda_2|^k), \quad \forall i, j \in \mathcal{X},\ k \in \mathbb{N} \end{aligned}$$

applies, where v_{li} denotes the i-th element of the l-th eigenvector of $\boldsymbol{G}$ and u_{lj} states the j-th element of the l-th eigenvector of $\boldsymbol{G}$, which proves the theorem. ∎

The theorem states a geometric bound for the convergence rate. This implies that a sequence exists, which geometrically converges for large values $k \in \mathbb{N}$ as $c\beta^k$, where $0 \leq \beta \leq 1$ and $c \in \mathbb{R}_{>0}$. Hence, $g_{ij}^k, \forall i, j \in \mathcal{X}$ converges geometrically fast towards π_j.

If the ergodic Markov chain is also reversible, a more precise estimate of the convergence rate can be found.

Theorem A.3 (Convergence rate of a reversible and ergodic Markov chain [145]). *Let $\{X(k, \omega)\}$, $(k = 0, \ldots, N)$ be a reversible and ergodic Markov chain with state space $\mathcal{X}$, transition matrix $\boldsymbol{G}$ and stationary distribution $\boldsymbol{\pi} \in \mathbb{R}^N$ and let $\lambda_1 > |\lambda_2| \geq \ldots \geq |\lambda_N|$ be the eigenvalues of $\boldsymbol{G}$ ordered according to the size of their magnitudes. Then*

$$|g_{ij}^k - \pi_j| \leq \frac{1}{\sqrt{\min_{i \in \mathcal{X}} \pi_i}} |\lambda_2|^k, \quad \forall i, j \in \mathcal{X},\ k \in \mathbb{N} \tag{A.16}$$

applies.

Proof. [145] Set $\boldsymbol{D} = \operatorname{diag}(\sqrt{\pi_i})$, $(i \in \mathcal{X})$ and $\boldsymbol{A} = \boldsymbol{D}\boldsymbol{G}\boldsymbol{D}^{-1}$. Then

$$\boldsymbol{A} = \boldsymbol{D}\boldsymbol{G}\boldsymbol{D}^{-1} = \begin{pmatrix} g_{11}\frac{\sqrt{\pi_1}}{\sqrt{\pi_1}} & \cdots & g_{1N}\frac{\sqrt{\pi_1}}{\sqrt{\pi_N}} \\ \vdots & \ddots & \vdots \\ g_{N1}\frac{\sqrt{\pi_N}}{\sqrt{\pi_1}} & \cdots & g_{NN}\frac{\sqrt{\pi_N}}{\sqrt{\pi_N}} \end{pmatrix} = \left(g_{ij}\frac{\sqrt{\pi_i}}{\sqrt{\pi_j}}\right), \quad i,j \in \mathcal{X}.$$

Since $\boldsymbol{G}$ is assumed to be reversible, it follows from (A.8)

$$\begin{aligned} g_{ij}\,\pi_i = g_{ji}\,\pi_j &\Rightarrow g_{ij}\frac{\pi_i}{\sqrt{\pi_j}} = g_{ji}\frac{\pi_j}{\sqrt{\pi_j}} = g_{ji}\sqrt{\pi_j} \\ &\Rightarrow g_{ij}\frac{\sqrt{\pi_i}}{\sqrt{\pi_j}} = g_{ji}\frac{\sqrt{\pi_j}}{\sqrt{\pi_i}}, \quad \forall i,j \in \mathcal{X}. \end{aligned}$$

Hence, $\boldsymbol{A}$ is a symmetrical matrix and the eigenvectors are orthogonal. It follows that the eigenvectors $\boldsymbol{w}_1,\ldots,\boldsymbol{w}_N$ of $\boldsymbol{A}$ form an orthonormal basis in $\mathbb{R}^N$ and they are linearly independent. It results from (A.2) that the right eigenvectors are equal to the left eigenvectors and all eigenvalues of $\boldsymbol{A}$ are real.

Let λ_i be the eigenvalue of eigenvector $\boldsymbol{w}_i$ of $\boldsymbol{A}$. Inserting in (A.1) leads to

$$\begin{aligned} \boldsymbol{A}\boldsymbol{w}_i = \lambda_i\boldsymbol{w}_i &\Rightarrow \boldsymbol{D}\boldsymbol{G}\boldsymbol{D}^{-1}\boldsymbol{w}_i = \lambda_i\boldsymbol{w}_i \\ &\Rightarrow \boldsymbol{D}\boldsymbol{G}\boldsymbol{w}_i = \boldsymbol{D}\lambda_i\boldsymbol{w}_i. \end{aligned} \tag{A.17}$$

With $\boldsymbol{v}_i = \boldsymbol{D}\boldsymbol{w}_i$ it follows from (A.17) that $\boldsymbol{G}\boldsymbol{v}_i = \lambda_i\boldsymbol{v}_i$ and hence the eigenvalues of $\boldsymbol{G}$ and $\boldsymbol{A}$ are equal. Therefore, $\boldsymbol{A}$ can be written in spectral representation as

$$\boldsymbol{G}^k = (\boldsymbol{D}^{-1}\boldsymbol{D}\boldsymbol{G}\boldsymbol{D}^{-1}\boldsymbol{D})^k = (\boldsymbol{D}^{-1}\boldsymbol{A}\boldsymbol{D})^k = \sum_{l=1}^{N}\lambda_l\boldsymbol{D}^{-1}\boldsymbol{w}_l\boldsymbol{w}_l^{\mathrm{T}}\boldsymbol{D}. \tag{A.18}$$

As $\boldsymbol{v}_1,\ldots,\boldsymbol{v}_N$ are the right eigenvectors of $\boldsymbol{G}$ and from Corollary A.1 it results that $\lambda_1 = 1$ and $\boldsymbol{v}_1 = \boldsymbol{\pi}$ holds, it follows

$$\boldsymbol{w}_1 = \boldsymbol{D}^{-1}\boldsymbol{v}_1 = \boldsymbol{D}^{-1}\boldsymbol{\pi} = \begin{pmatrix}\sqrt{\pi_1} & \cdots & \sqrt{\pi_N}\end{pmatrix}^{\mathrm{T}}$$

and the first addend $(l = 1)$ of (A.18) is given by

$$\lambda_1^k\boldsymbol{D}^{-1}\boldsymbol{w}_1\boldsymbol{w}_1^{\mathrm{T}}\boldsymbol{D} = \begin{pmatrix} \pi_1 & \cdots & \pi_N \\ \vdots & \ddots & \vdots \\ \pi_1 & \cdots & \pi_N \end{pmatrix}.$$

With $\boldsymbol{w}_l = \begin{pmatrix} w_{l,1} & \cdots & w_{l,N}\end{pmatrix}^{\mathrm{T}}$ it follows for all other summands $l \geq 2 \in \mathcal{X}$ of (A.18)

$$\lambda_l^k\boldsymbol{D}^{-1}\boldsymbol{w}_l\boldsymbol{w}_l^{\mathrm{T}}\boldsymbol{D} = \begin{pmatrix} \lambda_l^k\frac{\sqrt{\pi_1}}{\sqrt{\pi_1}}w_{l,1}w_{l,1} & \cdots & \lambda_l^k\frac{\sqrt{\pi_N}}{\sqrt{\pi_1}}w_{l,1}w_{l,N} \\ \vdots & \ddots & \vdots \\ \lambda_l^k\frac{\sqrt{\pi_1}}{\sqrt{\pi_N}}w_{l,N}w_{l,1} & \cdots & \lambda_l^k\frac{\sqrt{\pi_N}}{\sqrt{\pi_N}}w_{l,N}w_{l,N} \end{pmatrix} = \left(\lambda_l^k\frac{\sqrt{\pi_j}}{\sqrt{\pi_i}}w_{l,i}w_{l,j}\right)$$

with $i, j \in \mathcal{X}$. Overall it results

$$g_{ij}^k = \pi_j + \frac{\sqrt{\pi_j}}{\sqrt{\pi_i}} \sum_{l=2}^{N} \lambda_l^k w_{l,i} w_{l,j}$$

for arbitrary $i, j \in \mathcal{X}$ and $k \in \mathbb{N}$. With Corollary A.1 and Definition (A.10) it follows

$$\begin{aligned} |g_{ij}^k - \pi_j| &\leq \frac{\sqrt{\pi_j}}{\sqrt{\pi_i}} |\lambda_2|^k \sum_{l=2}^{N} |w_{l,i}||w_{l,j}| \leq \frac{1}{\sqrt{\pi_i}} |\lambda_2|^k \sum_{l=2}^{N} |w_{l,i}||w_{l,j}| \\ &\leq \frac{1}{\sqrt{\min_{i \in \mathcal{X}} \pi_i}} |\lambda_2|^k \sum_{l=2}^{N} |w_{l,i}||w_{l,j}| = \frac{\sum_{l=2}^{N} |w_{l,i}||w_{l,j}|}{\sqrt{\min_{i \in \mathcal{X}} \pi_i}} |\lambda_2|^k. \end{aligned}$$

With the Cauchy-Schwarz inequality

$$\sum_{l=2}^{N} |w_{l,i}||w_{l,j}| \leq \left(\sum_{l=1}^{N} w_{l,i}^2 \right)^{\frac{1}{2}} \left(\sum_{l=1}^{N} w_{l,j}^2 \right)^{\frac{1}{2}}$$

holds, where the right side is equal to one because the eigenvectors $\boldsymbol{w}_i$ are orthonormal to each other. In summary, it results

$$|g_{ij}^k - \pi_j| \leq \frac{\sum_{l=2}^{N} |w_{l,i}||w_{l,j}|}{\sqrt{\min_{i \in \mathcal{X}} \pi_i}} |\lambda_2|^k \leq \frac{1}{\sqrt{\min_{i \in \mathcal{X}} \pi_i}} |\lambda_2|^k, \quad \forall i, j \in \mathcal{X},\ k \in \mathbb{N},$$

which proves the theorem. ■

A.6 Hidden Markov model

A Hidden Markov model (HMM) is a Markov model in which the system that is modelled is assumed to be a Markov process with unobservable *hidden* states. The state transitions are not observable from the outside and are described by the function G as *transition model* by

$$\text{Prob}(X(k+1, \omega) = x_{k+1}) = \sum_{x_k \in \mathcal{X}} G(x_{k+1}|x_k) \cdot \text{Prob}(X(k, \omega) = x_k).$$

This equation models the so called *background process*. From the outside only the output w is observable, which is determined with the function H as

$$\text{Prob}(W(k, \omega) = w_k) = \sum_{x_k \in \mathcal{X}} H(w_k|x_k) \cdot \text{Prob}(X(k, \omega) = x_k).$$

This so called *sensor model* describes the probability distribution, which generates the output w for a given state probability distribution. The transition model and the sensor model are coupled by the probability distribution of the current state of the HMM.

A.7 Continuous probability theory

For the determination of the packet loss probability of the wireless channel some results of the continuous probability theory are required, which are taken from [139] and stated in this section.

Continuous random variable. A *continuous random variable* X is defined by a nonnegative function $f_X(x) : \mathbb{R} \to \mathbb{R}$ for all real $x \in (-\infty, \infty)$, called the probability density function. $f_X(x)$ has the following properties:

- $$f_X(x) \geq 0, \quad \forall x \in \mathbb{R}$$
- $$\int_{-\infty}^{\infty} f_X(x)\,\mathrm{d}x = 1.$$

Probability density function. A *probability density function* (PDF) is a function, which value at any given sample in the sample space Ω can be interpreted as a relative likelihood that the value of the random variable equals that sample. The probability density function $f_X(x)$ of a continuous random variable is given by the integral

$$\mathrm{Prob}(a \leq X \leq b) = \int_a^b f_X(x)\,\mathrm{d}x.$$

It is the probability of X to be in the interval $[a, b]$. Since X has to take on a specific value, $f_X(x)$ has to satisfy

$$1 = \mathrm{Prob}(X \in (-\infty, \infty)) = \int_{-\infty}^{\infty} f_X(x)\,\mathrm{d}x.$$

Cumulative distribution function. The *cumulative distribution function* (CDF) of a random variable evaluated at x is the probability that X has a value less than or equal to x. The function is given by

$$F_X = \mathrm{Prob}(X \leq x).$$

The CDF of a continuous random variable X can be expressed as the integral of its PDF as

$$F_X(x) = \int_{-\infty}^{x} f_X(x)\,\mathrm{d}x.$$

The probability that X is in the semi-closed interval $(a, b]$ can be determined with

$$\mathrm{Prob}(a < X \leq b) = F_X(b) - F_X(a).$$

The PDF of a continuous random variable can be determined from the CDF by differentiating as

$$f_X(x) = \frac{\mathrm{d}\,F_X(x)}{\mathrm{d}\,x}.$$

Expected value. The *expected value* $E(X)$ of a continuous random variable X is defined by

$$E(X) = \int_{-\infty}^{\infty} x \cdot f_X(x)\,\mathrm{d}x.$$

If the continuous random variable is uniformly distributed over an interval $[a, b]$, the expected value $E(X)$ is the midpoint of this interval.

Variance. The *variance* $V(X)$ of a continuous random variable X is defined by

$$V(X) = E(X^2) - (E(X))^2 = \int_{-\infty}^{\infty} (x - E(X))^2 \cdot f_X(x)\,\mathrm{d}x$$

and measures the expected square of the deviation of a continuous random variable from its expected value.

Proofs

B

B.1 Proof of Theorem 3.1

In the proof of Theorem 3.1 it is necessary to determine whether a point is located left or right of a given line. This can be verified with Lemma B.1.

Lemma B.1 (Location of a point relative to a line). *Let the points* $\boldsymbol{p} = \begin{pmatrix} p_x & p_y \end{pmatrix}^T$, $\boldsymbol{q} = \begin{pmatrix} q_x & q_y \end{pmatrix}^T$ *and* $\boldsymbol{r} = \begin{pmatrix} r_x & r_y \end{pmatrix}^T$ *be given. The directed line* g *passes through* $\boldsymbol{p}$ *and* $\boldsymbol{q}$ *and* $\boldsymbol{r}$ *does not lie not on* g. *Then with the determinant* $\det(\boldsymbol{A})$ *of the matrix*

$$\boldsymbol{A} = \begin{pmatrix} 1 & p_x & p_y \\ 1 & q_x & q_y \\ 1 & r_x & r_y \end{pmatrix}$$

the relative location of $\boldsymbol{r}$ *to* g *can be determined as:*

$$\det(\boldsymbol{A}) \text{is} \begin{cases} \text{negative:} & \boldsymbol{r} \text{ is right of } g \\ \text{zero:} & \boldsymbol{r} \text{ is on } g \\ \text{positive:} & \boldsymbol{r} \text{ is left of } g. \end{cases}$$

Proof. The cross product can be defined in the 2D case analogous to the 3D case. The cross product $(\boldsymbol{r} - \boldsymbol{p}) \times \boldsymbol{n}_{pq}$ with $\boldsymbol{n}_{pq}$ as the normal vector to the vector $(\boldsymbol{p} - \boldsymbol{q})$ is equal to 0 if and only if $\boldsymbol{r}$ is on g. Otherwise the sign of $\det(\boldsymbol{A})$ is positive or negative depending on $\boldsymbol{r}$ is left of g or $\boldsymbol{r}$ is right of g. ∎

For the proof of Theorem 3.1 the control points need to be determined, which satisfy the boundary requirements given in Assumption 1.4 on p. 17. As the validation is only necessary for the x-direction and the y-direction, the z-direction is neglected in the proof for simplicity. Inserting (1.12) in (3.6) and setting $t_\mathrm{t} = t^i_\mathrm{tc}$ the control points are given by

$$\begin{aligned}
\boldsymbol{b}^i_\mathrm{x} &= \begin{pmatrix} b^i_{0,\mathrm{x}} \\ b^i_{1,\mathrm{x}} \\ b^i_{2,\mathrm{x}} \\ b^i_{3,\mathrm{x}} \\ b^i_{4,\mathrm{x}} \end{pmatrix} = \begin{pmatrix} r \\ r \\ r \\ r \\ r \end{pmatrix}, \qquad
\boldsymbol{b}^i_\mathrm{x} = \begin{pmatrix} b^i_{5,\mathrm{x}} \\ b^i_{6,\mathrm{x}} \\ b^i_{7,\mathrm{x}} \\ b^i_{8,\mathrm{x}} \\ b^i_{9,\mathrm{x}} \end{pmatrix} = \begin{pmatrix} r\frac{4\,t^i_\mathrm{tc}}{9}\omega \\ r\frac{3\,t^i_\mathrm{tc}}{9}\omega \\ r\frac{2\,t^i_\mathrm{tc}}{9}\omega \\ r\frac{t^i_\mathrm{tc}}{9}\omega \\ 0 \end{pmatrix} \\
\boldsymbol{b}^i_\mathrm{y} &= \begin{pmatrix} b^i_{0,\mathrm{y}} \\ b^i_{1,\mathrm{y}} \\ b^i_{2,\mathrm{y}} \\ b^i_{3,\mathrm{y}} \\ b^i_{4,\mathrm{y}} \end{pmatrix} = \begin{pmatrix} 0 \\ r\frac{t^i_\mathrm{tc}}{9}\omega \\ r\frac{2\,t^i_\mathrm{tc}}{9}\omega \\ r\frac{3\,t^i_\mathrm{tc}}{9}\omega \\ r\frac{4\,t^i_\mathrm{tc}}{9}\omega \end{pmatrix}, \qquad
\boldsymbol{b}^i_\mathrm{y} = \begin{pmatrix} b^i_{5,\mathrm{y}} \\ b^i_{6,\mathrm{y}} \\ b^i_{7,\mathrm{y}} \\ b^i_{8,\mathrm{y}} \\ b^i_{9,\mathrm{y}} \end{pmatrix} = \begin{pmatrix} r \\ r \\ r \\ r \\ r \end{pmatrix}.
\end{aligned} \tag{B.1}$$

Proof. The convex hull formed by the control points (B.1) is shown in Fig. B.1 in the xy-plane.

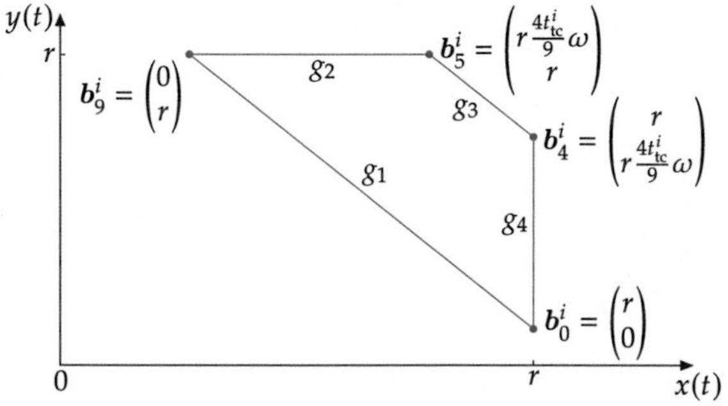

Fig. B.1: Boundary of the convex hull by g_1, g_2, g_3 and g_4.

The vertices of the convex hull are formed by the control points $\boldsymbol{b}^i_0 = \begin{pmatrix} b^i_{0,\mathrm{x}} & b^i_{0,\mathrm{y}} \end{pmatrix}^\mathrm{T}$, $\boldsymbol{b}^i_4 = \begin{pmatrix} b^i_{4,\mathrm{x}} & b^i_{4,\mathrm{y}} \end{pmatrix}^\mathrm{T}$, $\boldsymbol{b}^i_5 = \begin{pmatrix} b^i_{5,\mathrm{x}} & b^i_{5,\mathrm{y}} \end{pmatrix}^\mathrm{T}$ and $\boldsymbol{b}^i_9 = \begin{pmatrix} b^i_{9,\mathrm{x}} & b^i_{9,\mathrm{y}} \end{pmatrix}^\mathrm{T}$. Hence, the convex hull is limited by the four line segments g_1, g_2, g_3 and g_4. By forming the matrices $\boldsymbol{A}_1$, $\boldsymbol{A}_2$, $\boldsymbol{A}_3$ and $\boldsymbol{A}_4$ as

$$g_1: \quad \boldsymbol{A}_1 = \begin{pmatrix} 1 & b^i_{0,x} & b^i_{0,y} \\ 1 & b^i_{9,x} & b^i_{9,y} \\ 1 & s_x & s_y \end{pmatrix} \qquad g_2: \quad \boldsymbol{A}_2 = \begin{pmatrix} 1 & b^i_{9,x} & b^i_{9,y} \\ 1 & b^i_{5,x} & b^i_{5,y} \\ 1 & s_x & s_y \end{pmatrix}$$

$$g_3: \quad \boldsymbol{A}_3 = \begin{pmatrix} 1 & b^i_{4,x} & b^i_{4,y} \\ 1 & b^i_{5,x} & b^i_{5,y} \\ 1 & s_x & s_y \end{pmatrix} \qquad g_4: \quad \boldsymbol{A}_4 = \begin{pmatrix} 1 & b^i_{0,x} & b^i_{0,y} \\ 1 & b^i_{4,x} & b^i_{4,y} \\ 1 & s_x & s_y \end{pmatrix}$$

it can be proved that the control points (3.29) fulfilling the boundary requirements (3.28) are inside the convex hull. Due to the symmetrical arrangement of the points

$$\boldsymbol{b}^i = \begin{pmatrix} b^i_{l,x} \\ b^i_{l,y} \end{pmatrix}, \; l = 0, \ldots, 4 \text{ and } \boldsymbol{b}^i = \begin{pmatrix} b^i_{k,x} \\ b_{k,y} \end{pmatrix}, \; k = 5, \ldots, 9$$

it is sufficient to check only the points $\boldsymbol{b}^i$ on validity. A substitution of $\boldsymbol{s} = \begin{pmatrix} s_x \\ s_y \end{pmatrix}$ with $b^i_0 = \begin{pmatrix} b^i_{0,x} \\ b^i_{0,y} \end{pmatrix}$, $b^i_1 = \begin{pmatrix} b^i_{1,x} \\ b^i_{1,y} \end{pmatrix}$, $b^i_2 = \begin{pmatrix} b^i_{2,x} \\ b^i_{2,y} \end{pmatrix}$, $b^i_3 = \begin{pmatrix} b^i_{3,x} \\ b^i_{3,y} \end{pmatrix}$ and $b_4 = \begin{pmatrix} b^i_{4,x} \\ b^i_{4,y} \end{pmatrix}$ and an analysis of the determinants $\det(\boldsymbol{A}_1)$, $\det(\boldsymbol{A}_2)$, $\det(\boldsymbol{A}_3)$ and $\det(\boldsymbol{A}_4)$ for each control point gives that for

$$0 \leq t^i_{tc} \leq \frac{9}{4\,\omega}$$

the control points are inside the convex hull, which proves the theorem. ■

B.2 Proof of Theorem 4.1

For the proof of Theorem 4.1 the worst-case scenario is considered, that the agent moves with its maximum speed at a time instant as $||\boldsymbol{v}(t_k)|| = v_{\max}$, which also covers the cases with $||\boldsymbol{v}(t_k)|| < v_{\max}$. By rotating the coordinate system, initial speeds in any direction can be mapped to an initial speed $\begin{pmatrix} 0 & v_y(t_k) & 0 \end{pmatrix}^{\mathrm{T}}$ in y-direction only. The final speed $\dot{\boldsymbol{w}}(t_{k+1})$ at the end point $\boldsymbol{w}(t_{k+1})$ is assumed to be orthogonal to the ellipsoid to obtain the greatest distance. Without loss of generality, $t_k = 0$ is applied by a time shift. For the proof the set (4.4) is rewritten in terms of the end point $\boldsymbol{w}(t_{k+1})$ as

$$\mathcal{P}(t) = \left\{ \begin{pmatrix} w_\mathrm{x}(t) \\ w_\mathrm{y}(t) \\ w_\mathrm{z}(t) \end{pmatrix} \in \mathbb{R}^3 : \frac{(w_\mathrm{x}(t_{k+1}) - \bar{x}(t,t_k))^2}{r_\mathrm{x}^2(t,t_k)} + \frac{(w_\mathrm{y}(t_{k+1}) - \bar{y}(t,t_k))^2}{r_\mathrm{y}^2(t,t_k)} + \frac{(w_\mathrm{z}(t_{k+1}) - \bar{z}(t,t_k))^2}{r_\mathrm{z}^2(t,t_k)} - 1 \leq 0 \right\}.$$

Proof. The proof of Theorem 4.1 unfolds in three steps.

1): First, the z-component of (4.4) is excluded and it is proven that (4.4) describes a circle in the 2D space.

2): The factor e of the movement of the centre is proven.

3): Eqn. (4.6) describing the factor f of the radius is proven.

4): It is proven that (4.4) describes an ellipsoid in the 3D space.

1): The constraint (1.10) on the speed and Assumption 1.2 as

$$||\boldsymbol{v}(t)|| = ||\dot{\boldsymbol{w}}(t)|| \leq v_\mathrm{max}$$

can be transformed into

$$\dot{w}_\mathrm{x}^2(t) + \dot{w}_\mathrm{y}^2(t) \leq v_\mathrm{max}^2 \tag{B.2}$$

by considering the single speed components. With (2.10), (2.11) and (3.6), Assumption 1.4 and $\boldsymbol{v}(t_k) = \begin{pmatrix} 0 & v_\mathrm{y}(t_k) \end{pmatrix}^\mathrm{T}$ the Bézier curves for $\dot{w}_\mathrm{x}(t)$ and $\dot{w}_\mathrm{y}(t)$ are determined. A transformation of these Bézier curves into the monomial base with (2.18) yields:

$$\begin{aligned} \dot{w}_\mathrm{x}(t) &= w_\mathrm{x}(t_{k+1}) \cdot \tilde{a}_1(t,t_\mathrm{t}) + \dot{w}_\mathrm{x}(t_{k+1}) \cdot a_2(t,t_\mathrm{t}) \\ \dot{w}_\mathrm{y}(t) &= w_\mathrm{y}(t_{k+1}) \cdot \tilde{a}_1(t,t_\mathrm{t}) + \dot{w}_\mathrm{y}(t_{k+1}) \cdot a_2(t,t_\mathrm{t}) + v_{\mathrm{L},y}(t_k) \cdot b(t,t_\mathrm{t}) \end{aligned}$$

with the functions $\tilde{a}_1(t,t_\mathrm{t}) = \frac{1}{t_\mathrm{t}} a_1(t,t_\mathrm{t})$, $a_2(t,t_\mathrm{t})$ and $b(t,t_\mathrm{t})$ given by (4.7), (4.8), (4.9) and $w_\mathrm{x}(t_{k+1})$, $w_\mathrm{y}(t_{k+1})$ state the end points and $\dot{w}_\mathrm{x}(t_{k+1})$, $\dot{w}_\mathrm{y}(t_{k+1})$ are the speeds in x-direction and y-direction at the end points. Hence, eqn. (B.2) can be rewritten as

$$\begin{aligned} &(w_\mathrm{x}(t_{k+1}) \cdot \tilde{a}_1(t,t_\mathrm{t}) + \dot{w}_\mathrm{x}(t_{k+1}) \cdot a_2(t,t_\mathrm{t}))^2 \\ &\quad + \left(w_\mathrm{y}(t_{k+1}) \cdot \tilde{a}_1(t,t_\mathrm{t}) + \dot{w}_\mathrm{y}(t_{k+1}) \cdot a_2(t,t_\mathrm{t}) + v_\mathrm{y}(t_k) \cdot b(t,t_\mathrm{t})\right)^2 \leq v_\mathrm{max}^2. \end{aligned} \tag{B.3}$$

With the substitutions

$$\begin{aligned} x(t_{k+1}) &= w_\mathrm{x}(t_{k+1}), \quad \bar{x}(t,t_\mathrm{t}) = \dot{w}_\mathrm{x}(t_{k+1}) \cdot a_2(t,t_\mathrm{t}) \\ y(t_{k+1}) &= w_\mathrm{y}(t_{k+1}), \quad \bar{y}(t,t_\mathrm{t}) = \dot{w}_\mathrm{y}(t_{k+1}) \cdot a_2(t,t_\mathrm{t}) + v_\mathrm{y}(t_k) \cdot b(t,t_\mathrm{t}) \end{aligned}$$

eqn. (B.3) can be stated as

$$(x(t_{k+1}) \cdot \tilde{a}_1(t,t_\mathrm{t}) + \bar{x}(t,t_\mathrm{t}))^2 + (y(t_{k+1}) \cdot \tilde{a}_1(t,t_\mathrm{t}) + \bar{y}(t,t_\mathrm{t}))^2 \leq v_\mathrm{max}^2,$$

which describes a circle. Since the prefactors of $x(t_{k+1})$ and $y(t_{k+1})$ are equal, eqn. (B.3) is proved to be a circle.

2): A point on the circle is determined to obtain the factor e of the movement of the centre of the circle.

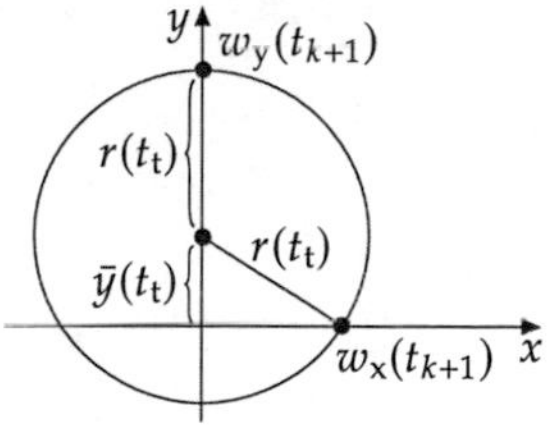

Fig. B.2: Description of the circle.

With

$$v_y(t_k) = v_{max} \tag{B.4}$$

and a constant movement of the agent in y-direction in the transition time interval t_t without a speed component in x-direction the maximum distance is covered. The conditions at the end point on the circle with the corresponding speed apply as

$$\begin{aligned} w_x(t_{k+1}) &= 0, & \dot{w}_x(t_{k+1}) &= 0, \\ w_y(t_{k+1}) &= v_{max} \cdot t_t, & \dot{w}_y(t_{k+1}) &= v_{max}. \end{aligned} \tag{B.5}$$

Figure B.2 illustrates that $w_y(t_{k+1}) = r(t_t) + \bar{y}(t_t)$ holds. With $r(t_t) = f \cdot v_{max} \cdot t_t$ and $\bar{y}(t_t) = e \cdot v_y(t_k) \cdot t_t$, (B.4) and (B.5) it follows

$$v_{max} \cdot t_t = f \cdot v_{max} \cdot t_t + e \cdot v_{max} \cdot t_t.$$

Hence, the factor of the movement of the centre of the circle in dependence on the radius is

$$e = 1 - f.$$

3): For the proof of (4.6) a second point on the circle is determined. According to Fig. B.2

$$r^2(t_t) = \left(w_y(t_{k+1}) - r(t_t)\right)^2 + w_x^2(t_{k+1})$$

holds. With (B.5) it results

$$(f \cdot v_{max} \cdot t_t)^2 = ((1-f) \cdot v_{max} \cdot t_t)^2 + w_x^2(t_{k+1}).$$

Hence, the point $w_x(t_{k+1})$ can be computed as

$$w_x(t_{k+1}) = \sqrt{2f-1} \cdot v_{max} \cdot t_t.$$

The end point with the corresponding final speed is given by

$$\begin{aligned} w_x(t_{k+1}) &= \sqrt{2f-1} \cdot v_{max} \cdot t_t, & \dot{w}_x(t_{k+1}) &= v_{max}, \\ w_y(t_{k+1}) &= 0, & \dot{w}_y(t_{k+1}) &= 0. \end{aligned} \tag{B.6}$$

Inserting (B.6) into (B.3) with the substitution $s = \frac{t}{t_t}$, with $s \in [0,1]$ yields

$$\left(\sqrt{2f-1} \cdot v_{max} \cdot t_t \cdot \tilde{a}_1(s) + v_{max} \cdot a_2(s)\right)^2 + (v_y(t_k) \cdot b(s))^2 \leq v_{max}^2.$$

With (B.4) and (4.7) it follows

$$v_{max}^2 \cdot \left(\sqrt{2f-1} \cdot a_1(s) + a_2(s)\right)^2 + v_{max}^2 \cdot b^2(s) \leq v_{max}^2$$

and

$$g_f(s) = \left(\sqrt{2f-1} \cdot a_1(s) + a_2(s)\right)^2 + b^2(s) \leq 1.$$

The family of functions $g_f(s)$ has an absolute maximum in the ε-neighbourhood $U[0.6]$ as defined in Definition 4.2 depending on f, which results in the estimate:

$$\begin{aligned} &\left(\sqrt{2f-1} \cdot a_1(s) + a_2(s)\right)^2 + b^2(s) \\ &\leq \left(\sqrt{2f-1} \cdot \max_{U[0.6]} a_1(s) + \max_{U[0.6]} a_2(s)\right)^2 + \left(\min_{U[0.6]} b(s)\right)^2 \leq 1. \end{aligned}$$

The function $a_1(s)$ is positive in the interval $[0,1]$. The function $a_2(s)$ is negative in the ε-neighbourhood $U[0.6]$. Hence, with $\max_{U[0.6]} a_2(s)$ the smallest value is subtracted from $a_1(s)$, which results in the maximum of the sum $\sqrt{2f-1} \cdot a_1(s) + a_2(s)$. The function $b(s)$ is negative in the ε-neighbourhood $U[0.6]$. Since the function $b(s)$ is squared, the minimum of $b(s)$ corresponds to the maximum of the squared function $b^2(s)$. With root extraction it follows

$$\sqrt{2f-1} \cdot \max_{U[0.6]} a_1(s) + \max_{U[0.6]} a_2(s) \leq \sqrt{1 - \left(\min_{U[0.6]} b(s)\right)^2}.$$

After algebraic manipulations the result is (4.6).

4): Due to Assumption 1.2 the agent moves with a maximum speed v_{max}. Since the radius in z-direction increases with v_{max} the positions of the agent in z-direction are included in the set. Therefore, according to Definition 4.1 the set (4.4) describes an ellipsoid, which proves the theorem. ■

B.3 Proof of Theorem 7.1

For the proof of Theorem 7.1 only the requirement (1.1) is considered. As for the requirement (1.2) the distance between $\bar{s}$ and the threshold is equal to requirement (1.1), the proof is similar with only small modifications. For the proof the worst-case scenario that both agents move directly towards each other with their maximum speed $v_{\max}$ is assumed. This scenario includes every other possible movement. Hence, the proof of the collision-free movement is valid for every possible movement of the agents.

Every possible intersection of the agents in the 3D space that leads to a violation of control aim (A1) can be mapped into the 2D space by a coordinate transformation. In addition, the coordinate system can be transformed in such a way that the agents move along the y-axis. For simplicity and without loss of generality the movement of the agents in the 3D space is considered where $\boldsymbol{p}_{\mathrm{G}}(t_{\mathrm{s1}}) = (0 \;\; 0 \;\; 0)^{\mathrm{T}}$ should apply by a coordinate transformation and the stand-on agent has an arbitrary position in $\mathbb{R}^2$ as $\boldsymbol{p}_{\mathrm{S}}(t_{\mathrm{s1}}) = (0 \;\; -y_{\mathrm{S}}(t_{\mathrm{s1}}) \;\; 0)^{\mathrm{T}}$. Due to the coordinate transformation the proof holds for every possible movement of the agents. For the specified movement of the agents, the scaling factors $\boldsymbol{\mu} = (1 \;\; 0 \;\; 0)^{\mathrm{T}}$ and $\boldsymbol{\nu} = (0 \;\; 1 \;\; 0)^{\mathrm{T}}$ apply. Hence, it needs to be proven that the end point (7.8) appears to be

$$\boldsymbol{w}_{\mathrm{G}}(t_{\mathrm{e1}}) = \left(x_{\mathrm{G}}(t_{\mathrm{s1}}) + \tfrac{4}{9}\, t_{\mathrm{t,min}}\, v_{G} \quad y_{\mathrm{G}}(t_{\mathrm{s1}}) + \tfrac{4}{9}\, t_{\mathrm{t,min}}\, v_{\mathrm{G,max}} \quad 0\right)^{\mathrm{T}}. \tag{B.7}$$

The worst-case scenario of a possible collision is exemplarily shown in Fig. B.3. The agents move along the y-axis directly towards each other with their maximum speeds $v_{\max}$. The reactive trajectory together with the control points is shown in the upper part of the figure. The position of the stand-on agent is illustrated as a blue dot. The possible movement of the agent in the time interval $t_{\mathrm{t,min}}$ is included in the set $\mathcal{P}_{\mathrm{S}}(t, t_k)$. As it can be seen, the set $\mathcal{P}_{\mathrm{S}}(t, t_k)$ does not intersect with the convex hull of the trajectory of the give-way agent. Hence, no collision occurs.

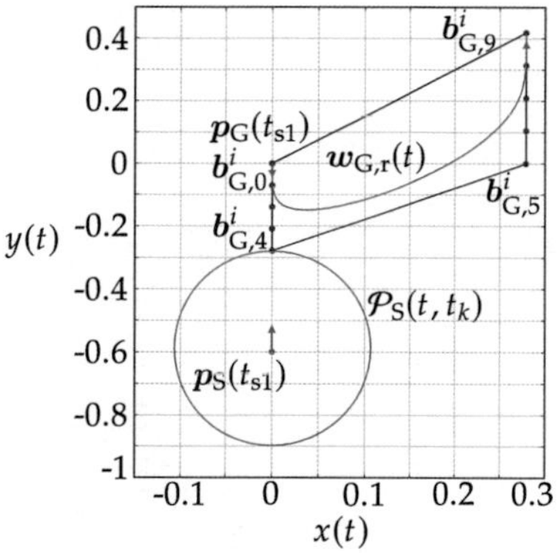

Fig. B.3: Worst-case scenario of a possible collision with $v_{\max} = 1\frac{\mathrm{m}}{\mathrm{s}}$, $v_{\mathrm{G,max}} = 1.5\frac{\mathrm{m}}{\mathrm{s}}$, $t_{\mathrm{t,min}} = 0.8\,\mathrm{s}$.

Proof. The idea of the proof is to show that with an event generation with condition (7.7) the set $\mathcal{P}_{\mathrm{S}}(t, t_k)$ does not intersect the convex hull of the reactive trajectory of the give-way agent. The proof unfolds in four steps.

1): It is proven that the points $\boldsymbol{b}^i_{\mathrm{G},4}$ and $\boldsymbol{b}^i_{\mathrm{G},5}$ are candidates for being closest to the stand-on agent and that $\boldsymbol{b}^i_{\mathrm{G},4}$ is closest to the stand-on agent.

2): The stand-on agent does not intersect the convex hull of the reactive trajectory if the events are generated with $\bar{e}_{\mathrm{G}}$ as (7.7).

3): It is proven that the distance between the agents is at least $\underline{s} + \varepsilon_{\mathrm{d}}\,\bar{e}_{\mathrm{G}}$ after the time interval $t_{\mathrm{t,min}}$.

4): It is shown that in the worst-case scenario all other intersection situations are included.

1): Due to Assumption 1.4 the control points $\boldsymbol{b}^i_{\mathrm{G},0} \dots \boldsymbol{b}^i_{\mathrm{G},4}$ and $\boldsymbol{b}^i_{\mathrm{G},5} \dots \boldsymbol{b}^i_{\mathrm{G},9}$ are located along a line. As $\dot{w}_{\mathrm{G}}(t_{\mathrm{s1}}) = -v_{\mathrm{max}}$ and $\dot{w}_{\mathrm{G}}(t_{\mathrm{e1}}) = v_{\mathrm{G,max}}$ holds, it can be seen with (3.7) that $\boldsymbol{b}^i_{\mathrm{G},4}$ and $\boldsymbol{b}^i_{\mathrm{G},5}$ are candidates for being closest to the stand-on agent. The locations of the control points are shown in Fig. B.3. With Assumption 1.4, $\boldsymbol{p}_{\mathrm{G}}(t_{\mathrm{s1}}) = \boldsymbol{0}$ and eqns. (3.7) and (B.7) the locations of the control points in $\mathbb{R}^2$ are given by

$$\boldsymbol{b}^i_{\mathrm{G},4} = \begin{pmatrix} 0 \\ -\frac{4}{9} t_{\mathrm{t,min}}\, v_{\mathrm{max}} \end{pmatrix}$$

$$\boldsymbol{b}^i_{\mathrm{G},5} = \begin{pmatrix} \frac{4}{9} t_{\mathrm{t,min}}\, v_{\mathrm{max}} \\ \frac{4}{9} t_{\mathrm{t,min}}\, v_{\mathrm{G,max}} - \frac{4}{9} t_{\mathrm{t,min}}\, v_{\mathrm{G,max}} \end{pmatrix} = \begin{pmatrix} \frac{4}{9} t_{\mathrm{t,min}}\, v_{\mathrm{max}} \\ 0 \end{pmatrix}.$$

As the stand-on agent is assumed to move on the y-axis the control point $\boldsymbol{b}^i_{\mathrm{G},4}$ is closest to the stand-on agent.

2): The stand-on agent covers the distance

$$d_{\mathrm{S}} = t_{\mathrm{t,min}}\, v_{\mathrm{max}}$$

in the time interval $t_{\mathrm{t,min}}$. The distance between $\boldsymbol{p}_{\mathrm{G}}(t_{\mathrm{s1}})$ and $\boldsymbol{p}_{\mathrm{S}}(t_{\mathrm{s1}})$ is given by $\bar{e}_{\mathrm{G}}$. It results:

$$\begin{aligned} \bar{e}_{\mathrm{G}} &\geq \frac{13}{9} t_{\mathrm{t,min}}\, v_{\mathrm{max}} \\ &= \left(\frac{9}{9} + \frac{4}{9}\right) t_{\mathrm{t,min}}\, v_{\mathrm{max}} \\ &= \underbrace{t_{\mathrm{t,min}}\, v_{\mathrm{max}}}_{d_{\mathrm{S}}} + \underbrace{\frac{4}{9} t_{\mathrm{t,min}}\, v_{\mathrm{max}}}_{|b^i_{\mathrm{G},4,x}|} . \end{aligned}$$

Hence, after the time interval $t_{\mathrm{t,min}}$ the stand-on agent touches the control point $\boldsymbol{b}^i_{\mathrm{G},4}$ but does not intersect with the convex hull, which proves the first part of the theorem.

3): As the stand-on agent is at most on the position of the control point $\boldsymbol{b}^i_{\mathrm{G},4}$ the scaling factor ε_d of the distance to the position $\boldsymbol{w}_\mathrm{G}(t_\mathrm{e1}) = (w_{\mathrm{G,x}}(t_\mathrm{e1}) \quad w_{\mathrm{G,y}}(t_\mathrm{e1}))^\mathrm{T}$ is with Assumptions 1.2 and 1.4 given by

$$\begin{aligned}
\varepsilon_\mathrm{d}\,\bar{e}_\mathrm{G} &= \sqrt{\left(|b^i_{\mathrm{G},4,\mathrm{x}}| + |w_{\mathrm{G,x}}(t_\mathrm{e1})|\right)^2 + \left(|b^i_{\mathrm{G},4,\mathrm{y}}| + |w_{\mathrm{G,y}}(t_\mathrm{e1})|\right)^2} \\
&= \sqrt{\left(\tfrac{4}{9} t_\mathrm{t,min}\, v_\mathrm{max}\right)^2 + \left(\tfrac{4}{9} t_\mathrm{t,min}\, v_\mathrm{max} + \tfrac{4}{9} t_\mathrm{t,min}\, v_\mathrm{G,max}\right)^2} \\
&= \sqrt{\left(\tfrac{4}{9} t_\mathrm{t,min}\, v_\mathrm{max}\right)^2 + \left(\tfrac{4}{9} t_\mathrm{t,min}\, v_\mathrm{max} + \tfrac{4}{9} t_\mathrm{t,min}\, 1.5\, v_\mathrm{max}\right)^2} \\
&= \sqrt{\tfrac{116}{81} t^2_\mathrm{t,min}\, v^2_\mathrm{max}} \\
&= \tfrac{\sqrt{116}}{9} t_\mathrm{t,min}\, v_\mathrm{max}.
\end{aligned}$$

Inserting $\bar{e}_\mathrm{G}$ gives

$$\varepsilon_\mathrm{d}\,\frac{13}{9} t_\mathrm{t,min}\, v_\mathrm{max} = \frac{\sqrt{116}}{9} t_\mathrm{t,min}\, v_\mathrm{max} \tag{B.8}$$

and results in

$$\varepsilon_\mathrm{d} = 0.8285.$$

Hence, with the safety distance $\underline{s}$ the distance between the agents after the time interval $t_\mathrm{t,min}$ is given by

$$d_\mathrm{SG} = \underline{s} + \varepsilon_\mathrm{d}\,\bar{e}_\mathrm{G}.$$

4): If the give-way agent has a speed with a positive x-component and a negative y-component as

$$v_\mathrm{max} = \left\| \begin{pmatrix} v_\mathrm{x,max} \\ -v_\mathrm{y,max} \end{pmatrix} \right\|,$$

shown in Fig. B.4 the location of the control point $\boldsymbol{b}^i_{\mathrm{G},4}$ is given by

$$\boldsymbol{b}^i_{\mathrm{G},4} = \begin{pmatrix} \tfrac{4}{9} t_\mathrm{t,min}\, v_\mathrm{x,max} \\ -\tfrac{4}{9} t_\mathrm{t,min}\, v_\mathrm{y,max} \end{pmatrix}. \tag{B.9}$$

At time instant t_s the position of the stand-on agent is given by

$$\boldsymbol{p}_\mathrm{S}(t_\mathrm{s1}) = \begin{pmatrix} 0 \\ -\bar{e}_\mathrm{G} \end{pmatrix} = \begin{pmatrix} 0 \\ -\tfrac{13}{9} t_\mathrm{t,min}\, v_\mathrm{max} \end{pmatrix}. \tag{B.10}$$

The distance between (B.9) and (B.10) can be calculated as

$$d_\mathrm{Sb} = ||\boldsymbol{p}_\mathrm{S}(t_\mathrm{s1}) - \boldsymbol{b}^i_{\mathrm{G},4}|| = \sqrt{\left(-\tfrac{4}{9} t_\mathrm{t,min}\, v_\mathrm{x,max}\right)^2 + \left(-\tfrac{13}{9} t_\mathrm{t,min}\, v_\mathrm{max} + \tfrac{4}{9} t_\mathrm{t,min}\, v_\mathrm{y,max}\right)^2}.$$

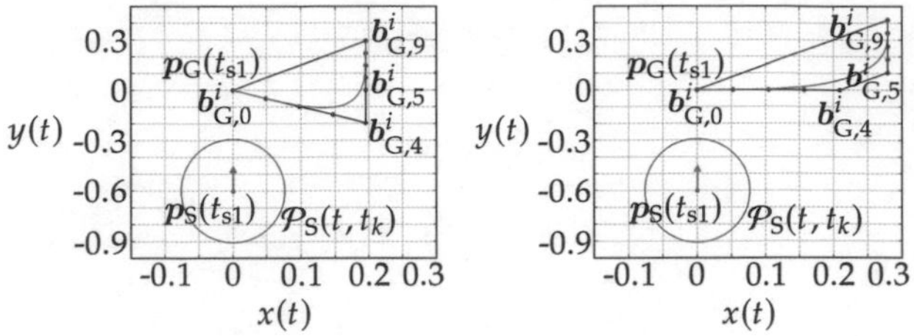

Fig. B.4: Scenario of a possible collision with speeds of the give-way agent given by $v_{x,max} > 0$, $v_{y,max} < 0$ (left) and $v_{x,max} > 0$, $v_{y,max} = 0$ (right).

With $v_{x,max} = \sqrt{v_{max}^2 - v_{y,max}^2}$ it results

$$\begin{aligned} d_{Sb} &= \sqrt{\left(-\tfrac{4}{9}t_{t,min}\sqrt{v_{max}^2 - v_{y,max}^2}\right)^2 + \left(-\tfrac{13}{9}t_{t,min}\,v_{max} + \tfrac{4}{9}t_{t,min}\,v_{y,max}\right)^2} \\ &= \sqrt{\tfrac{185}{81}t_{t,min}^2\,v_{max}^2 - \tfrac{104}{81}t_{t,min}^2\,v_{max}\,v_{y,max}} \\ &> \sqrt{\tfrac{185}{81}t_{t,min}^2\,v_{max}^2 - \tfrac{104}{81}t_{t,min}^2\,v_{max}^2} \quad \text{with } v_{y,max} < v_{max} \\ &= t_{t,min}\,v_{max} = d_S. \end{aligned}$$

It can be concluded that no risk of collision occurs if the give-way agent has a speed with a positive x-component and a positive y-component as the control point $b^i_{G,0}$ is closest to the stand-on agent, which proves the theorem. ■

B.4 Proof of Theorem 7.2

For the proof the worst-case scenarios are considered. As in the proof of Theorem 7.1 the worst-case considerations include all possible movements of the agents. Hence, with the derivation of the thresholds for these cases for each possible change of speed or change of height of the stand-on agent the control aim (A1) for a circular movement of the agents is satisfied when events are generated with condition (7.17). The following two worst-case scenarios can occur that are considered in the proof:

1. Both agents move with the speed $v_{S,max}$ and $\bar{P}_S$ reduces its speed to $v_{S,min}$ in the shortest possible time span. Then, $\bar{P}_G$ has to reduce its speed to $v_{G,min}$ as fast as possible.

2. Both agents move with the speed $v_{S,min}$ and $\bar{P}_S$ accelerates as fast as possible to $v_{S,max}$. Then, $\bar{P}_G$ has to accelerate to $v_{G,max}$ as quickly as possible.

Proof. The proof unfolds in four steps.

1): The first part of the threshold $\bar{e}_\Phi$ is proven that holds for the first worst-case scenario.
2): For the second worst-case scenario the second part of $\bar{e}_\Phi$ is proven.
3): The threshold $\bar{e}_\Phi$ is proven
4): The threshold $\bar{e}_z$ is derived.

1): In the first worst-case scenario the agent $\bar{P}_G$ has to accelerate. During the change of speed, the agent $\bar{P}_S$ continues moving and the phase difference changes further. Hence, the threshold needs to be determined to be identical to the phase difference added after the generation of event $\bar{e}_{G2}$.

The minimum time span $t_{t,min,a}$ to accelerate from $v_G(t) = v_{S,min}$ to $v_G(t) = v_{G,max}$ is determined with Algorithm 2 on p. 56. The phase that is covered during $t_{t,min,a}$ is given by $\Delta\Phi_a$. Hence, the phase difference added after the event e_{G2} for the first worst-case scenario is given by

$$\Delta\Phi_{add,a} = \left| \frac{v_{S,max} \cdot t_{t,min,a}}{2\pi r} \cdot 360° - \Delta\Phi_a \right| . \tag{B.11}$$

2): Similar considerations apply for the second worst-case scenario. Here, the minimum time span $t_{t,min,b}$ to reduce the speed from $v_G(t) = v_{S,max}$ to $v_G(t) = v_{G,min}$ is determined with Algorithm 2. The phase that is covered during $t_{t,min,b}$ is given by $\Delta\Phi_b$. The added phase difference after triggering event e_{G2} is given by

$$\Delta\Phi_{add,b} = \left| \frac{v_{S,min} \cdot t_{t,min,b}}{2\pi r} \cdot 360° - \Delta\Phi_b \right| . \tag{B.12}$$

3): In order to derive a feasible threshold for every case the maximum of the values $\Delta\Phi_{add,a}$ and $\Delta\Phi_{add,b}$ is taken as

$$\bar{e}_\Phi = \max\left(\Delta\Phi_{add,a}, \Delta\Phi_{add,b}\right),$$

which proves the first part of the theorem.

4): The choice of the threshold $\bar{e}_z$ defines how large the deviation of the height difference from the set point $\bar{z}$ is allowed to be before communication is invoked. As no specific tolerance range is defined, the threshold can be chosen to a value within the interval

$$0 < \bar{e}_z < \bar{z}$$

where $\bar{z}$ is defined in (1.7), which proves the theorem. ■

B.5 Proof of Lemma 7.1

Proof. The frequency of the data communication depends upon the speed difference $||v_{SG}(t)||$ of both agents. Due to Assumption 1.2 on p. 17 this difference is bounded by

$$||v_{SG}(t)|| = ||v_S(t) - v_G(t)|| \leq v_{SG,max}, \quad t \geq 0. \tag{B.13}$$

with $v_{\mathrm{SG,max}}$ given by (7.31). For the proof only communication invoked by condition (I) of (7.4) is considered. Invocation of communication with condition (II) of (7.4) leads to the same result. The desired distance between the positions of the stand-on agent and the give-way agent according to (7.11) is given by

$$||\boldsymbol{p}_{\mathrm{G}}(t) - \boldsymbol{p}_{\mathrm{S}}(t)|| = \underline{s} + 2\,\bar{e}_{\mathrm{G}}.$$

The distance between both agents for the generation of the communication event $e_{\mathrm{G}0}$ is

$$||\boldsymbol{p}_{\mathrm{G}}(t) - \boldsymbol{p}_{\mathrm{S}}(t)|| = \underline{s} + \bar{e}_{\mathrm{G}}.$$

Hence, communication is invoked at time instant t if the agents have covered the relative distance

$$\left\| \int_{t_k}^{t} \boldsymbol{v}_{\mathrm{SG}}(\tau)\mathrm{d}\tau \right\| = \bar{e}_{\mathrm{G}}$$

to one another after the last event at time instant t_k. The minimum time span $t = t_k + \tilde{t}$ for which this condition is satisfied leads to the minimum time span between two consecutive events:

$$t_{\mathrm{sep}} = \arg\min_{\tilde{t}} \left\{ \left\| \int_{t_k}^{t_k+\tilde{t}} \boldsymbol{v}_{\mathrm{SG}}(\tau)\mathrm{d}\tau \right\| = \bar{e}_{\mathrm{G}} \right\}$$
$$\text{s.t. } ||\boldsymbol{v}_{\mathrm{SG}}(t)|| \leq v_{\mathrm{SG,max}}, \quad t_k \leq t \leq t_k + \tilde{t}.$$

As

$$\arg\min_{\tilde{t}} \left\{ \left\| \int_{t_k}^{t_k+\tilde{t}} \boldsymbol{v}_{\mathrm{SG}}(\tau)\mathrm{d}\tau \right\| = \bar{e}_{\mathrm{G}} \right\} \geq \arg\min_{\tilde{t}} \left\{ \int_{t_k}^{t_k+\tilde{t}} ||\boldsymbol{v}_{\mathrm{SG}}(\tau)||\mathrm{d}\tau = \bar{e}_{\mathrm{G}} \right\}$$
$$= \frac{\bar{e}_{\mathrm{G}}}{v_{\mathrm{SG,max}}}$$

holds, one get

$$t_{\mathrm{sep}} \geq \tilde{t} = \frac{\bar{e}_{\mathrm{G}}}{v_{\mathrm{SG,max}}},$$

which proves the lemma. ■

B.6 Proof of Lemma 7.2

For the proof without loss of generality the coordinate transformations stated in Appendix B.3 are considered. In addition again the agents are assumed to move towards one another along the y-axis. The proof is valid for every possible movement of the agents.

Proof. For the proof only the event generation with condition (I) of (7.4) and (7.15) is considered. The proof for the event generation with condition (II) of (7.4) and (7.15) is identical. Without loss of generality $p_G(t_{s1}) = (0 \quad 0 \quad 0)^T$ is assumed by a coordinate transformation and the stand-on agent has the position $p_S(t_{s1}) = (0 \quad -\bar{e}_G \quad 0)^T$. The proof considers the worst-case scenario that both agents approach one other with their maximum speed and change their movement to the opposite directions directly after the event time instant t_k, as shown in Fig. B.5. At t_k the distance

$$d_{SG} = \underline{s} + \bar{e}_G$$

between the agents applies. According to (B.7) the give-way agent plans a reactive trajectory to reach the point

$$w_{G,r}(t_{e1}) = \begin{pmatrix} x_G(t_{s1}) + \frac{4}{9}\,\varepsilon_t\, t_{t,min}\, v_{max} & y_G(t_{s1}) + \frac{6}{9}\,\varepsilon_t\, t_{t,min}\, v_{max} & 0 \end{pmatrix}^T$$

after the time span $t_{com} = \varepsilon_t\, t_{t,min}$ given by (7.33). As the stand-on agent has similar dynamics to the give-way agent it moves to the point

$$w_S(t_{e1}) = \begin{pmatrix} x_S(t_{s1}) - \frac{4}{9}\,\varepsilon_t\, t_{t,min}\, v_{max} & y_S(t_{s1}) - \frac{4}{9}\,\varepsilon_t\, t_{t,min}\, v_{max} & 0 \end{pmatrix}^T$$

within the time span t_{com} in the worst-case scenario, as shown in Fig. B.5.

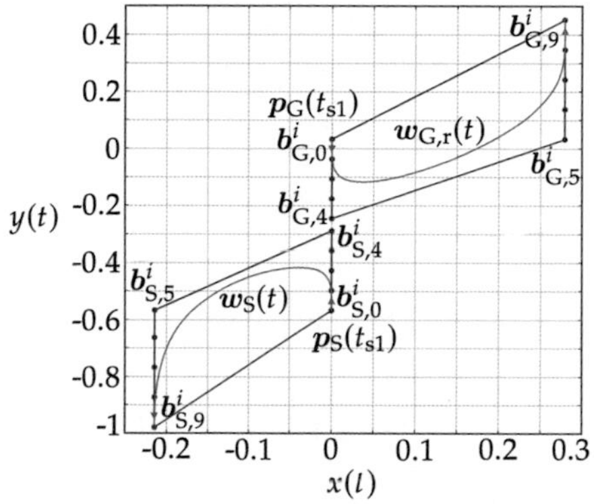

Fig. B.5: Worst-case scenario for the determination of the parameter t_{com}.

After the time span t_{com} the distance between the agents in the 2D space is given by

$$\begin{aligned} d_{SG} &= \underline{s} + \sqrt{\left(|w_{G,r,x}(t_{e1})| + |w_{S,x}(t_{e1})|\right)^2 + \left(|w_{G,r,y}(t_{e1})| + |w_{S,y}(t_{e1})|\right)^2} \\ &= \underline{s}\sqrt{\left(\left|\tfrac{4}{9}\varepsilon_t\, t_{t,min}\, v_{max}\right| + \left|-\tfrac{4}{9}\varepsilon_t\, t_{t,min}\, v_{max}\right|\right)^2 + \left(\left|\tfrac{6}{9}\varepsilon_t\, t_{t,min}\, v_{max}\right| + \left|-\tfrac{17}{9}\varepsilon_t\, t_{t,min}\, v_{max}\right|\right)^2} \\ &= \underline{s}\sqrt{\tfrac{593}{81}\,\varepsilon_t^2\, t_{t,min}^2\, v_{max}^2}. \end{aligned}$$

Using $\bar{e}_G$ (7.7) and requirement (7.32) results in

$$\underline{s} + 2\,\bar{e}_G \overset{!}{=} d_{SG} = \underline{s} + \frac{\sqrt{593}}{9}\varepsilon_t\, t_{t,min}\, v_{max}$$

and

$$2\frac{13}{9} t_{t,min}\, v_{max} = \frac{\sqrt{593}}{9}\varepsilon_t\, t_{t,min}\, v_{max}.$$

Hence,

$$\varepsilon_t = 1.07$$

results, which proves the lemma. ■

B.7 Proof of Lemma 7.3

Proof. The proof is similar to the proof of Lemma 7.1 and unfolds in two steps. First, the minimum time span between two events for the communication triggered with conditions (I) and (II) of (7.17) is proven. Second, the minimum time span between two events e_{G0} triggered by condition (III) of (7.17) is proven.

1) The proof considers only communication triggered by condition (I) of (7.17), because communication caused by condition (II) leads to the same results. The desired phase difference between the agents on a circle is specified by

$$\Phi_G(t) - \Phi_S(t) = \underline{s} + 2\,\bar{e}_\Phi.$$

An event e_{G0} is generated if for the phase difference

$$\Phi_G(t) - \Phi_S(t) = \underline{s} + \bar{e}_\Phi$$

applies, which yields that communication is invoked at a time instant t if the agents covered the relative phase

$$\int_{t_k}^{t} \frac{v_{SG}(\tau)}{2\,\pi\, r}\mathrm{d}\tau = \bar{e}_\Phi$$

after the last event time instant t_k. $v_{SG}(t)$ is the speed difference given by (B.13) and r is the radius of the circle. Then, the minimum time span $t = t_k + \tilde{t}$ is determined to be

$$t_{\Phi,sep} = \arg\min_{\tilde{t}} \left\{ \int_{t_k}^{t_k+\tilde{t}} \frac{v_{SG}(\tau)}{2\,\pi\, r}\mathrm{d}\tau = \bar{e}_\Phi \right\} = \frac{\bar{e}_\Phi\, 2\,\pi\, r}{v_{SG,max}}$$
$$\text{s.t. } v_{SG}(t) \le v_{SG,max}, \quad t_k \le t \le t_k + \tilde{t}.$$

2) Considering the desired height difference to be

$$|z_G(t) - z_S(t)| = \bar{z}$$

and the height difference at which event e_{G0} is triggered to be

$$|z_{\mathrm{G}}(t) - z_{\mathrm{S}}(t)| = \bar{z} - \bar{e}_{\mathrm{z}}$$

communication is invoked at time instant t if the agents have covered the relative height

$$\left| \int_{t_k}^{t} v_{\mathrm{SG,z}}(\tau) \mathrm{d}\tau \right| = \bar{e}_{\mathrm{z}}$$

after the last event time instant t_k with $v_{\mathrm{SG,z}}(t)$ as the speed difference of the agents in z-direction. Hence, the minimum time span $t = t_k + \tilde{t}$ satisfying the condition is given by

$$\begin{aligned} t_{\mathrm{z,sep}} &= \arg\min_{\tilde{t}} \left\{ \left| \int_{t_k}^{t_k+\tilde{t}} v_{\mathrm{SG,z}}(\tau) \mathrm{d}\tau \right| = \bar{e}_{\mathrm{z}} \right\} \\ &\text{s.t. } v_{\mathrm{SG,z}}(t) \leq v_{\mathrm{SG,z,max}}, \quad t_k \leq t \leq t_k + \tilde{t}. \end{aligned}$$

As

$$\begin{aligned} \arg\min_{\tilde{t}} \left\{ \left| \int_{t_k}^{t_k+\tilde{t}} v_{\mathrm{SG,z}}(\tau) \mathrm{d}\tau \right| = \bar{e}_{\mathrm{z}} \right\} &\geq \arg\min_{\tilde{t}} \left\{ \int_{t_k}^{t_k+\tilde{t}} |v_{\mathrm{SG,z}}(\tau)| \mathrm{d}\tau = \bar{e}_{\mathrm{z}} \right\} \\ &= \frac{\bar{e}_{\mathrm{z}}}{v_{\mathrm{SG,z,max}}} \end{aligned}$$

applies,

$$t_{\mathrm{z,sep}} \geq \tilde{t} = \frac{\bar{e}_{\mathrm{z}}}{v_{\mathrm{SG,z,max}}}$$

results and the minimum time span between two events e_{G0} is given by

$$t_{\mathrm{c,sep}} = \min\left(t_{\Phi,\mathrm{sep}}, t_{\mathrm{z,sep}}\right),$$

which proves the lemma. ■

B.8 Proof of Lemma 7.4

Proof. At first, for the proof the first worst-case scenario is considered. Both agents move with their minimum speeds $v_{\mathrm{S,min}}$ and $v_{\mathrm{G,min}}$ so that the phase difference $\Phi_{\mathrm{diff}}(t) = 180°$ will be recovered since $v_{\mathrm{G,min}} < v_{\mathrm{S,min}}$ applies. If $\bar{P}_{\mathrm{S}}$ accelerates to its maximum speed $v_{\mathrm{S,max}}$ without a reaction of $\bar{P}_{\mathrm{G}}$ a violation of the control aim at the upper limit threatens. Hence, the maximum time between two consecutive event time instants t_k, t_{k+1} must only be so large that the give-way agent has enough time to react to the change of speed of the stand-on agent. In the worst case, the phase difference $\Phi_{\mathrm{diff}}(t) = 180°$ is recovered at a time instant $t_e = t_k + \tilde{t}_I + t_{t,min}$, where $\tilde{t}_I$ states the time span when $\bar{P}_S$ moves with $v_{S,max}$ while $\bar{P}_G$ still moves with $v_{G,min}$ before reacting on the change of speed and $t_{t,min}$ states the time span $\bar{P}_G$ needs to accelerate from $v_{G,min}$ to $v_{S,max}$. At an event time instant t_k the phase difference is given by

$$\Phi_{diff}(t) = \underline{s} + \bar{e}_\Phi. \tag{B.14}$$

During the time span $\tilde{t}_I$ the phase difference

$$\Delta\Phi_{diff,1} = \frac{(v_{S,max} - v_{G,min}) \cdot \tilde{t}_I}{2\pi r} \cdot 360° \tag{B.15}$$

arises and during $t_{t,min}$ the phase difference

$$\Delta\Phi_{diff,2} = \frac{t_{t,min}\, v_{S,max}}{2\pi r} \cdot 360° - \Delta\Phi_G \tag{B.16}$$

occurs. $\Delta\Phi_G$ states the phase $\bar{P}_G$ covers during $t_{t,min}$. At $t = t_e$ the equation

$$\Phi_{diff}(t_e) = \Phi_{diff}(t_k) + \Delta\Phi_{diff,1} + \Delta\Phi_{diff,2} \stackrel{!}{=} 180° \tag{B.17}$$

should apply. Inserting (B.14), (B.15) and (B.16) in (B.17) and solving for $\tilde{t}_I$ gives

$$\tilde{t}_I = \frac{\frac{\gamma - \bar{e}_\Phi + \Delta\Phi_G}{360°} \cdot 2\pi r - t_{t,min} \cdot v_{S,max}}{v_{S,max} - v_{G,min}}.$$

With similar considerations the time $\tilde{t}_{II}$ is obtained for the second worst-case scenario:

$$\tilde{t}_{II} = \frac{\frac{-\gamma + \bar{e}_\Phi + \Delta\Phi_G}{360°} \cdot 2\pi r - t_{t,min} \cdot v_{S,min}}{v_{S,min} - v_{G,max}}.$$

For the third worst-case scenario similar considerations are applied using $\bar{e}_z$ leading to

$$\tilde{t}_{III} = \frac{\bar{e}_z - v_{G,max} \cdot t_{t,min}}{v_{S,max}}.$$

Then, the parameter t_{com} results to be

$$t_{com} = \min\left(\tilde{t}_I, \tilde{t}_{II}, \tilde{t}_{III}\right),$$

which proves the lemma. ■

B.9 Proof of Theorem 7.5

Proof. By fulfilling Assumptions 1.2, 1.4 and 6.1 and eqn. (7.36) the control aim (A1) is initially guaranteed. The proof unfolds in three steps, which correspond to the three tasks of the event-based control unit and is composed of the proofs of the parts of the method.

1): It is proven that the give-way agent has knowledge about the future positions of the stand-on agent.

2): The event threshold is proven to be appropriate for an event generation so that a proactive trajectory $w_{G,p}(t)$ or a reactive trajectory $w_{G,r}(t)$ can be planned.

3): It is proven that the trajectory planning method leads to the satisfaction of the control aims (A1) – (A3) with satisfaction of the constraints (C1) and (C2).

1): By planning the trajectories based on Bézier curves, the requirement for Theorem 4.1 is fulfilled, which ensures the inclusion of all future positions of the stand-on agent according to the proof in Appendix B.2.

2): As the optimisation problem (3.18) is convex, with Algorithm 2 the global minimum $t_{t,min}$ of the transition time interval is obtained. The event threshold is determined in a way that there is sufficient space for the give-way agent to perform an avoidance manoeuvre accordingly, proven in Appendix B.3.

3): By executing Algorithm 14 on p. 168 trajectories for the give-way agent are planned in a way to satisfy control aim (A1) at any time, which proves the theorem. ■

B.10 Proof of Theorem 9.1

Proof. The proof unfolds in two steps.

1): It is proven that (9.15) includes the movement of the agent in the time delay-free case ($\tau_k = 0$). The proof is equal to the proof of Theorem 4.1 in Section B.2. Here, also the factors e and f are derived.

2): It is proven that considering τ_k ensures the movement of the agent during the time delay to be included by (9.15). During the time span $t_{r,k} - \tau_k \leq t \leq t_{r,k}$ the movement of the agent is included in $\mathcal{P}(t_{r,k}, t_{r,k}, \tau_k)$ with

$$\begin{aligned}
\bar{x}(t_{r,k}, t_{r,k}, \tau_k) &= x(t_{c,k}) + v_x(t_{c,k}) \cdot e \cdot \tau_k \\
\bar{y}(t_{r,k}, t_{r,k}, \tau_k) &= y(t_{c,k}) + v_y(t_{c,k}) \cdot e \cdot \tau_k \\
\bar{z}(t_{r,k}, t_{r,k}, \tau_k) &= z(t_{c,k}) + v_z(t_{c,k}) \cdot e \cdot \tau_k \\
r_x(t_{r,k}, t_{r,k}, \tau_k) &= r_y(t_{r,k}, t_{r,k}, \tau_k) = f \cdot v_{max} \cdot \tau_k \\
r_z(t_{r,k}, t_{r,k}, \tau_k) &= v_{max} \cdot \tau_k
\end{aligned}$$

according to the first part of the proof. In the time span $t > t_{r,k}$ the movement is included in $\mathcal{P}(t, t_{r,k}, \tau_k)$ with

$$\begin{aligned}
\bar{x}(t, t_{r,k}, \tau_k) &= x(t_{c,k}) + v_x(t_{c,k}) \cdot e \cdot (t - t_{r,k} + \tau_k)\\
\bar{y}(t, t_{r,k}, \tau_k) &= y(t_{c,k}) + v_y(t_{c,k}) \cdot e \cdot (t - t_{r,k} + \tau_k)\\
\bar{z}(t, t_{r,k}, \tau_k) &= z(t_{c,k}) + v_z(t_{c,k}) \cdot e \cdot (t - t_{r,k} + \tau_k)\\
r_x(t, t_{r,k}, \tau_k) &= r_y(t, t_{r,k}, \tau_k) = f \cdot v_{\max} \cdot (t - t_{r,k} + \tau_k)\\
r_z(t, t_{r,k}, \tau_k) &= v_{\max} \cdot (t - t_{r,k} + \tau_k),
\end{aligned}$$

where $p(t) \in \mathcal{P}(t, t_{r,k}, \tau_k)$, which proves the theorem. ■

B.11 Proof of Theorem 9.2

Proof. The proof is similar to the proof of Theorem 9.1 and unfolds in two steps.
1): It is proven that (9.17) includes the movement of the agent in the time delay-free case ($\tau_k = 0$). The proof is equal to the proof of Theorem 4.2.
2): It is proven that considering τ_k ensures the circular movement of the agent during the time delay to be included by (9.17). During the time span $t_{r,k} - \tau_k \leq t \leq t_{r,k}$ the circular movement of the agent is included in $\mathcal{P}_c(t_{r,k}, t_{r,k}, \tau_k)$ with

$$\begin{aligned}
\Delta\Phi(t_{r,k}, t_{r,k}, \tau_k) &= [\Phi_{\min}(t_{r,k}, t_{r,k}, \tau_k), \Phi_{\max}(t_{r,k}, t_{r,k}, \tau_k)]\\
\Delta z(t_{r,k}, t_{r,k}, \tau_k) &= [z_{\min}(t_{r,k}, t_{r,k}, \tau_k), z_{\max}(t_{r,k}, t_{r,k}, \tau_k)]
\end{aligned}$$

where

$$\begin{aligned}
\Phi_{\min}(t_{r,k}, t_{r,k}, \tau_k) &= \bmod\left(\Phi_S(t_{c,k}) + \tfrac{v_{\min}}{2\pi r} \cdot \tau_k \cdot 360°, 360°\right)\\
\Phi_{\max}(t_{r,k}, t_{r,k}, \tau_k) &= \bmod\left(\Phi_S(t_{c,k}) + \tfrac{v_{\max}}{2\pi r} \cdot \tau_k \cdot 360°, 360°\right)\\
z_{\min}(t_{r,k}, t_{r,k}, \tau_k) &= \min\left(z_1(t_{r,k}, t_{r,k}, \tau_k), z_2(t_{r,k}, t_{r,k}, \tau_k)\right)\\
z_{\max}(t_{r,k}, t_{r,k}, \tau_k) &= \max\left(z_1(t_{r,k}, t_{r,k}, \tau_k), z_2(t_{r,k}, t_{r,k}, \tau_k)\right)
\end{aligned}$$

and

$$\begin{aligned}
z_1(t_{r,k}, t_{r,k}, \tau_k) &= |z_S(t_{c,k}) + v_{\max} \cdot \tau_k|\\
z_2(t_{r,k}, t_{r,k}, \tau_k) &= |z_S(t_{c,k}) - v_{\max} \cdot \tau_k| .
\end{aligned}$$

In the time span $t > t_{r,k}$ the movement is included in $\mathcal{P}_c(t, t_{r,k}, \tau_k)$ with

$$\begin{aligned}
\Delta\Phi(t, t_{r,k}, \tau_k) &= [\Phi_{\min}(t, t_{r,k}, \tau_k), \Phi_{\max}(t, t_{r,k}, \tau_k)]\\
\Delta z(t, t_{r,k}, \tau_k) &= [z_{\min}(t, t_{r,k}, \tau_k), z_{\max}(t, t_{r,k}, \tau_k)]
\end{aligned}$$

where

$$\begin{aligned}
\Phi_{\min}(t, t_{\mathrm{r},k}, \tau_k) &= \bmod\ \left(\Phi_{\mathrm{S}}(t_{\mathrm{c},k}) + \tfrac{v_{\min}}{2\pi r} \cdot (t - t_{\mathrm{r},k} + \tau_k) \cdot 360°, 360°\right) \\
\Phi_{\max}(t, t_{\mathrm{r},k}, \tau_k) &= \bmod\ \left(\Phi_{\mathrm{S}}(t_{\mathrm{c},k}) + \tfrac{v_{\max}}{2\pi r} \cdot (t - t_{\mathrm{r},k} + \tau_k) \cdot 360°, 360°\right) \\
z_{\min}(t, t_{\mathrm{r},k}, \tau_k) &= \min\left(z_1(t, t_{\mathrm{r},k}, \tau_k), z_2(t, t_{\mathrm{r},k}, \tau_k)\right) \\
z_{\max}(t, t_{\mathrm{r},k}, \tau_k) &= \max\left(z_1(t, t_{\mathrm{r},k}, \tau_k), z_2(t, t_{\mathrm{r},k}, \tau_k)\right)
\end{aligned}$$

and

$$\begin{aligned}
z_1(t, t_{\mathrm{r},k}, \tau_k) &= |z_{\mathrm{S}}(t_{\mathrm{c},k}) + v_{\max} \cdot (t - t_{\mathrm{r},k} + \tau_k)| \\
z_2(t, t_{\mathrm{r},k}, \tau_k) &= |z_{\mathrm{S}}(t_{\mathrm{c},k}) - v_{\max} \cdot (t - t_{\mathrm{r},k} + \tau_k)| \,.
\end{aligned}$$

It holds $p(t) \in \mathcal{P}_{\mathrm{c}}(t, t_{\mathrm{r},k}, \tau_k)$, which proves the theorem. ■

B.12 Proof of Lemma 9.1

Proof. The proof unfolds in two steps in which the event conditions of (9.24) are considered.

1): The stand-on agent is able to cover the distance $v_{\mathrm{S,max}} \cdot 2 \cdot \tilde{\tau}_{\max}(d(t_{k+1}))$ in the time span $2\,\tilde{\tau}_{\max}(d(t_{k+1}))$ by moving with its maximum speed. The first information transfer of the stand-on agent at $t = 0$ and every communication after a change of $\boldsymbol{w}_{\mathrm{G}}(t)$ or $\boldsymbol{w}_{\mathrm{S}}(t)$ is a reaction on a request of the give-way agent at a communication time instant $t_{\mathrm{c},k+1}$. The transmission of $\boldsymbol{R}_{\mathrm{G}}(t_{\mathrm{c},k+1})$ and $\boldsymbol{S}_{\mathrm{S}}(t_{\mathrm{c},k+1})$ lasts at most the time delay $2\,\tilde{\tau}_{\max}(d(t_{k+1}))$. The event e_{G0} is generated with a condition that is extended by the delay threshold (9.27) that leads to an earlier event generation. This causes that information is received at an event time instant t_{k+1}, where the agents have at least the distance $\underline{s} + \bar{e}_{\mathrm{G}}$ or $\bar{s} - \bar{e}_{\mathrm{G}}$ between them.

2): In the time span $\tilde{\tau}_{\max}(d(t_{k+j}))$ the stand-on agent is able to move the distance $v_{\mathrm{S,max}} \cdot \tilde{\tau}_{\max}(d(t_{k+j}))$ with its maximum speed. The stand-on agent sends its information automatically at communication time instants $t_{\mathrm{c},k+j}$, $(j = 2, \ldots, N)$, which takes at most the time span $\tilde{\tau}_{\max}(d(t_{k+j}))$. As events are generated with a condition that is extended by the delay threshold (9.28), the information is received at event time instants t_{k+j}. Again the agents have at least the distance $\underline{s} + \bar{e}_{\mathrm{G}}$ or $\bar{s} - \bar{e}_{\mathrm{G}}$ between one another, which proves the theorem. ■

B.13 Proof of Lemma 9.2

Proof. The proof unfolds in two steps in which the event conditions of (9.33) are considered. The proof is similar to the proof of Lemma 9.1.

1): The stand-on agent is able to cover the phase $\frac{v_{\mathrm{S,max}}}{r} \cdot 2 \cdot \tilde{\tau}_{\max}(d(t_{k+1}))$ or the height $v_{\mathrm{S,max}} \cdot 2 \cdot \tilde{\tau}_{\max}(d(t_{k+1}))$ in the time span $2\,\tilde{\tau}_{\max}(d(t_{k+1}))$ by moving with its maximum speed.

The first information transfer of the stand-on agent at $t = 0$ and every communication after a change of $\boldsymbol{w}_{\mathrm{G}}(t)$ or $\boldsymbol{w}_{\mathrm{S}}(t)$ is a reaction on a request of the give-way agent at a communication time instant $t_{\mathrm{c},k+1}$. The transmission of $\boldsymbol{R}_{\mathrm{G}}(t_{\mathrm{c},k+1})$ and $\boldsymbol{S}_{\mathrm{S}}(t_{\mathrm{c},k+1})$ lasts at most the time delay $2\,\tilde{\tau}_{\max}(d(t_{k+1}))$. The event e_{G0} is generated with a condition that is extended by the delay thresholds (9.34) and (9.35), which lead to an earlier event generation. This causes that information is received at an event time instant t_{k+1}, where the agents have at least the phase difference $\Phi_{\mathrm{diff}}(t) \in [\underline{s}, \bar{s}]$ and the height difference $z_{\mathrm{diff}}(t) = \bar{z}$ between them.

2): In the time span $\tilde{\tau}_{\max}(d(t_{k+j}))$ the stand-on agent is able to cover the phase $\frac{v_{\mathrm{S,max}}}{r} \cdot \tilde{\tau}_{\max}(d(t_{k+1}))$ or the height $v_{\mathrm{S,max}} \cdot \tilde{\tau}_{\max}(d(t_{k+1}))$ with its maximum speed. The stand-on agent sends its information automatically at communication time instants $t_{\mathrm{c},k+j}$, $(j = 2, \ldots, N)$, which takes at most the time span $\tilde{\tau}_{\max}(d(t_{k+j}))$. As events are generated with a condition that is extended by the delay thresholds (9.36) and (9.37), the information is received at event time instants t_{k+j}. Again the agents have at least the phase difference $\Phi_{\mathrm{diff}}(t) \in [\underline{s}, \bar{s}]$ and the height difference $z_{\mathrm{diff}}(t) = \bar{z}$ between one another, which proves the theorem. ■

B.14 Proof of Theorem 9.5

Proof. By fulfilling Assumptions 1.2, 1.4 and 8.1 and eqn. (9.45) the control aim (A1) is initially guaranteed. The proof unfolds in four steps.

1): It is proven that the give-way agent has knowledge about the future positions of the stand-on agent.

2): The event threshold is proven to be appropriate for an event generation so that a proactive trajectory $\boldsymbol{w}_{\mathrm{G,p}}(t)$, a reactive trajectory $\boldsymbol{w}_{\mathrm{G,r}}(t)$ or an avoidance trajectory $\boldsymbol{w}_{\mathrm{G,a}}(t)$ can be planned.

3): It is proven that the trajectory planning method guarantees the satisfaction of requirement (1.1) of control aim (A1) and fulfils requirement (1.2) of (A1) with probability (9.46) with satisfaction of the constraints (C1) – (C3).

1): By planning the trajectories based on Bézier curves, the requirement for Theorem 9.1 is fulfilled, which ensures the inclusion of all future positions of the stand-on agent according to the proof in Appendix B.10.

2): As the optimisation problem (3.18) is convex, with Algorithm 2 on p. 56 the global minimum $t_{\mathrm{t,min}}$ of the transition time interval is obtained. The event threshold is determined in a way that there is sufficient space for the give-way agent to perform an avoidance manoeuvre according to the scenario to be controlled proven in Appendix B.3. The delay thresholds are determined in a way that data are received by the give-way agent when the agents are sufficiently separated, as proven in Appendix B.12.

3): If no packet losses occur by the execution of the Algorithm 14 on p. 168 with the modifications stated in Section 9.8 trajectories for the give-way agent are planned in a way to satisfy control aim (A1) at any time.

After a packet loss at $t = t_k$ the newly sent information arrives at the give-way agent after the time delay $2\,\tilde{\tau}_{\max}(d(\tilde{t}_i))$. During this time span the stand-on agent is able to cover a maximum distance d_S defined in (9.41) towards the give-way agent. The give-way agent increases the distance to the stand-on agent by d_S in this time span according to (9.39) with respect to the closest possible position (9.40) of the stand-on agent at $t = t_k$. This means that the distance at $t = t_k$ between the agents is maintained or larger after the time span $2\,\tilde{\tau}_{\max}(d(\tilde{t}_i))$ compared to the distance between the agents before the packet loss occurs. Hence, requirement (1.1) of (A1) is satisfied at any time.

In case of a packet loss the give-way agent does not react with respect to requirement (1.2) of (A1). Hence, a violation of (1.2) is possible with probability $p_{\mathrm{e,max}}$, because the packet loss probability of the network is bounded from above by $p_{\mathrm{e,max}}$ (Assumption 8.1). The give-way agent receives data from the stand-on agent with probability greater than or equal to $1 - p_{\mathrm{e,max}}$. According to the first part of the proof, requirement (1.2) is satisfied if the give-way agent receives data in time. As not every relative movement of the stand-on agent and the give-way agent leads to a violation of (1.2), the probability of fulfilment of (1.2) is given by (9.46), which proves the theorem. ∎

B.15 Proof of Proposition 10.1

Proof. The proof of the flatness property unfolds in three steps to proof the three conditions:

1) Proof of condition 1: The state $\boldsymbol{x}(t)$ defined in (10.4) is directly composed of $\boldsymbol{y}_{\mathrm{f}}(t)$ and $\dot{\boldsymbol{y}}_{\mathrm{f}}(t)$ except for $\phi(t)$, $\vartheta(t)$ and their derivatives and it applies

$$\begin{aligned}
x_3(t) &= f(\dot{\boldsymbol{y}}_{\mathrm{f}}(t)), & x_9(t) &= f(\dot{\boldsymbol{y}}_{\mathrm{f}}(t)) \\
x_6(t) &= f(\boldsymbol{y}_{\mathrm{f}}(t)), & x_{10}(t) &= f(\boldsymbol{y}_{\mathrm{f}}(t)) \\
x_7(t) &= f(\dot{\boldsymbol{y}}_{\mathrm{f}}(t)), & x_{11}(t) &= f(\boldsymbol{y}_{\mathrm{f}}(t)) \\
x_8(t) &= f(\dot{\boldsymbol{y}}_{\mathrm{f}}(t)), & x_{12}(t) &= f(\boldsymbol{y}_{\mathrm{f}}(t)).
\end{aligned} \tag{B.18}$$

By rearranging the first row and the second row of (10.7) the parametrisation of $\phi(t)$ and $\vartheta(t)$ in terms of the flat output is given by

$$\phi(t) = \frac{k_2(t)\cos(\psi(t)) - k_1(t)\sin(\psi(t))}{a_{\mathrm{z}}(t)} \tag{B.19}$$

$$\vartheta(t) = \frac{-k_1(t)\cos(\psi(t)) - k_2(t)\sin(\psi(t))}{a_{\mathrm{z}}(t)} \tag{B.20}$$

with

$$a_z(t) = g - \ddot{z}(t) - \tfrac{c_z}{m}\dot{z}(t)$$
$$k_1(t) = \left(\ddot{x}(t) + \dot{x}(t)\tfrac{c_x}{m}\right), \quad k_2(t) = \left(\ddot{y}(t) + \dot{y}(t)\tfrac{c_y}{m}\right).$$

Hence, $\phi(t)$ and $\vartheta(t)$ are functions of $\boldsymbol{y}_f(t)$ and its derivatives as

$$\begin{aligned} x_4(t) &= f(\boldsymbol{y}_f(t), \dot{\boldsymbol{y}}_f(t), \ddot{\boldsymbol{y}}_f(t)) \\ x_5(t) &= f(\boldsymbol{y}_f(t), \dot{\boldsymbol{y}}_f(t), \ddot{\boldsymbol{y}}_f(t)). \end{aligned} \tag{B.21}$$

While the parametrisation of $\dot{\phi}(t)$ and $\dot{\vartheta}(t)$ is represented by

$$\dot{\phi} = \frac{-k_1\,\dot{\psi}\cos(\psi) - k_2\,\dot{\psi}\sin(\psi) - \dot{k}_1\sin(\psi) + \dot{k}_2\cos(\psi)}{a_z} - \frac{\dot{a}_z\,(k_2\cos(\psi) - k_1\sin(\psi))}{a_z^2} \tag{B.22}$$

$$\dot{\vartheta} = \frac{k_1\,\dot{\psi}\sin(\psi) - k_2\,\dot{\psi}\cos(\psi) - \dot{k}_1\cos(\psi) + \dot{k}_2\sin(\psi)}{a_z} - \frac{\dot{a}_z\,(k_1\cos(\psi) - k_2\sin(\psi))}{a_z^2}, \tag{B.23}$$

in which the time dependencies are omitted for simplicity. $\dot{\phi}(t)$ and $\dot{\vartheta}(t)$ can be stated as

$$\begin{aligned} x_1(t) &= f(\boldsymbol{y}_f(t), \dot{\boldsymbol{y}}_f(t), \ddot{\boldsymbol{y}}_f(t), \boldsymbol{y}_f^{(3)}(t)) \\ x_2(t) &= f(\boldsymbol{y}_f(t), \dot{\boldsymbol{y}}_f(t), \ddot{\boldsymbol{y}}_f(t), \boldsymbol{y}_f^{(3)}(t)). \end{aligned} \tag{B.24}$$

Hence, with (B.18), (B.21) and (B.24) the state $\boldsymbol{x}(t)$ is posed as

$$\boldsymbol{x}(t) = \boldsymbol{\Psi}_1\left(\boldsymbol{y}_f(t), \dot{\boldsymbol{y}}_f(t), \ddot{\boldsymbol{y}}_f(t), \boldsymbol{y}_f^{(3)}(t)\right).$$

2) Proof of condition 2: The parametrisation of the control input (10.14) in terms of the flat output is given by

$$\alpha_\phi(t) = \ddot{\phi}(t) \tag{B.25}$$
$$\alpha_\vartheta(t) = \ddot{\vartheta}(t) \tag{B.26}$$
$$\alpha_\psi(t) = \ddot{\psi}(t) \tag{B.27}$$
$$a_z(t) = g - \ddot{z}(t) - \tfrac{c_z}{m}\dot{z}(t). \tag{B.28}$$

With (B.25) and

$$\begin{aligned} \ddot{\phi} =& \frac{-2\,\dot{a}_z\left(-k_1\,\dot{\psi}\cos(\psi) - k_2\,\dot{\psi}\cos(\psi) - \dot{k}_1\sin(\psi) + \dot{k}_2\cos(\psi)\right)}{a_z^2} \\ &+ \frac{-2\,\dot{k}_1\,\dot{\psi}\cos(\psi) - 2\,\dot{k}_2\,\dot{\psi}\sin(\psi) - \ddot{k}_1\sin(\psi) + \ddot{k}_2\cos(\psi)}{a_z} \\ &+ \frac{k_1\left(\ddot{\psi}\cos(\psi) - \dot{\psi}^2\sin(\psi)\right) + k_2\left(-\ddot{\psi}\sin(\psi) - \dot{\psi}^2\cos(\psi)\right)}{a_z} \\ &+ \left(\frac{2\,\dot{a}_z^2}{a_z^3} - \frac{\ddot{a}_z}{a_z^2}\right)(k_2\cos(\psi) - k_1\sin(\psi)) \end{aligned} \tag{B.29}$$

it follows

$$\tilde{u}_1(t) = f(y_\mathrm{f}(t), \dot{y}_\mathrm{f}(t), \ldots, y_\mathrm{f}^{(4)}(t)). \tag{B.30}$$

For sake of simplicity the time dependencies are dropped again. Using (B.26) and

$$\begin{aligned}\ddot{\vartheta} =& \frac{-2\,\dot{a}_\mathrm{z}\left(k_1\,\dot{\psi}\sin(\psi) - k_2\,\dot{\psi}\cos(\psi) - \dot{k}_1\cos(\psi) - \dot{k}_2\sin(\psi)\right)}{a_\mathrm{z}^2} \\ &+ \frac{2\,\dot{k}_1\,\dot{\psi}\sin(\psi) - 2\,\dot{k}_2\,\dot{\psi}\cos(\psi) - \ddot{k}_1\cos(\psi) - \ddot{k}_2\sin(\psi)}{a_\mathrm{z}} \\ &+ \frac{k_1\left(\ddot{\psi}\sin(\psi) + \dot{\psi}^2\cos(\psi)\right) - k_2\left(\ddot{\psi}\cos(\psi) - \dot{\psi}^2\sin(\psi)\right)}{a_\mathrm{z}} \\ &+ \left(\frac{2\,\dot{a}_\mathrm{z}^2}{a_\mathrm{z}^3} - \frac{\ddot{a}_\mathrm{z}}{a_\mathrm{z}^2}\right)\left(-k_2\sin(\psi) - k_1\cos(\psi)\right).\end{aligned} \tag{B.31}$$

results in

$$\tilde{u}_2(t) = f(y_\mathrm{f}(t), \dot{y}_\mathrm{f}(t), \ldots, y_\mathrm{f}^{(4)}(t)). \tag{B.32}$$

Use of (B.27) and (B.28) leads to

$$\begin{aligned}\tilde{u}_3(t) &= f(\dot{y}_\mathrm{f}(t), \ddot{y}_\mathrm{f}(t)) \\ \tilde{u}_4(t) &= f(\dot{y}_\mathrm{f}(t), \ddot{y}_\mathrm{f}(t)).\end{aligned} \tag{B.33}$$

Hence, with (B.30), (B.32) and (B.33) the input is represented as

$$\tilde{\boldsymbol{u}}(t) = \begin{pmatrix}\alpha_\phi(t)\\ \alpha_\vartheta(t)\\ \alpha_\psi(t)\\ a_\mathrm{z}(t)\end{pmatrix} = \boldsymbol{\Psi}_2\left(\boldsymbol{y}_\mathrm{f}(t), \dot{\boldsymbol{y}}_\mathrm{f}(t), \ldots, \boldsymbol{y}_\mathrm{f}^{(4)}(t)\right).$$

3) Proof of condition 3: With the input (10.14) and the flat output (10.19) it follows

$$\dim(\boldsymbol{y}_\mathrm{f}(t)) = \dim(\tilde{\boldsymbol{u}}(t)) = 4.$$

It results that with Definition 10.1 the nonlinear system (10.6), (10.7) is flat with respect to the output (10.19). ■

C
List of Symbols

The following convention is generally used in this thesis: Scalars are denoted by lowercase italic letters (e.g. a). Vectors are denoted by lowercase bold italic letters (e.g. $\boldsymbol{x}$). Matrices are represented by uppercase bold italic letters (e.g. $\boldsymbol{A}$). An agent and its components are denoted by uppercase italic letters (e.g. $\bar{P}_{\mathrm{G}}$). The initial condition of a system or the initial value of a signal is indicated with 0 as part of its index (e.g. $\boldsymbol{x}_0$). The indices 'S' and 'G' state values of the stand-on agent or the give-way agent (e.g. $\boldsymbol{p}_{\mathrm{S}}(t)$, $\boldsymbol{w}_{\mathrm{G}}(t)$). A signal always includes the dependence of the time (e.g. $\boldsymbol{x}(t)$). It is possible that this thesis violates the described convention to improve the readability, whenever this happens, the reader is informed about the changes that are made.

The following Tables C.1 to C.5 lists all symbols and operators that are used in this thesis.

Table C.1: Notations

Symbol	Description
$\bar{P}$	Controlled agent
P	Physical object
A	Event-based control unit
G	Prediction unit
E	Event generator
T	Trajectory planning unit
D	Delay estimator
C^*	Flatness-based two-degrees-of-freedom controller
S_{A},S_{B}	Start position and end position of the movement of the agents
$P_i(t)$	Intermediate points at which two quarter circles are connected
$\mathrm{dist}(a,b)$	Euclidean distance between a and b

Table C.1: Notations (continued)

Symbol	Description
$\mathrm{dist}_{\mathrm{min}}(a,b)$	Minimum euclidean distance between a and b
$\mathrm{dist}_{\mathrm{max}}(a,b)$	Maximum euclidean distance between a and b
SNR	Signal-to-noise ratio
AWGN	Additive white Gaussian noise channel
LOS	Line-of sight path
QoS	Quality-of-service parameters of a communication channel
ACK	Acknowledgement message

Table C.2: Scalars

Symbol	Description
$x(t)$	x-position of an agent
$y(t)$	y-position of an agent
$z(t)$	z-position of an agent
z_{l},z_{h}	Lower and upper height of a circular movement
r	Radius of the circular movement
$v(t)$	Speed of an agent
$\omega(t)$	Angular speed of an agent
$s(t)$	Spatial separation between two agents
$\underline{s}$,$\bar{s}$	Lower and upper bound on the spatial separation
$d(t)$	Distance between two agents
$\Phi(\boldsymbol{p}(t))$	Phase of an agent
$\Phi_{\mathrm{diff}}(t)$	Phase difference between two agents
$z_{\mathrm{diff}}(t)$	Height difference between two agents

Table C.2: Scalars (continued)

Symbol	Description
$\Delta\Phi(t, t_k)$	Set of possible future phases of an agent
$\Delta z(t, t_k)$	Set of possible future heights of an agent
$B_k^m(u)$	Bernstein polynomials on the interval $u \in [\bar{t}_i, \bar{t}_{i+1}]$
$\tilde{B}_k^m(u)$	Bernstein polynomials on the interval $u \in [0, 1]$
t	Current time instant
$t_{\mathrm{start}}, t_{\mathrm{end}}$	Start time instant and end time instant of the movement of an agent
$t_{\mathrm{s}}, t_{\mathrm{e}}$	Start time instant and end time instant of a proactive, a reactive or an avoidance trajectory
t_{ret}	Time instant for the return to the initial trajectory
t_k	Event time instant for communication
$t_{\mathrm{c},k}$	Communication time instant
$t_{\mathrm{r},k}$	Time instant of reception of information
$\bar{t}_i$	Time instant at which two pieces of a trajectory are connected
$\tilde{t}_i$	Event time instant for the network estimation
t_{EG}	Event time instant for a change of the trajectory
$t_{\mathrm{sep}}, t_{\mathrm{c,sep}}$	Time span between two communication events
t_{com}	Time span between two communication events in the worst-case scenario
$t_{\mathrm{min,com}}, t_{\mathrm{c,min,com}}$	Minimum time span between two communication events
t_{t}	Transition time interval
t_{tc}	Transition time interval for a circular movement
τ_k	Time delay
τ_{c}	Computational delay
$\tau_{\mathrm{c,max}}$	Upper bound of the computational delay

Table C.2: Scalars (continued)

Symbol	Description
τ_{n}	Transmission delay of the communication network
$\tilde{\tau}_{\mathrm{n,max}}(d(\tilde{t}_i))$	Estimated transmission delay
$\tilde{\tau}_{\mathrm{max}}(d(\tilde{t}_i))$	Estimated time delay
e	Events, indices state different types of events
$\bar{e}$	Event threshold, indices state different types of thresholds
$\bar{d}$	Event threshold on the distance
R	Receiver of an agent
$R(d(\tilde{t}_i))$	Data rate of a communication channel
S_{rs}	Receiver sensitivity
P_0	Transmission power
$P_{\mathrm{r}}(d(\tilde{t}_i))$	Received signal power
$\Omega(d(\tilde{t}_i))$	Mean total received signal power
$x(s)$	Sequence of N packets
$x_{\mathrm{p}}(s)$	Received observation sequence
$C(d(\tilde{t}_i))$	Channel capacity
$p_{\mathrm{g}}(d(\tilde{t}_i)), p_{\mathrm{b}}$	Packet loss probability of the states 'good' and 'bad' of the Markov model
$p_{\mathrm{e}}(d(\tilde{t}_i))$	Mean packet loss probability
$p_{\mathrm{gg}}, p_{\mathrm{gb}}, p_{\mathrm{bg}}, p_{\mathrm{bb}}$	Transition probabilities of the Markov model
$\hat{p}_{\mathrm{gg}}, \hat{p}_{\mathrm{gb}}, \hat{p}_{\mathrm{bg}}, \hat{p}_{\mathrm{bb}}$	Estimated transition probabilities of the Markov model
$n(t)$	Rotor speeds of an UAV
$\underline{n}, \bar{n}$	Lower and upper bound on the rotor speeds
$\phi(t)$	Roll angle

Table C.2: Scalars (continued)

Symbol	Description
$\underline{\phi}, \bar{\phi}$	Lower and upper bound on the roll angle
$\vartheta(t)$	Pitch angle
$\underline{\vartheta}, \bar{\vartheta}$	Lower and upper bound on the pitch angle
$\psi(t)$	Yaw angle

Table C.3: Vectors

Symbol	Description
$\boldsymbol{w}(t)$	Trajectory of an agent
$\boldsymbol{w}^i(t)$	Piece of a trajectory of an agent
$\boldsymbol{p}(t)$	Position of an agent
$\boldsymbol{u}(t)$	Control input
$\boldsymbol{v}(t)$	Speed of an agent
$\boldsymbol{b}$	Control points of a Bézier curve
$\boldsymbol{b}^i$	Control points of a piece of a Bézier curve
$\boldsymbol{r}(t_k)$	Request of an agent for an ideal communication network

Table C.4: Matrices

Symbol	Description
$\boldsymbol{W}(t)$	Trajectory and its first four derivatives
$\boldsymbol{R}(t_{\mathrm{c},k})$	Request of an agent for an unreliable communication network
$\boldsymbol{S}(t_{\mathrm{c},k})$	Information including the trajectory of an agent
$\boldsymbol{G}_{\mathrm{M}}$	Transition matrix of a Markov model

Table C.5: Calligraphic letters

Symbol	Description
$\mathcal{P}(t, t_{\mathrm{r},k}, \tau_k)$	Set of possible future positions of an agent
$\mathcal{P}_{\mathrm{d}}(t_{\mathrm{c},k}, t_{\mathrm{c},k}, \tau_k)$	Set of possible future positions of an agent during τ_k
$\mathcal{P}_{\mathrm{O}}(\bar{t}_i)$	List of static obstacles
$\mathcal{K}(d(\tilde{t}_i))$	Ratio of the received signal power from the LOS path to the received signal power from the scattered paths